Nanofabrication and its Application in Renewable Energy

RSC Nanoscience & Nanotechnology

Editor-in-Chief:
Paul O'Brien FRS, *University of Manchester, UK*

Series Editors:
Ralph Nuzzo, *University of Illinois at Urbana-Champaign, USA*
Joao Rocha, *University of Aveiro, Portugal*
Xiaogang Liu, *National University of Singapore, Singapore*

Honorary Series Editor:
Sir Harry Kroto FRS, *University of Sussex, UK*

How to obtain future titles on publication:
A standing order plan is available for this series. A standing order will bring delivery of each new volume immediately on publication.

For further information please contact:
Book Sales Department, Royal Society of Chemistry, Thomas Graham House, Science Park, Milton Road, Cambridge, CB4 0WF, UK
Telephone: +44 (0)1223 420066, Fax: +44 (0)1223 420247
Email: booksales@rsc.org
Visit our website at www.rsc.org/books

Nanofabrication and its Application in Renewable Energy

Edited by

Gang Zhang
Department of Electronic, Peking University, Beijing, P. R. China
Email: zhanggang@pku.edu.cn

and

Navin Manjooran
Siemens AG, Princeton, USA
Email: jose@vt.edu

ROYAL SOCIETY
OF CHEMISTRY

RSC Nanoscience & Nanotechnology No. 32

ISBN: 978-1-84973-640-4
ISSN: 1757-7136

A catalogue record for this book is available from the British Library

Published by The Royal Society of Chemistry,
Thomas Graham House, Science Park, Milton Road,
Cambridge CB4 0WF, UK

Registered Charity Number 207890

For further information see our web site at www.rsc.org

Preface

We are living in interesting times where technology has been progressing at a significant pace in the last few decades. One of the major areas that will fulfill an important role in this technological progress is the science of nanotechnology. Experts predict that nanotechnology has the potential to play a critical role in this global technology advancement.

With the increase of the global population and the growth in the number of metro cities, there is a continuing greater need for energy. In order to ensure we have a clean environment and that we generate energy safely, renewable energy is very important. Renewable energy comprises of energy generated primarily from renewable sources such as wind and solar.

This book highlights the various nanotechnology innovations in renewable energy by world renowned experts. The aim of this book is to provide an introduction for both theorists and experimentalists to the current technology and looks at the applications of nanostructures in renewable energy, and the associated research topics. The book should also be useful for graduate-level students who want to explore this new field of research. With content relevant to both academic and commercial viewpoints, the book will interest researchers and postgraduates as well as consultants in the renewable energy industry.

The single chapters have been written by internationally recognized experts and provide in-depth introductions to the directions of their research. Along these lines, our intention was to embed research on new energy materials into a wider context of nanotechnology research. We thus hope that this book may serve as a catalyst both to fuse existing nanotechnology and to inspire new tools in the rapidly growing area of new energy material research.

I finally remark that the various points of view expressed in the single chapters may not always be in full agreement with each other. As editors,

RSC Nanoscience & Nanotechnology No. 32
Nanofabrication and its Application in Renewable Energy
Edited by Gang Zhang and Navin Manjooran
© The Royal Society of Chemistry 2014
Published by the Royal Society of Chemistry, www.rsc.org

we do not necessarily aim to achieve a complete consensus among all the authors, as differences in opinions are typical for a very active field of research such as the one presented in this book.

We are most grateful to Royal Society of Chemistry, for the invitation to edit this book, and for kind and efficient assistance in the project. We finally wish to thank all book chapter authors for sharing their expertise in this multi-author monograph. Their strong efforts and enthusiasm for this project were indispensable for bringing it to success.

Gang Zhang
Navin Manjooran

Contents

RSC Nanoscience & Nanotechnology No. 32
Nanofabrication and its Application in Renewable Energy
Edited by Gang Zhang and Navin Manjooran
Published by the Royal Society of Chemistry, www.rsc.org

Author Biographies

Prof. Dr Gang Zhang

 Dr Zhang is senior scientist and group manager at the Institute of High Performance Computing, A*STAR, Singapore. Before joining IHPC, he was a full professor at Peking University, China. He received his BS and PhD in Physics from Tsinghua University in 1998 and 2002, respectively. From 2002 to 2004 he was a SMF research fellow at National University of Singapore (NUS) and from 2005 to 2006 at Stanford University. He joined the Institute of Microelectronics, A*STAR, Singapore, as a senior research engineer in 2006 and Department of Electronics, Peking University in 2010 as a full professor. He is a world recognized expert in the electrical and thermal properties simulation of nano–materials. He has developed several novel approaches for molecular dynamic, and quantum chemistry simulations. He has authored and co-authored more than 100 publications in peer-reviewed international journals and conferences, including 1 review article in *Reviews of Modern Physics*, 1 in *Nature Communications*, 6 in *Nano letters and Nano Today*, and more than 10 invited reviews and book chapters. His research has gained him a number of international recognition and media highlights. He was awarded with an Outstanding Ph.D. thesis Award in Tsinghua University (2002), Singapore Millennium Foundation Fellowship (2002–2004), and IME Excellence Award (2008). Dr Zhang's research achievements are also matched with his competency in teaching. Several of his PhD students received the Outstanding Awards both in China and Singapore.

Prof. Dr Navin Manjooran

Prof. Dr Navin Manjooran serves as the Global Director (CRM), Energy for Siemens AG. In this role, he provides leadership from the corporate side for the entire Siemens energy portfolio; including expanding international collaborations, developing the global energy strategy and ensuring Siemens stays a trend-setter for advanced technology and innovation. He also serves as an Adjunct/ Distinguished Professor at Virginia Tech in Materials Science and Engineering (Nanotechnology). Dr Manjooran holds a Bachelors degree in Engineering from the National Institute of Technology (Warangal, India), a Masters degree in Materials Science and Engineering from the University of Florida (Gainesville, USA), and a Doctorate in Materials Science and Engineering from Virginia Tech (Blacksburg, USA) all with the highest honors. He also has an MBA from the prestigious University of Chicago, Booth School of Business. Navin has 11 patents/ disclosures, 5 books, 37 publications and 51 presentations at national/international conferences. He has received several awards including the following prestigious awards: TMS Young Leader (among the top 5 leaders internationally under the age of 35), ASM International Leadership Award, Siemens Performance Award, Siemens Collaboration Award, VT Outstanding Leader Award, VT Paul Torgerson Doctoral Research Excellence Award, VT Outstanding Service Commendation Award, VT Outstanding Student Award, NITW Sri Kabadi Subalu Medal and the NITW Alumni Association Gold Medal (Valedictorian and best student at National Institute of Technology, Warangal, India). Navin is a Member of the US Technology Advisory Board. He has also served on the Virginia Tech University Board of Visitors and the University of Chicago, Board of Trustees. He was also a Co-Chair for the Virginia Tech Capital Campaign aimed at raising USD 1 billion. In his spare time, Navin enjoys playing tennis and golf.

CHAPTER 1

Fabrication Techniques of Graphene Nanostructures

XINRAN WANG* AND YI SHI

National Laboratory of Solid State Microstructures and School of Electronic Science and Engineering, Nanjing University, Nanjing 210093, P. R. China
*Email: xrwang@nju.edu.cn

1.1 Introduction to Graphene

Carbon is one of the most studied elements in the periodic table. The versatility of chemical bonds enables many carbon allotropes. In three-dimensional bulk form, carbon can exist as diamonds and graphite, which comprise of sp^3 and sp^2 covalent bonds, respectively. In the 1980s and 1990s, another two types of carbon allotropes, the zero-dimensional fullerene[1] and one-dimensional carbon nanotubes,[2] were discovered (Figure 1.1(a)).[3] These nanomaterials, with fascinating physical and chemical properties, have driven an enormous amount of research in many areas.[4] However, the two-dimensional counterpart of carbon allotrope was still missing until 2004, when a single layer of graphite, or graphene, was successfully isolated on a substrate.[5] In this section, we give a brief introduction to graphene. We do not intend to derive the properties of graphene from the lattice and band structures. Readers who are interested in those aspects are encouraged to read the excellent reviews that are available.[6,7]

RSC Nanoscience & Nanotechnology No. 32
Nanofabrication and its Application in Renewable Energy
Edited by Gang Zhang and Navin Manjooran
© The Royal Society of Chemistry 2014
Published by the Royal Society of Chemistry, www.rsc.org

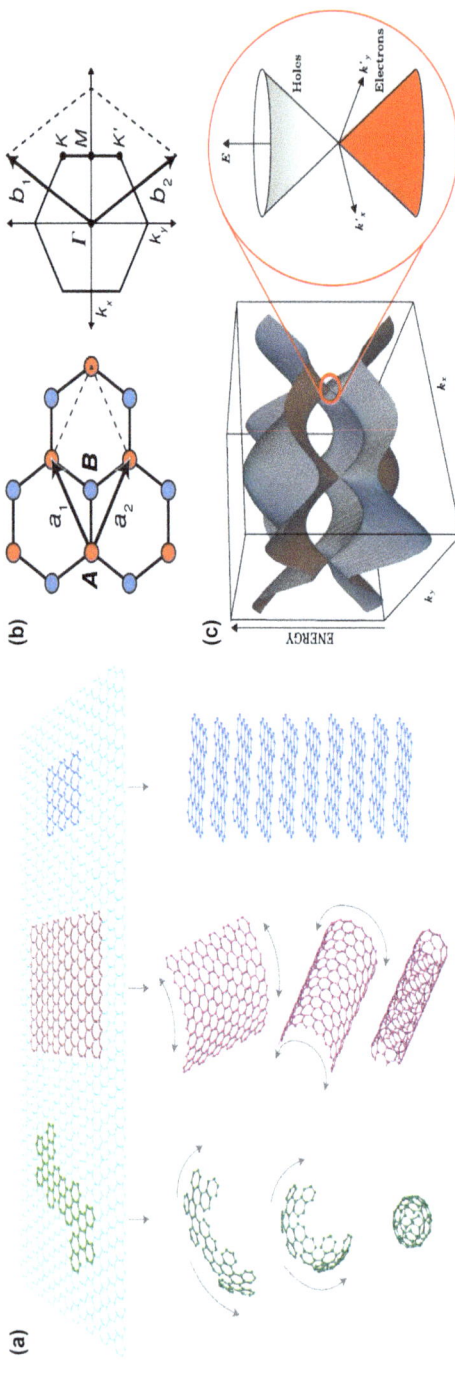

Figure 1.1   (a) Schematic drawing of the relationship between graphene and other carbon materials. Graphene is a 2D building material for carbon materials of all other dimensions. It can be wrapped up into 0D buckyballs, rolled into 1D nanotubes or stacked into 3D graphite. Adapted from Ref. 3. The lattice structure of graphene with the yellow region as the unit cell. (b) (Left) Hexagonal crystal structure of graphene with lattice parameters a_1 and a_2 and inequivalent atomic positions A and B of the diatomic basis shown in red and blue. (Right) Reciprocal lattice and Brillouin zone with reciprocal lattice parameters b_1 and b_2, showing Dirac points K and K' at the Brillouin zone corners, as well as M, the midpoint of the BZ edge and Γ, the center of the Brillouin zone. Adapted from Ref. 90. (c) (Left) Electronic band structure of graphene. (Right) Enlargement of the band structure near the Dirac cone. Adapted from Ref. 91.

1.1.1 Lattice and Band Structure

Graphene is composed of a honeycomb lattice of carbon atoms (Figures 1.1(b) and 1.2). Structurally, graphene is related to many carbon allotropes. For example, carbon nanotubes can be formed by rolling graphene along certain axes, and graphite can be formed by stacking graphene vertically[3] (Figure 1.1(a)). The structure can be seen as a triangular lattice with two equivalent atoms in each unit cell. The unit vectors $\boldsymbol{a_{1,2}} = \frac{a}{2}\left(3, \pm\sqrt{3}\right)$, where $a = 1.42\text{Å}$ is the carbon–carbon distance (Figure 1.1(b)). The reciprocal lattice is also triangular, with unit vectors $\boldsymbol{b_{1,2}} = \frac{2\pi}{3a}\left(1, \pm\sqrt{3}\right)$. The shape of the Brillouin zone is rotated $90°$ compared to the lattice unit cell. In each Brillouin zone, there are two inequivalent corners, $\boldsymbol{K}$ and $\boldsymbol{K'}$, called Dirac points. This is related to the presence of a pseudospin degree of freedom or two independent sublattices in graphene.[6,7]

The tight-binding band structure of graphene was calculated in 1947 by Wallace,[8] which can be described by eqn (1).

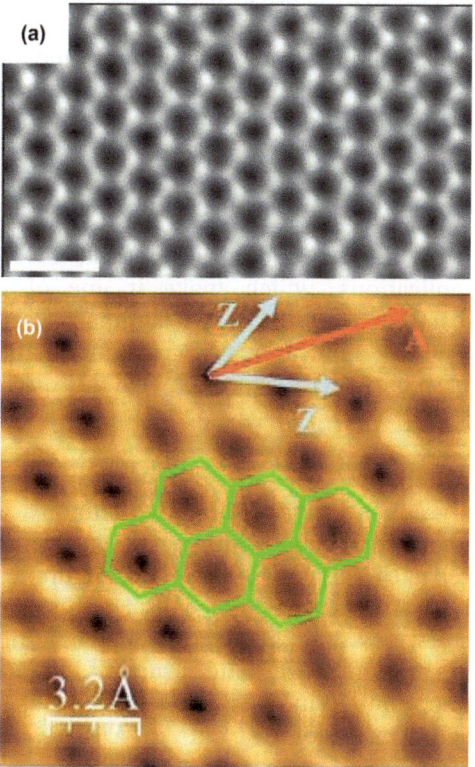

Figure 1.2 Lattice structure of graphene under (a) TEM and (b) STM showing the hexagonal lattice.
Adapted from Refs. 92 and 93.

$$E(\boldsymbol{k}) = E_F \pm t\sqrt{1 + 4\cos\left(\frac{\sqrt{3}k_x a}{2}\right)\cos\left(\frac{k_y a}{2}\right) + 4\cos^2\left(\frac{k_y a}{2}\right)}, \qquad (1)$$

here E_F is the Fermi energy, t is the nearest neighbor hopping integral. The $\pm$ signs represent conduction and valance bands respectively. Figure 1(c) is the plot near the first Brillouin zone. The band structure of graphene is drastically different from that of conventional semiconductors. The most remarkable feature is the linear dispersion relation near the Dirac points, which resembles ultrarelativistic particles and can be described by the massless Dirac equation.[6,7] Therefore, the electrons in graphene are often called massless Dirac fermions. This can be seen quantitatively by expanding eqn (1) near the Dirac points to give eqn (2).

$$E(\boldsymbol{k}) = \pm h\nu_F \boldsymbol{k} \qquad (2)$$

where $\nu_F = 3ta/2h$ is the graphene Fermi velocity. Take $t = 2.5$ eV, $\nu_F \approx 10^8$ cm s^{-1}, about 1/300 of the speed of light.[6] Due to the unique dispersion relation, the density of states of graphene is linear with energy and vanishes at the Dirac points. This is again in contrast to conventional 2D materials with a constant density of states. The unique band structure of graphene has led to many intriguing phenomena like half integer quantum Hall effect,[9,10] Klein tunneling,[11] and electron focusing.[12]

1.1.2 Properties and Potential Applications

Graphene has many unique electrical, optical, mechanical properties and potential applications, which make it one of the most studied materials in the past ten years. Below we summarize some of the physical properties and potential applications of graphene, many of which require the fabrication of nanostructures, which will be discussed later in the chapter.

1. High carrier mobility
   Electrons in graphene have well-defined chirality, which is related to the two component pseudospin degree of freedom. Such chirality prevents intervalley backscattering and therefore leads to intrinsically high mobility.[6] Another advantage of graphene, compared to III-V semiconductors,[13] is that the electron and hole mobility are equally high, due to the symmetric band structure. Theories have predicted that the phonon limited mobility can be as high as 2×10^5 cm^2 Vs^{-1} at a carrier concentration of 10^{12} cm^{-2}.[14] A similar mobility value has already been measured experimentally.[15,16]
   The ultrahigh mobility makes graphene a potential candidate in electronic applications, including radio frequency and logic transistors.[3,17-19] Although the absence of a bandgap places a major

obstacle for logic applications, many ways of opening a bandgap have been proposed and tested, which will be the focus of discussion in this chapter.

2. Conductive and transparent

   Due to the absence of backscattering, the resistivity of pristine graphene is expected to be $\approx 1 \times 10^{-6}$ Ωcm, which is lower than the most conductive metal, silver, at room temperature.[20] Optically, graphene is quite transparent over the visible spectra, absorbing 2.3% of the light. However, considering that graphene only has one atomic layer, such absorption is quite dramatic. The absorption is expressed by a simple formula $\pi\alpha$, where $\alpha \approx 1/137$ is the fine-structure constant.[21] Such an optical feature originates from the unique conical band structure of graphene near the Dirac points.

   With a combination of high conductivity and transparency, graphene is expected to find applications in transparent conducting films.[17] As a possible candidate to replace the increasingly costly indium tin oxide (ITO) based films, graphene is made entirely of carbon. Another advantage is the extreme mechanical strength and flexibility.[22] Potential near term applications include touch screen display, e-paper and organic light-emitting diodes.[17]

3. Wide-band and tunable optical absorption

   Graphene has a tunable optical absorption. Since the density of states is vanishing at the Dirac points, the Fermi energy can be tuned by carrier density (normally through electrostatic gating). The Pauli blockade ensures that optical absorption is only possible for energy greater than $2E_F$.[23] Therefore, graphene is expected to respond to wide-band signals spanning microwave to ultraviolet, including the commonly used fiber-optic communication band at 1.55 μm. Combined with the superior carrier mobility, possible applications include ultrafast photodetectors, modulators, terahertz wave detectors and tunable fiber mode-locked lasers.[17,24]

4. Large specific surface area

   Since every atom of graphene is on its surface, graphene has one of the largest specific surface areas among all materials, 2630 m^2 g^{-1}. Therefore graphene is expected to have applications in sensors: it has been shown that detection of single molecule is possible with graphene.[25] Combined with high conductivity and transparency, graphene also has great potential in energy related applications. For example, supercapacitor electrodes require large conductivity for high specific power and large surface area for high specific energy. Chemically exfoliated graphene is shown to possess both properties and have superior capacitance.[26,27] Another example is lithium ion batteries. Since (chemically-reduced) graphene has a higher conductivity and can accommodate more strain than many cathode materials, the addition of graphene into these active materials often increased the charging rate and afforded more stable cycling performances.[28,29]

1.2   Fabrication Techniques of Graphene Electronic Devices

Among the aforementioned applications, many require the fabrication of electronic devices, such as transistors, photodetectors and sensors. This section briefly reviews the most commonly employed forms of graphene device structures and their general fabrication processes.

The most commonly used device structures are summarized in Figure 1.3. Backgated devices, with silicon as a backgate and SiO_2 as the gate dielectrics (Figure 1.3(a)), are widely used in graphene research.[5] First graphene is transferred (*e.g.* by mechanical exfoliation) on to SiO_2/highly conducting silicon substrates. The normally used thickness of SiO_2 is ≈ 300 nm or 90 nm for easy identification of graphene under optical microscope.[5] Then optical or electron beam lithography is used to create patterns on graphene, followed by Ar or O_2 plasma etching to etch away the unnecessary parts (Figure 1.3(a)). Another lithography step is used to pattern the metal source/drain of the devices, which is done by evaporation or sputtering. In many cases thermal annealings are necessary to remove resistant residues and improve the contacts. In such a device, the backgate is used to globally adjust the carrier type and concentration (or equivalently, Fermi energy) in graphene. The mobility of the device can be readily measured by field effects (with two metal electrodes) or Hall measurements (with Hall bar structure as Figure 1.3(a)).[5] Other than SiO_2, other substrates such as hexagonal boron

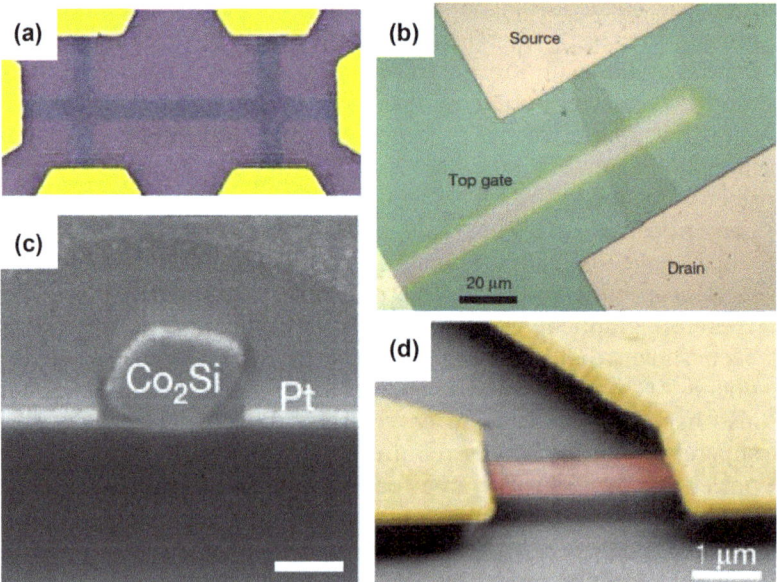

Figure 1.3   Various device structures of graphene. (a) backgated, (b) topgated, (c) self-aligned topgate and (d) suspended device.
Adapted from Refs. 9, 31, 34 and 36.

nitride (hBN) have also been used to improve the mobility of graphene, which will be discussed later.

To add extra control of graphene devices, the double gate structure is commonly used (Figure 1.3(b)). Such devices include a topgate on top of the backgated device to locally adjust the carrier concentration. Used in combination with a backgate, many interesting structures can be created such as p-n junctions[30] and tunnel barriers.[11] As the topgate and backgate can be tuned independently to control the vertical displacement field, the bandgap opening in AB stacked bilayer graphene can be demonstrated with such a device structure.[31] During the fabrication of topgate, it is challenging to deposit a uniform topgate dielectric layer with an atomic layer deposition since the plane of graphene lacks dangling bonds to start the nucleation.[32] Quite often a seeding layer is needed, such as a thin layer of aluminium, that is natively oxidized,[33] gas molecules,[30] or organic layers like perylene tetra-carboxylic acid (PTCA).[32]

A more challenging form of topgated device is shown in Figure 1.3(c), where the source/drain is self-aligned without any lithography step.[34] Such a structure is advantageous to probe the ultimate potential of graphene transistors because it minimizes the parasitics. During the fabrication, a core-shell nanowire is deposited on graphene. The conducting core acts as the topgate, while the insulating shell acts as the gate dielectrics. Then a thin layer of metal is evaporated as a source and drain. Due to the directional nature of evaporation, the source and drain fall on each side of the nanowire in a self-aligned manner. This way there is minimal gap between the source/drain and the gate-controlled region. With a similar structure, graphene radio frequency transistors with channel length as low as 100 nm are demonstrated, showing the record cut-off frequency of 427 GHz.[35]

Finally, a graphene device without any substrate is shown in Figure 1.3(d). The motivation to fabricate a suspended device is to remove any substrate effect that could affect the intrinsic properties of graphene, such as carrier scattering by phonons and charged impurities from the substrate.[36] In fact, the highest experimental mobility of graphene, 2×10^5 cm^2 Vs^{-1}, was measured in a suspended device at cryogenic temperature.[15] To obtain the suspended structure in Figure 1.3(d), first a backgated device with metal electrodes is fabricated. Then the structure is immersed in buffered oxide etch to remove the exposed SiO$_2$, including those underlying the graphene. The SiO$_2$ under the metal electrodes is only partially etched. Finally, the device is transferred to ethanol and dried with critical point drying to prevent any surface tension, which would otherwise collapse the suspended graphene.[15]

1.3 Transfer Techniques of Graphene

Recently rapid progress has been made in the area of chemical vapor deposition (CVD) synthesis of graphene on metal surfaces. It is now possible to grow graphene single crystals over a few millimeters (Figure 1.4(b))[37] and polycrystal films over meters.[38] The quality of CVD graphene can be

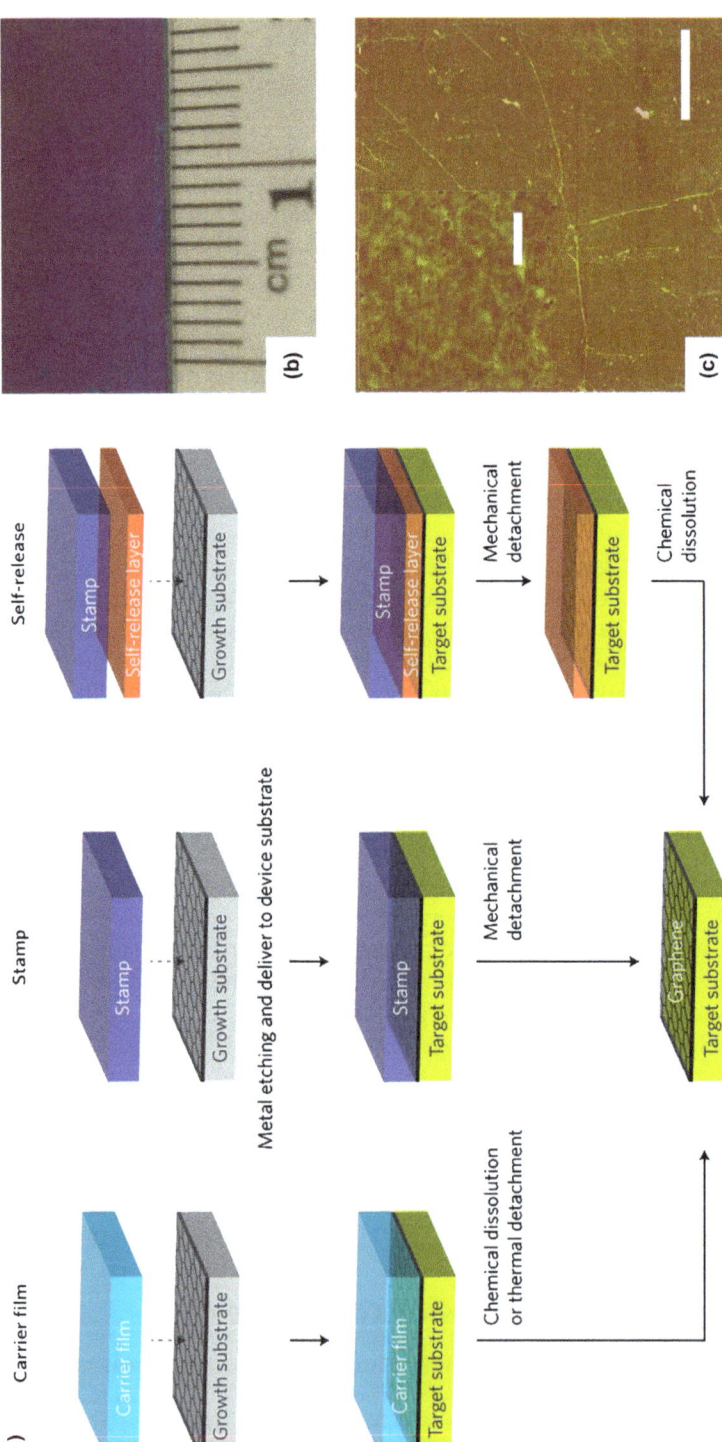

Figure 1.4 Transfer of CVD graphene. (a) Three different approaches for transferring CVD graphene with millimeters single-crystal domain. (b) An optical image of CVD graphene after transfer by the "self-release transfer method".
(c) AFM image of clean CVD graphene after transfer by the "self-release transfer method".
Adapted from Ref. 37, 39 and 43.

comparable to mechanically exfoliated samples in terms of carrier mobility, therefore more and more research groups now use CVD graphene to fabricate devices. During the fabrication process, one of the most critical steps is the transfer from the metal surface to the target substrates.[17] In this process, it is important not to damage graphene while maintaining a clean surface without any chemical residue. The transfer process is also important for the application of graphene. Below we review several available methods for graphene transfer.[39]

The most commonly used approach is the "carrier film" method (Figure 1.4(a)), where the graphene layer is attached to a carrier film during transfer, usually a thick film of PMMA polymer or a thermal release tape.[39] Here we take the most commonly used system as a detailed example: to transfer graphene from Cu foil to SiO_2 substrate with PMMA. In a typical procedure, first one side of the as-grown copper surface (covered by graphene) is spin-coated with PMMA and then baked to remove the solvents. The other side of the sample is etched by O_2 plasma to remove the back-side graphene and the overflowing PMMA. After that, the underlying Cu foil is etched away using aqueous copper etchant (such as $FeCl_3$ or CE-100 from Transene), resulting in free-standing PMMA/graphene membrane floating on the surface of the etchant.[40] The PMMA/graphene film is then treated with a modified RCA cleaning process to further clean the Cu residue:[41] rinsing with a volumetric mixture of $H_2O : H_2O_2 : HCl = 20 : 1 : 1$ for 15 mins, and then a volumetric mixture of $H_2O : H_2O_2 : NH_3H_2O = 20 : 1 : 1$ for 15 mins. After washing with deionized water, the PMMA/graphene layer is then transferred onto SiO_2 substrate. After soft baking at 150 °C to promote graphene–substrate adhesion, the PMMA support is dissolved in acetone to yield a graphene film on the substrate. We find that this process works very well, leaving only a trace amount of PMMA residue on the graphene surface.

The second way is the "stamp method", *i.e.* picking up graphene by an elastomeric material, such as polydimethylsiloxane (PDMS) (Figure 1.4(a)). Once the growth substrate is etched away, the stamp/graphene is delivered to the target substrate. The difference from the "carrier film" method is the way to remove the stamp. Here, the stamp is removed by mechanical detachment, leaving the graphene behind with the target substrate.[42] We note that the stamp method works only if the adhesion energy of the graphene–substrate interface is stronger than the graphene stamp, restricting its usefulness to flat and hydrophilic substrates.

The third approach is the "self-release transfer method", which is more or less a combination of the first and second methods (Figure 1.4(a)).[43] In this method, a specific polymer film (self-release layer) is first spun-cast over the graphene. An elastomeric stamp is then placed in conformal contact with the self-release layer. The growth metal substrate is etched away to leave the graphene/self-release layer on the elastomeric stamp. Then graphene is brought into contact with the target substrate by stamping and the stamp is removed mechanically. Finally, the self-release polymer is dissolved under mild conditions in a suitable solvent. A number of possible release polymers have been reported, including polystyrene (PS), poly(isobutylene) (PIB) and Teflon AF (poly[4,5-difluoro-2,2-bis(trifluoromethyl)-1,3-dioxole-co-tetra-fluoroethylene]). These belong to the class of aromatic hydrocarbon polymers, aliphatic hydrocarbon polymers and fluorocarbon polymers, respectively, which together provide sufficient diversity for solvents to be found that are orthogonal to (that is, do not swell or dissolve) features on the destination substrate. This method was found to afford much cleaner graphene than the PMMA carrier film (Figure 1.4(c)). It also enables the possibility of transferring graphene onto a variety of fragile/soft substrates.[43]

Besides the transfer of CVD graphene, it is also worth discussing the transfer of mechanically exfoliated graphene onto other 2D materials. Recently, vertical heterostructures of graphene and other 2D materials have attracted a lot of attention for novel electronic and optoelectronic devices.[44] As an example, graphene on hBN has been shown to exhibit much higher mobility than on SiO_2 and many interesting quantum phenomena because hBN has little dangling bonds or roughness to disturb intrinsic graphene (Figure 1.5(c)).[16,45–48] We will discuss the transfer process for such heterostructures (Figure 1.5(a)).[16] Fabrication begins with the mechanical exfoliation of hBN crystals onto SiO_2/Si substrates. Graphene is exfoliated separately onto a polymer stack consisting of a water-soluble layer and PMMA on Si substrate. The PMMA thickness is precisely tuned to allow the identification of monolayer graphene by optical means. The substrate was floated on the surface of a deionized water bath. Once the water-soluble polymer has dissolved, the Si substrate sinks to the bottom of the bath, leaving the extremely hydrophobic PMMA floating on the surface. The PMMA/graphene membrane is adhered to a glass transfer slide, which is clamped onto the arm of a micromanipulator mounted under an optical

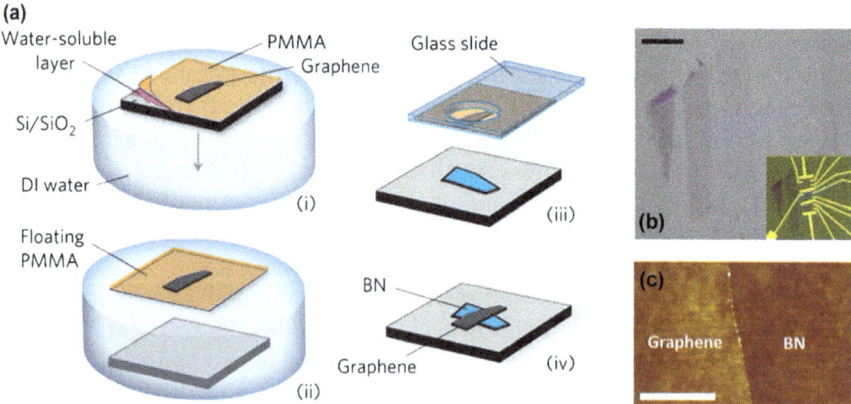

Figure 1.5   Transfer of graphene on to hBN substrate. (a) Schematic drawing of the transfer process. (b) Optical image of graphene transferred on hBN. Inset shows the same piece after device fabrication. (c) AFM of graphene on hBN showing ultraflat surface.
Adapted from Ref. 16.

microscope. The graphene is precisely aligned to the target hBN under the microscope and brought into contact (Figure 1.5(b)). During transfer, the target substrate is heated to 110 °C in an effort to drive off any water adsorbed on the surface of the graphene or hBN flakes, as well as to promote good adhesion of the PMMA to the target substrate. Once transferred, the PMMA is dissolved in acetone, resulting in the vertical heterojunctions with graphene and hBN.[16] Finally metal electrodes can be made by EBL (electron beam lithography) and evaporation. More complex stacks with multiple-layer devices can be transferred sequentially in a layer-by-layer fashion.[49]

1.4 Fabrication Techniques of Graphene Nanostructures

Like many other materials, the nanostructures of graphene can exhibit very different properties from their mother material. They attract a lot of attention not only in fundamental science but also in many applications. For example, solution processed graphene sheets (also referred to as chemically-reduced graphene) usually come in micrometer size or even smaller, in order to be solvent soluble and synthesized in bulk quantities,[50] which is necessary for applications such as batteries and supercapacitors.[26–29]

This section focuses on the fabrication techniques of a particular class of nanostructures: graphene nanoribbons (GNRs). Although graphene is a promising candidate for next generation electronics, the unique electronic structure is, however, not ideal for logic device applications. The primary concern is the absence of a bandgap, so at finite temperature, electrons are thermally excited to the conduction band. As a result, graphene transistors usually have an on/off ratio lower than 10 at room temperature, far below the requirement for logic applications. Among several approaches to create a gap in graphene, quantum confinement by nanostructures is most attractive due to the size and tunability of bandgap. The simplest quantum confined structure of graphene is the one-dimensional GNR.

Theoretical investigations of GNRs started as early as 1996. Nakada *et al.* used tight-binding calculations to show that the band structure depends highly on the orientation and width of a GNR.[51] In particular, GNRs with armchair edges (Figure 1.6(a)) could be either metallic or semiconducting, depending on the number of repeating unit cells across the width direction (Figure 1.6(b)), whereas GNRs with zigzag edges (Figure 1.6(d)) are always metallic due to localized edge states (Figure 1.6(e)). This first order picture is further refined by first-principle calculations, which include edge termination groups and other subtle edge effects.[52,53] It was shown that hydrogen terminated armchair and zigzag GNRs always have non-zero direct bandgaps, albeit for different reasons. For armchair ribbons, the bandgap opens up due to the combinational effect of quantum confinement and shortened interatomic distance (thus increased hopping integral) at the edges. For zigzag edges, the bandgap opens up because of a staggered sublattice

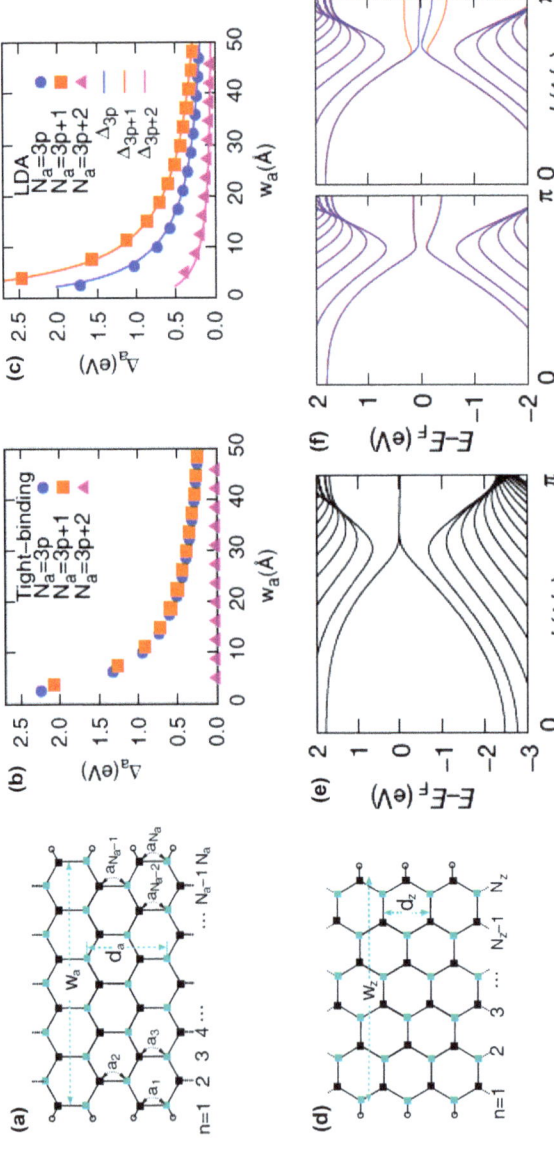

Figure 1.6 (a) Schematic of a 11-AGNR. The empty circles denote hydrogen atoms passivating the edge carbon atoms, and the black and gray rectangles represent atomic sites belonging to different sublattice in the graphene structure. The 1D unit cell distance and ribbon width are represented by d_a and w_a, respectively. The carbon-carbon distances on the nth dimer line are denoted by a_n. (b) The variation of band gaps of N_a-AGNRs as a function of width (w_a) obtained from TB calculations with $t = 2.70$ (eV). (c) The variation of band gaps of N_a-AGNRs as a function of width (w_a) obtained from first-principles calculations (symbols). The solid lines in (c) are from analytical expressions in Ref. 52. (d) Schematic of a 6-ZGNR. The empty circles and rectangles follow the same convention described in (a). The 1D unit cell distance and the ribbon width are denoted by d_z and w_z, respectively. (e) The spin-unpolarized band structure of a 16-ZGNR. And the Fermi energy (E_F) is set to zero. (f) From left to right, the spin-resolved band structures of a 16-ZGNR with an external electric field of 0 and 0.05 VÅ^{-1}, respectively. The red and blue lines denote bands of a-spin and b-spin states, respectively. And the Fermi energy (E_F) is set to zero. (a)–(d), adapted from Ref. 52. (e)–(f), adapted from Ref. 54.

potential on the hexagonal lattice due to edge magnetization. For both cases, the gaps are roughly inversely proportional to the width of the GNRs (Figure 1.6(c)).[52] More interestingly, the localized state in zigzag GNRs could be antiferromagnetic (*i.e.* with opposite spin directions). In such a system, one could achieve half metals with an electric field across the width direction, as one can close the gap for one spin while increasing the gap for the other (Figure 1.6(f)).[54]

Although GNRs offer an appealing approach to engineer the bandgap, they are non-trivial to synthesize experimentally. As a rule of thumb, $E_g \approx 1/w(\text{nm})$ eV in GNRs (Figure 1.6(c)).[55] So in order to achieve a 1 eV gap, the width of the ribbon has to be of the order of 1 nm, which is extremely challenging to make. Another issue is the edge structure. Most of the appealing properties of GNRs rely on atomically well-defined edges. However, most experimental techniques to produce GNRs do not have such capability, with the only exception being the bottom-up synthesis from aromatic precursors (Figure 1.9(c)).[56] Next we review the major approaches to synthesize GNRs and related nanostructures.

1.4.1 Top-down Etching of Graphene Nanostructures

One of the most common ways to make GNRs is by top-down plasma etching. In such a process, one uses resists,[57,58] metals[59] or nanowires[60] as masks and plasma to etch away the exposed graphene regions (Figure 1.7(a)). Han *et al.* spin coat negative-tone resist hydrogen silsesquioxane (HSQ) on graphene and expose line patterns by EBL.[57] After development, only the exposed parts remain, which protect the underlying GNRs from being etched. In this way, they fabricate GNRs down to sub-20 nm (Figure 1.7(b)). However, since it is difficult to remove the exposed HSQ, careful characterization of GNR width is not possible.

In a different approach, Wang *et al.* used positive tone resist poly(methyl methacrylate (PMMA) to fabricate GNR arrays.[59] They first exposed a single pixel line on PMMA, followed by a cold development to ensure the sharp edge profile for the ≈20 nm-wide trenches. Then a thin layer of Al was evaporated as the etching mask (Figure 1.7(a)). After a brief plasma etching to remove the unprotected graphene, Al lines are removed by dilute KOH solution. The GNRs obtained by this method can be as narrow as ≈20 nm (Figure 1.7(c)), limited by the proximity effect during EBL.

For device applications, it is highly desirable to make dense arrays of GNRs to achieve high drive current. However, the density is limited by resolution in optical lithography and by scaling up in EBL. Jiao *et al.* developed a process that used self-assembled diblock copolymer as the etching mask to fabricate large scale dense GNR arrays.[61] First, a poly(styrene-*b*-dimethylsiloxane) (PS-PDMS) block copolymer is spin-coated onto a gold film on a Si substrate. The substrate is then solvent-annealed in toluene vapor to allow the BCP to self-assemble into a monolayer of parallel cylinders of PDMS in a PS matrix with a PDMS top layer. After that, CHF_3 and O_2

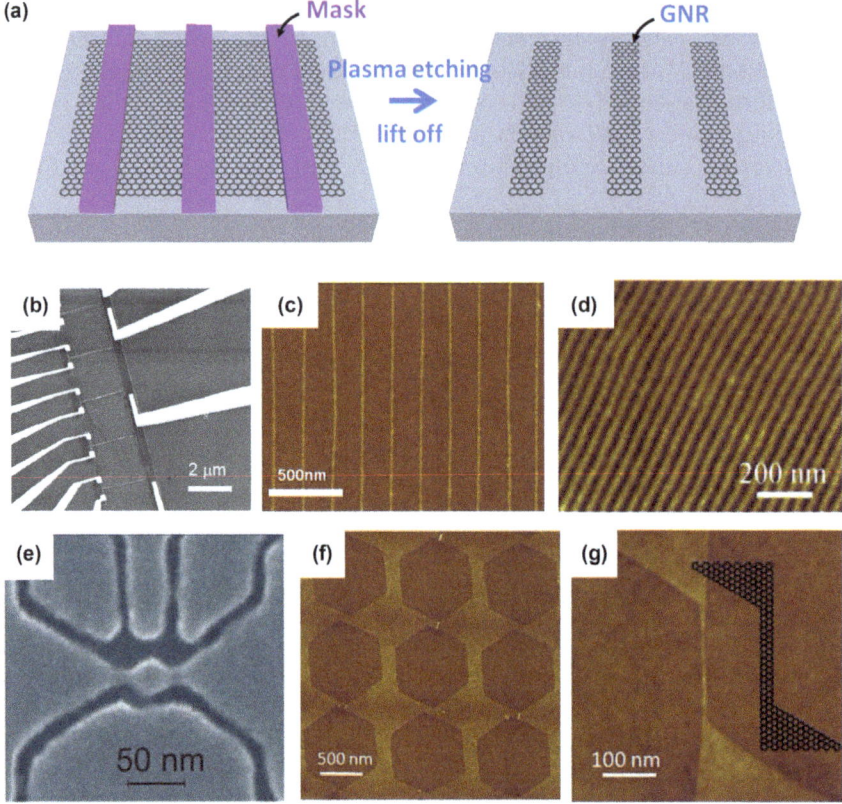

Figure 1.7   Graphene nanostructures made by plasma etching. (a) Schematics of
plasma etching process for GNRs. (b)–(g) Various graphene nanostructures including GNRs and quantum dots fabricated by different approaches.
(a), (c), adapted from Ref. 59. (b), adapted from Ref. 57. (d), adapted
from Ref. 61. (e), adapted from Ref. 62. (f), (g), adapted from Ref. 64.

plasma are used to remove the top PDMS surface layer and PS matrix, respectively, leaving parallel oxidized PDMS lines on the gold film. The gold
film is peeled-off from the underlying substrate by Scotch tape with a square
window. After dissolving the supporting gold film in a KI/I_2 solution, the
densely packed, oxidized PDMS lines are placed on top of graphene as the
etching mask. A drop of ethanol is dried on the mask to promote contact
between the oxidized PDMS film and graphene. Finally, a mild Ar plasma is
applied to transfer the pattern from the mask to the underlying graphene.
The oxidized PDMS patterns are successfully transferred to GNRs, with
12 nm width and 35 nm pitch (Figure 1.7(d)).

Other materials can also act as masks for physical etching. Bai *et al.* used
nanowires as etching masks and was able to push the narrowest ribbon
width to ≈ 8 nm.[60] However, due to the difficulty in controlling the positions
of the nanowire mask, this method cannot be scaled up. In addition to

GNRs, more complex nanostructures can be fabricated as well. Figure 1.7(e) shows a graphene quantum dot (QD) surrounded by source, drain and several gates, all made on graphene.[62] This particular structure was fabricated by EBL with PMMA resist and plasma etching. QDs of ≈ 100 nm in size act as single-electron transistors. These devices demonstrate the possibility of molecular-scale electronics based on graphene.

The plasma etching approaches already discussed all require an etching mask. However, this is not always true. In some cases, the etching can selectively start from the defects/edges of graphene, and proceed along certain crystallographic directions to give nanostructures, without any mask. For example, Wang *et al.* develop a gas-phase chemistry to narrow graphene and GNRs by heating them in a mixture of NH_3 and trace amount of O_2 to 800 °C.[59] Interestingly, such a process could only etch graphene from the edges without creating defects in the basal plane, due to the higher chemical reactivity of the edges. The etching rate could be as low as 1 nm min^{-1}, making it possible to control the width of GNRs in the nanometer region. Using a combined two-step approach (EBL + gas-phase chemistry), GNRs down to 4 nm (limited by the roughness of the starting GNRs) are successfully fabricated, which is the record of all lithographic GNRs.[63] Another example is H_2 plasma etching. Shi *et al.* found that graphene that had undergone a mild remote H_2 plasma at 450 °C could be etched from defects and could proceed in zigzag directions, giving rise to hexagonal holes with zigzag edges (Figure 1.7(f)).[64] Further small holes are patterned on the graphene as the seed for H_2 plasma etching. As the neighboring hexagons approach each other, GNRs with zigzag edges can be readily fabricated (Figure 1.7(g)). Such an example highlights the advantage of combining physical and chemical etching to obtain graphene nanostructures.

Another graphene nanostructure that is closely related to GNR is the nanomesh. A graphene nanomesh is a network of GNRs interconnected in a 2D manner (Figure 1.8), equivalent of a densely packed GNR array.[65] Similar to GNRs, the bandgap is determined by the narrowest neck width. Several methods have been demonstrated to etch graphene nanomesh, including block copolymer lithography and anodic aluminium oxide (AAO) templates. In particular, Bai *et al.* used poly(styrene-*b*-methyl methacrylate) (P(S-*b*-MMA)) block copolymer to produce hexagonal patterns on graphene followed by plasma etching to create the nanomesh (Figure 1.8(a)).[65] First graphene is mechanically exfoliated on a SiO_2/Si substrate. Then a 10 nm thick silicon oxide (SiO_x) film is evaporated onto graphene as the protecting layer and also as the grafting substrate for the subsequent block copolymer nanopatterning. The P(S-*b*-MMA) block copolymer thin film is spun on the surface and annealed to form cylindrical PMMA domains that are packed hexagonally within the polystyrene matrix. The annealed film is exposed to 295 nm UV irradiation to degrade the PMMA domains, which are removed by glacial acid, leaving the polystyrene on SiO_x as an etching mask. CHF_3 plasma is then used to punch holes into the evaporated SiO_x to expose the underlying graphene layer. Additional O_2 plasma is used to completely etch

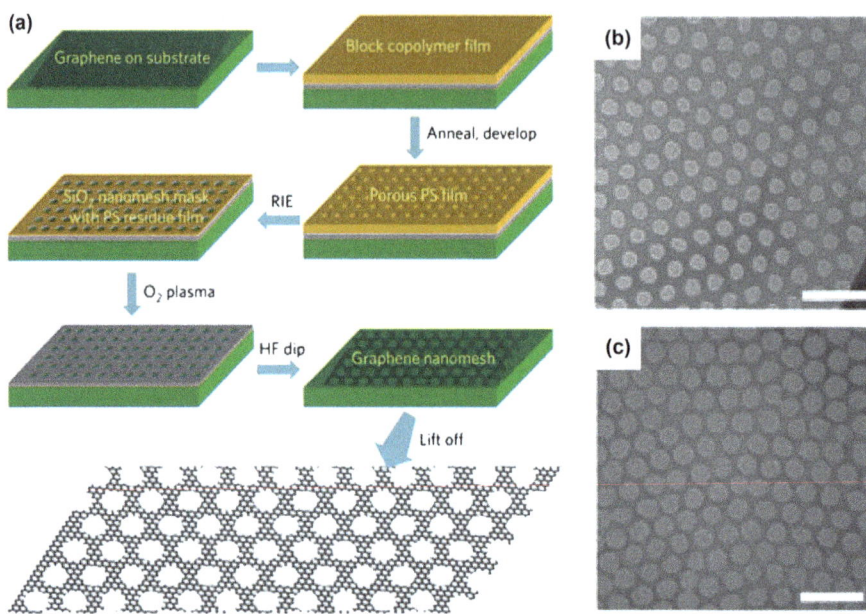

Figure 1.8   Fabrication of graphene nanomesh. (a) Schematics of the fabrication process by copolymer lithography. (b), (c) TEM images of graphene nanomesh with average neck width of 11.2 nm and 7.1 nm, respectively. Adapted from Ref. 65.

away the exposed region of graphene and the evaporated SiO_x is removed by HF (Figure 1.8(a)). The width of the neck can be engineered by the time of the final etching step to be $\approx$5–15 nm (Figure 1.8(b, c)).

In another approach, Zeng *et al.* used AAO as a template with hexagonally arranged holes to create similar nanomesh patterns on graphene, with a neck width as small as 15 nm.[66] They spin cast a thin layer of PMMA as an adhesion layer between graphene and the AAO template to achieve uniform mesh structure over a large area. O_2 plasma is used to etch the exposed PMMA and graphene. Finally, the AAO and PMMA are removed by NaOH and acetone, respectively. We note that although the two methods successfully demonstrated the advantages of graphene nanomesh, a technologically relevant approach, which requires low cost, minimal processing and the ability to produce sub-10 nm nanomesh over wafer scale, still remains to be developed.

1.4.2   Chemical Synthesis of Graphene Nanostructures

One of the biggest drawbacks of graphene nanostructures made by top-down etching is the poor edge quality. During the etching, it is inevitable to break the chemical bonds by the bombardment of energetic species, leaving behind rough edges with little control over the orientation (except the mild H₂ plasma discussed in the previous section). The rough edges can severely degrade the electronic properties, especially for sub-20 nm nanostructures. For example, edge defects can introduce a transport gap (different from the confinement gap) in etched GNRs.[67] At low temperatures, such a transport gap caused electron localization and hopping transport near the Dirac point, which was highly undesirable for device applications. On the other hand, chemistry could offer new possibilities in the fabrication of graphene nanostructures. With careful control, chemists could now produce graphene nanostructures down to true nanometer scale with well-defined edges, opening up a new paradigm to investigate the properties and device applications of these nanostructures.

An early chemical approach to synthesize GNRs is based on sonication, developed by Dai and colleagues (Figure 1.9(a)).[55] They start by exfoliating commercial expandable graphite at 1000 °C in forming gas to exfoliate the graphite into a number of few-layer graphenes or even single-layer graphene, as evidenced by the huge volume expansion. They then sonicate the resulting exfoliated material in a 1,2-dichloroethane (DCE) solution of poly(*m*-phenylenevinylene-co-2,5-dioctoxy-*p*-phenylenevinylene) (PmPV) (0.1 mg mL⁻¹) to disperse and break up the graphene into small pieces. A final centrifugation of the suspension is needed to remove large/thick graphene pieces and to retain the GNRs (together with small sheets) in the supernatant. Surprisingly, such a sonication approach could produce GNRs with very high aspect ratio over 1000. Semiconducting GNRs as narrow as ≈2 nm are synthesized for the first time (Figure 1.9(b)) with a bandgap of up to 400 meV. They also observe many other interesting nanostructures such as GNRs with varying width and bent several times. The bandgap of these GNRs is due to quantum confinement, rather than the transport gap associated with defects. Such a difference clearly highlights the advantage of chemical methods in producing high-quality graphene nanostructures.

The next example of a chemical approach represents an extreme case of bottom-up molecular assembly. Since graphene nanostructures can be considered as big aromatic molecules, it is possible to self-assemble them from various small aromatic molecules as basic building blocks. In particular, Cai *et al.* demonstrated a bottom-up synthesis of N = 7 armchair GNRs from the 10,109-dibromo-9,99-bianthryl precursor monomers *via* a two-step annealing on the Au(111) surface (Figure 1.9(c)).[56] In the first annealing, the dehalogenated molecules undergo a radical addition reaction to form linear polymer chains (Figure 1.9(d)). In the second annealing, a surface mediated cyclodehydrogenation reaction establishes the full aromatic GNRs. A scanning tunneling microscope image shows a perfect match between the final product and the simulation of N = 7 armchair GNRs. The GNR structure is atomically well-defined, and is completely determined by the precursor monomer. Other shapes of GNRs such as chevron-type can also be synthesized with different precursors. It provides a precise way of

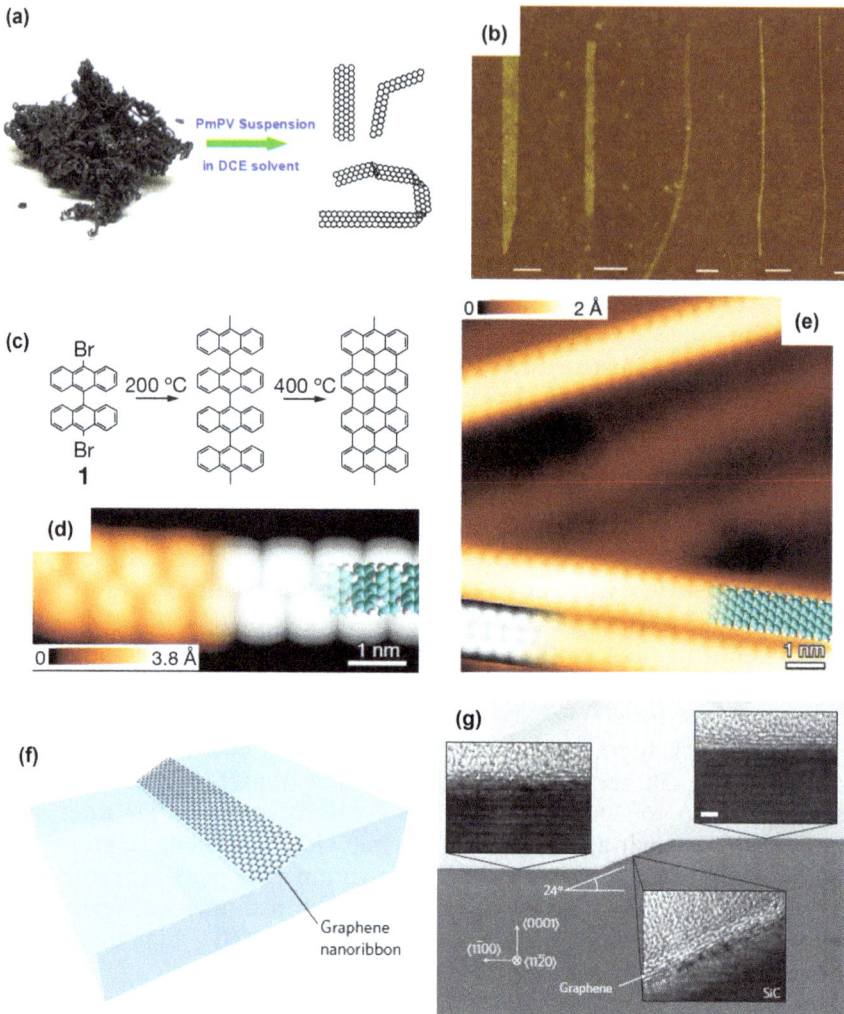

Figure 1.9   Chemically derived GNRs. (a) Schematic drawing of the fabrication of GNRs by sonication of expandable graphite. (b) AFM images of GNRs with different widths made by the sonication method. (c) Reaction scheme from precursor 1 to straight N = 7 GNRs. (d) STM image taken after surface assisted C–C coupling at 200 °C, but before the final cyclodehydrogenation step, showing a polyanthrylene chain, and DFT-based simulation of the STM image (right) with partially overlaid model of the polymer (blue, carbon; white, hydrogen). (e) High-resolution STM image with partly overlaid molecular model (blue) of the ribbon. At the bottom left is a DFT-based STM simulation of then = 7 ribbon shown as a grayscale image. (f) Schematic drawing of epitaxial GNR growth on patterned SiC step. (g) High-resolution TEM cross-sectional images of a step on (0001) confirm preferential growth on the ($1\bar{1}0n$) facet. Scale bar, 2 nm (and for all insets).
(a), (b), adapted from Ref. 55. (c)–(e), adapted from Ref. 56. (f), (g), adapted from Ref. 68.

engineering graphene nanostructures, which opens up the possibility of designing molecular devices from the bottom-up.

The previous two chemical approaches, although they can make narrow and precise nanostructures, have the intrinsic limit of scaling because neither method could control the position of the nanostructures. For real applications, it is highly desirable to place the target structure where we want it. In this respect, Sprinkle *et al.* developped a patterned growth of GNRs based on an epitaxial technique on SiC substrates.[68] They took advantage of the fact that epitaxial growth of graphene starts first along ($1\bar{1}0n$) facets. Controlled facets are achieved by the photolithographic definition of nickel lines as etching masks on a SiC substrate perpendicular to the ($1\bar{1}00$) direction. After a fluorine-based reactive ion etch, the exposed SiC is etched down with an etch depth of nanometer precision. After removal of the nickel mask and cleaning, the SiC is heated to 1200–1300 °C at an intermediate vacuum (10^{-4} torr) for 30 min to let the abrupt step relax to a ($1\bar{1}10n$) facet. Then the temperature is elevated to 1450 °C within 1.5 min and maintained for 10 min for graphene growth before cooling down (Figure 1.9(f)). Cross-sectional TEM images confirm the growth of GNRs on the tilted nanofacets, while the horizontal (0001) plane is only partially covered (Figure 1.9(g)). The 24° tilt angle indicates the nanofacet along the ($1\bar{1}08$) direction. The GNR width is determined by the etch depth. In their work, etch depths of 20 nm are readily achieved resulting in 40 nm-wide GNR arrays on wafer scale.

1.4.3 Graphene Nanoribbons from Unzipping Carbon Nanotubes

GNRs are structurally related to carbon nanotubes (CNTs), which can be viewed as rolled-up GNRs with little defects. The synthesis and separation of CNTs are actively researched for 20 years now, with great success to control the chirality and metallic/semiconducting nature.[69,70] Therefore, if one can develop a reliable approach to longitudinally unzip CNTs, synthesis of a large quantity of GNRs with well-defined chirality may be possible. Over the past few years, several methods have been successfully demonstrated in this direction, including metal nanoparticle cutting, oxidation and reduction, plasma etching and sonication (Figure 1.10). These different approaches are reviewed here.

Since CNTs are structurally very stable, one needs to introduce defects to open them. Kosynkin *et al.* reported a solution based oxidation process to open multiwalled CNTs (MWCNTs).[71] They suspend MWCNTs in concentrated H_2SO_4 for 1–12 hours and then treated them with 500 wt% $KMnO_4$. The mixture is stirred at room temperature for 1 hour and then heated to 55–70 °C for an additional hour. When all of the $KMnO_4$ have been consumed, the reaction is quenched by pouring over ice containing a small amount of H_2O_2. The solution is filtered over a polytetrafluoroethylene membrane, and the remaining solid is washed with acidic water followed by

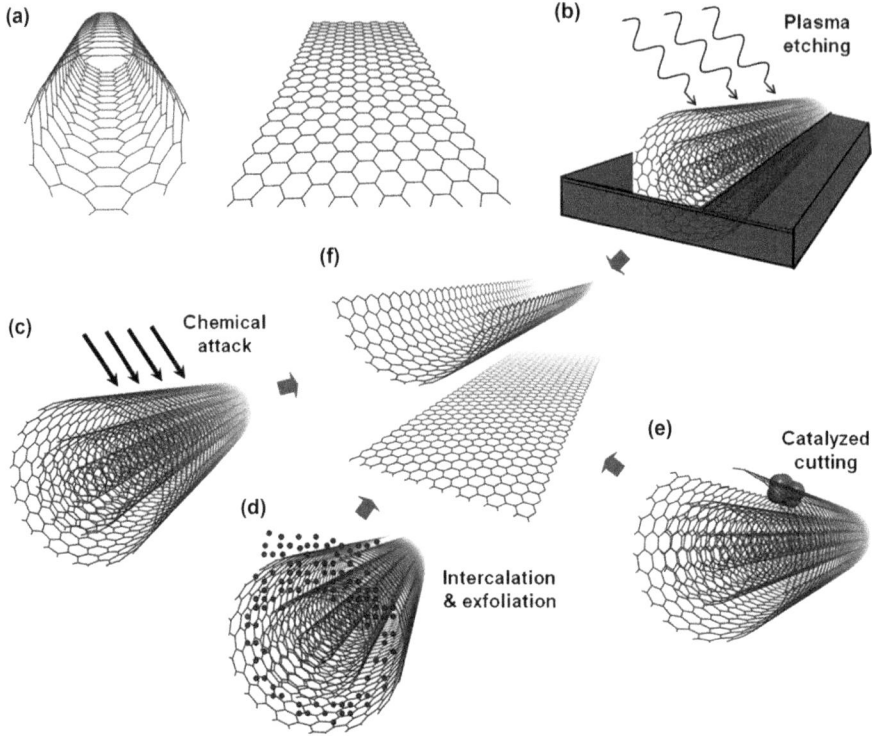

Figure 1.10   Schematic drawing of various methods to fabricate GNRs by unzipping CNTs.
Adapted from Ref. 94.

ethanol/ether. TEM and AFM characterizations both illustrate the formation
of GNRs, with little MWCNTs remaining (Figure 1.11(a, b)). Since the original MWCNTs are 40–80 nm in diameter, the resulting GNRs could be over
100 nm wide. Since a strong oxidation process is employed in the synthesis,
the raw GNRs are highly oxidized with many oxygen groups like carboxyls
and hydroxyls. These functionalities make the GNRs very insulating. In order
to improve the material quality, they further chemically reduce the as-made
GNRs in an aqueous solution of ammonium hydroxide and hydrazine
monohydrate. Attenuated-total-reflection infrared spectroscopy and X-ray
photoemission spectroscopy confirm the significant reduction of oxygen
content from 42% to 16%, which is still higher than the original MWCNTs,
most likely due to the edge carboxylic acid moieties.

Another unzipping approach based on plasma etching is developed in
parallel with the above work by Dai and colleagues.[72] Here the controllability
is crucial because CNTs are easily etched away by plasma. Elegantly, they
embed the MWCNTs in PMMA polymer matrix to partially protect the CNTs.
A mild Ar plasma is used to etch the MWCNT/PMMA matrix. The etch rate of
CNTs is significantly decreased by the PMMA making it possible to control

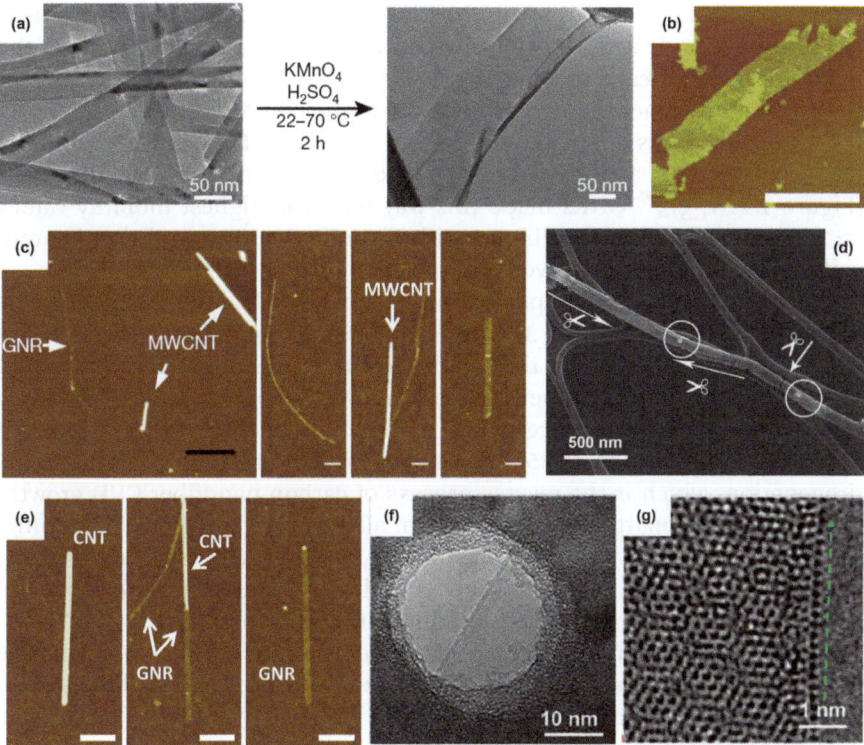

Figure 1.11 Fabrication of GNRs by unzipping CNTs. (a) TEM images of MWCNTs before and after unzipping by the oxidation process. The experimental conditions are shown. (b) AFM image of a GNR made by oxidation unzipping process. (c) AFM images of GNRs made by plasma etching after depositing on substrate. Some MWCNTs are also observed. (d) SEM image of MWCNTs cut by metal nanoparticles (highlighted by the circles). (e) AFM images of GNRs made by sonication unzipping of MWCNTs. (f) TEM image of a GNR made by sonication unzipping MWCNT with a smooth edge. (g) High-resolution TEM image of the edge.

(a), (b), adapted from Ref. 71. (c), adapted from Ref. 72. (d), adapted from Ref. 76. (e), adapted from Ref. 73. (f), (g), adapted from Ref. 95.

the layer-by-layer etching process. Most of the ribbons obtained are within a narrow distribution of 10–20 nm in width and 1–3 layers in thickness (Figure 1.11(c)). The quality of the GNRs is higher than those produced by oxidation-reduction because the plane of the GNRs is undamaged in this process. Raman measurements also confirm the high quality with low defect peak. The GNRs are narrow enough to afford FETs with an on/off ratio up to ≈ 100.

Further improvement of GNR quality is achieved by using a sonication based method by the same group (Figure 1.4(e)).[73] First, the MWCNTs are mildly calcined in air to etch away the impurities and oxidize the CNTs from the defects. Then a sonication step in DCE with PmPV surfactant is employed to longitudinally unzip the CNTs with high efficiency, which likely starts from the defects. Finally ultracentrifugation is used to remove the bulky pieces and leave most of the GNRs in the supernatant. The yield of GNRs is much higher than the sonication approach on expandable graphite. Most of the ribbons are below 20 nm and 1–3 layers (Figure 1.11(e)). A TEM study reveals that a portion of the GNRs have atomically smooth edges (Figure 1.11(f, g)).[95] GNRs made this way have the highest mobility values (≈ 1500 cm^2 Vs^{-1}) among ribbons with a similar width,[74] and the theoretically predicted edge states were observed by STM for the first time.[75]

CNTs could also be unzipped by metal nanoparticles. Elías *et al.* disperse MWCNTs in a solution of 3% wt CoCl$_2$ in methanol by high power tip sonication for 1 min.[76] After the solution is dried on Si substrate, it is heated to 500 °C for 1 hour in a tube furnace to allow a proper nucleation of the Co nanoparticles on the surface of the MWCNTs. The catalytic unzipping of MWCNTs is performed at 850 °C for 30 min in a mixture of Ar/H$_2$ by hydrogenation, which is the reverse process of carbon nanofiber CVD growth. Other metal nanoparticles such as Ni can also be used to achieve unzipping. Figure 1.11(d) clearly shows the cuts by the nanoparticles, creating partially open CNTs. However, cuts are observed both along and perpendicular to the axes of CNTs making it difficult to control the unzipping.

Having compared many physical and chemical approaches to synthesize graphene nanostructures, we can clearly see that each method has its own pros and cons. Generally, physical etching has the advantage of large scale patterning capability, which is required for device applications. However, the poor edge quality can seriously affect the electrical properties of GNRs, especially at nanometer regime. Chemical methods, on the other hand, can produce GNR with true nanometer width, high quality, and even well-defined edges. Yet they are not able to pattern the nanostructures for integration. At this moment, a technically scalable method to produce atomically well-defined graphene nanostructures still remains to be developed. This area is still under active research.

1.4.4   Crystallographic Cutting of Graphene

Metal nanoparticles can cut not only CNTs, but graphene under proper experimental conditions. In a H$_2$-rich environment, graphene can be hydrogenated in the presence of a metal nanoparticle catalyst, which has been reported by several groups.[77-79] Figure 1.12(a) shows the schematic of the reaction.[77] In a typical process, graphene is exfoliated on a SiO$_2$/Si substrate. An aqueous solution of NiCl$_2$ (2.4 mg mL^{-1}) is spun onto the substrate surface and then baked for 10 min at 90 °C on a hot plate to evaporate the H$_2$O. The NiCl$_2$ treated sample is then submitted to a two-step process under Ar : H$_2$ flow (850 : 150 sccm): annealing at 500 °C for 20 min to form Ni nanoparticles, and then cutting happens at 1000 °C.

After the cutting, trenches with well-defined angles are observed by AFM, indicating that the cutting is highly anisotropic (Figure 1.12(b, c)). Therefore

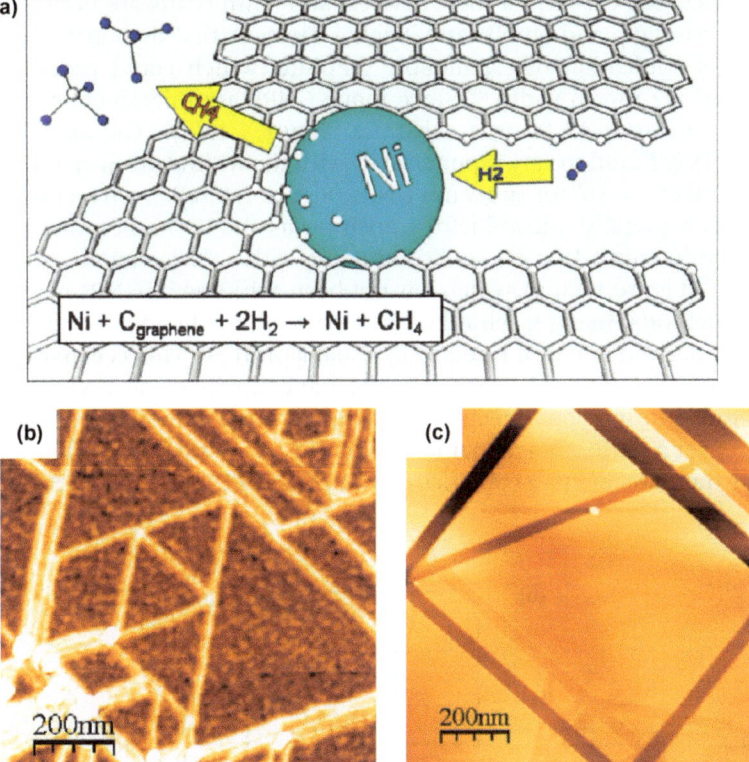

Figure 1.12 Anisotropic cutting of graphene. (a) Schematic drawing of graphene cutting by metal nanoparticles. (b), (c) AFM of graphene and graphite after cutting respectively.
Adapted from Ref. 77.

the cutting method could be used to synthesize graphene nanostructures with the ability to control edges. For single-layer graphene, the angles between the trenches are predominantly 60° and 120° (Figure 1.12(b)). Both TEM studies of Ni nanoparticle etching[80] and ab initio calculations,[78] suggest that the zigzag edges are more favorable compared to armchair edges. For graphite however, 30°, 90° and 150° angles appear as well, suggesting the existence of both zigzag and armchair edges. When two nanoparticles are close together, sub-10 nm GNRs could be obtained.

1.5 Electronic Devices Based on Graphene Nanostructures

In the previous section, we introduced various ways of making graphene nanostructures. Now let us briefly discuss one of the most important applications of these nanostructures – electronic devices.

As discussed earlier, for logic devices, sub-5 nm GNRs are necessary for a sizable bandgap and on/off ratio. Such a device is first demonstrated by the chemical sonication of expandable graphite, which could produce GNRs down to ≈ 2 nm in width.[55] Since many GNRs are micrometers long, it is possible to make FETs on a substrate (Figure 1.13(a)). The on/off ratio of GNRFETs is found to vary exponentially with the widths, from ≈ 1 for 50 nm-wide GNRs to $\approx 10^6$ for sub-5 nm GNRs (Figure 1.13(c)). Such an observation suggests a metal-semiconductor transition as the GNRs become narrower. All the sub-5 nm GNRs are semiconducting with a device on/off ratio higher than 10^5 (Figure 1.13(c)), which has not been achieved by GNRs made by any other methods. Using a Schottky barrier model, the bandgap of these GNRs are also derived and well fitted by $E_g \approx 0.8/w(\text{nm})$ eV. This is consistent with theoretical calculations. Therefore the bandgap in these GNRs is due to quantum confinement effects. The GNRs derived from chemical sonication can deliver a performance that is similar to CNT transistors.[81] Figure 1.13(b) shows the transfer characteristics of a 2 nm GNRFET with 10 nm SiO_2 as gate

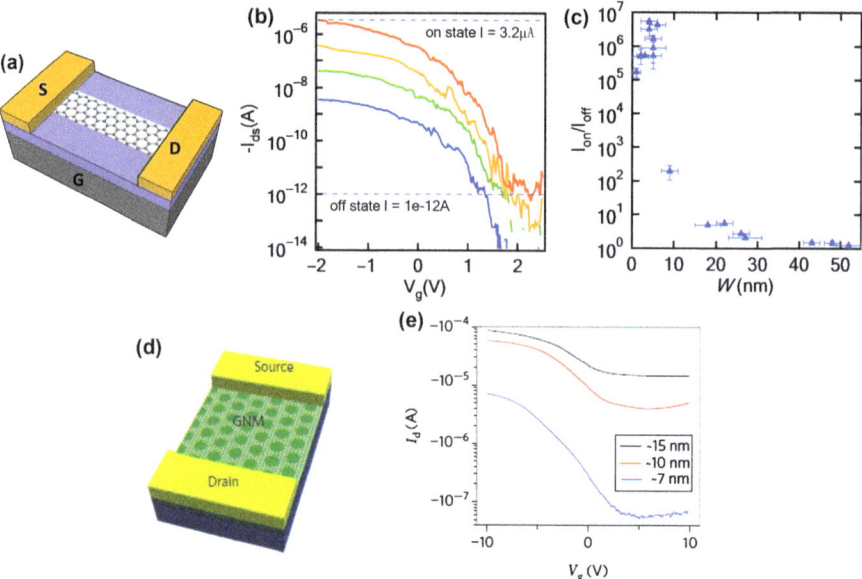

Figure 1.13 Electronic devices based on GNR and graphene nanomesh. (a) Schematics of backgated GNR FETs metal source/drain. Highly doped silicon is used as backgate. (b) Transfer characteristics (current vs gate voltage $I_{ds} - V_{gs}$) under various V_{ds} for a $w \approx 2 \pm 0.5$ nm, $L \approx 236$ nm GNRFET. (c) I_{on}/I_{off} ratios (under $V_{ds} = 0.5$ V) for GNRs of various ribbon widths. (d) Schematic of a graphene nanomesh FET. (e) Transfer characteristics at $V_d = -100$ mV for graphene nanomesh FET with different estimated neck widths of ≈ 15 nm, ≈ 10 nm and ≈ 7 nm. (a), (b), adapted from Ref. 81. (c), adapted from Ref. 55. (d), (e), adapted from Ref. 65.

dielectrics. The transistor delivers an over 10^6 on/off ratio and an ≈ 2 mA/μm on-state current at $V_{ds} = 1$ V. The mobility is limited by edge scattering to be less than 200 cm^2/Vs.

Similar to GNRs, the on/off ratio of nanomesh devices is also sensitive to the neck width. The highest on/off ratio of ≈ 100 is achieved at room temperature for the 7 nm-wide nanomesh, while that of the 15 nm nanomesh device is only less than 10 (Figure 1.14(d, e)).[65] However, a clear advantage of graphene nanomesh devices is the drive current. Since the density of the mesh is quite high, a single nanomesh device could deliver ≈ 100 times higher drive current than a single GNR device.

Other than logic transistor applications, graphene nanostructures are also interesting in quantum devices. For example, graphene QDs (Figure 1.7(e)) could act as single-electron transistors. Figure 1.14(a) plots the stability diagram for a graphene QD with diameter, D, <30 nm.[62] The Coulomb blockade diamonds are well-defined, indicating that the system is dominated by quantum confinement. The charge transport is suppressed when the energy levels are not in line with the Fermi energy of the leads. From the charging energy, ≈ 40 meV in Figure 1.14(a), D ≈ 15 nm can be estimated. Since graphene is chemically stable even down to a few benzene rings, such

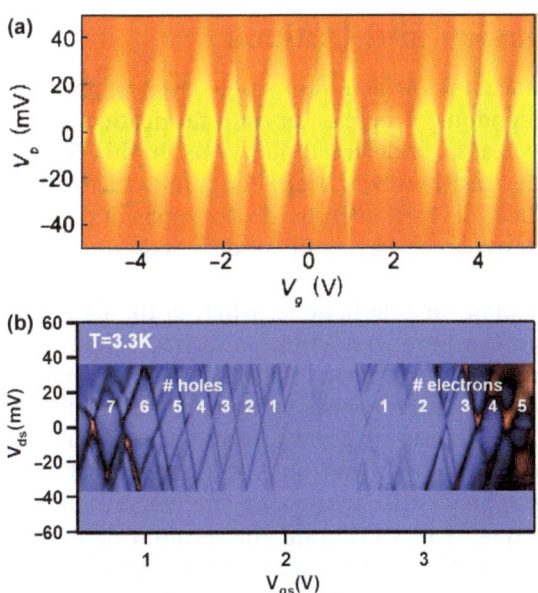

Figure 1.14   Quantum devices based on graphene nanostructures. (a) Coulomb blockade diamonds from a graphene QD with an estimated size of 15 nm. (b) Differential conductance near the bandgap of a high-quality GNR made by sonication unzipping MWCNTs, showing regular Coulomb blockade patterns and excited states. The number of electrons and holes in the QD are marked.
(a), adapted from Ref. 62. (b), adapted from Ref. 84.

single-electron transistors are suitable for the top-down approach to molecular electronics.

GNRs are also quantum confined systems, and therefore are interesting candidates for single-electron transistors or spintronic devices. As previously discussed, edges greatly affect the charge transport and device performances. Electron transport in plasma etched GNRs is dominated by defect states introduced by edge roughness.[67,82,83] At low temperature, those GNRs usually exhibit multiple QD behavior. However, dramatic change can be made by improving edges. For example, metallic behavior and signature of bandgap are observed in the high-quality GNRs derived from sonication unzipping CNTs, indicating that edge disorder is not dominant.[84] Compared to the etched ribbons, the unzipped ribbons act as a high-quality single QD at low temperature with well-defined Coulomb blockade, making it possible to count the number of electrons and holes (Figure 1.14(b)). Clear and regular excited states are in good agreement with theoretical calculations. Many other transport phenomena, such as the Fabry-Perot interference and Kondo effect, are also observed, suggesting the pristine nature of these GNRs despite the open edges. Therefore, these GNRs may have the potential as new types of quantum wires to explore the widely predicted magnetic edge states and realize novel spintronic devices.

1.6   Conclusion and Outlook

Since the first successful isolation of single-layer graphene in 2004, we have witnessed tremendous progress in the fabrication of graphene nanostructures, as well as their device applications. In this chapter, we only focus on electronic devices, however, graphene nanostructures are also very important for optical devices, such as phototransistors[24,85,86] and plasmonic devices.[87,88] Many areas are still under active research as we write this chapter.

The next question for graphene is: when is the first major application? Earlier this year, the European Union launched "Graphene" as one of the two Future Emerging Technology flagships. The purpose of this 10 year, 1000 million Euro project is to "take graphene and related layered materials from academic laboratories to society, revolutionize multiple industries and create economic growth and new jobs in Europe".[89] For industrial device applications, graphene still faces many challenges, such as the cost, controllability and quality of CVD graphene. From a nanofabrication point of view, we need to develop large-area, CMOS compatible processes that can pattern sub-10 nm features with controlled orientation, edge structure and passivation. In the end, we need to have uniform graphene nanostructures to ensure the device to device reproducibility for any applications. This is the ultimate goal of making graphene nanostructures. Although it appears very challenging, but 50 years ago, who would have known what it was going to be like today?

References

1. H. W. Kroto, A. W. Allaf and S. P. Balm, *Chem. Rev.*, 1991, **91**, 1213.
2. S. Iijima, *Nature*, 1991, **354**, 56.
3. A. K. Geim and K. S. Novoselov, *Nature Mater.*, 2007, **6**, 183.
4. Y. Gogotsi and V. Presser ed., *Carbon nanomaterials*. CRC Press, Taylor & Francis, 2010.
5. K. S. Novoselov, A. K. Geim, S. V. Morozov, D. Jiang, Y. Zhang, S. V. Dubonos, I. V. Grigorieva and A. A. Firsov, *Science*, 2004, **306**, 666.
6. S. D. Sarma, S. Adam, E. H. Hwang and E. Rossi, *Rev. Mod. Phys.*, 2011, **83**, 407.
7. A. H. C. Neto, F. Guinea, N. M. R. Peres, K. S. Novoselov and A. K. Geim, *Rev. Mod. Phys.*, 2009, **81**, 109.
8. P. R. Wallace, *Phys. Rev.*, 1947, **71**, 622.
9. K. S. A. Novoselov, A. K. Geim, S. V. Morozov, D. Jiang, M. I. Katsnelson, I. V. Grigorieva, S. V. Dubonos and A. A. Firsov, *Nature*, 2005, **438**, 197.
10. Y. Zhang, Y. W. Tan, H. L. Stormer and P. Kim, *Nature*, 2005, **438**, 201.
11. A. F. Young and P. Kim, *Nat. Phys.*, 2009, **5**, 222.
12. V. V. Cheianov, V. Fal'ko and B. L. Altshuler, *Science*, 2007, **315**, 1252.
13. J. A. del Alamo, *Nature*, 2011, **479**, 317.
14. E. H. Hwang and S. D. Sarma, *Phys. Rev. B*, 2008, 77, 115449.
15. K. I. Bolotin, K. J. Sikes, Z. Jiang, M. Klima, G. Fudenberg, J. Hone, P. Kim and H. L. Stormer, *Solid. State. Commun.*, 2008, **146**, 351.
16. C. R. Dean, A. F. Young, I. Meric, C. Lee, L. Wang, S. Sorgenfrei, K. Watanabe, T. Taniguchi, P. Kim, K. L. Shepard and J. Hone, *Nat. Nanotechnol.*, 2010, **5**, 722.
17. K. S. Novoselov, V. I. Fal, L. Colombo, P. R. Gellert, M. G. Schwab and K. Kim, *Nature*, 2012, **490**, 192.
18. W. Xin-Ran, S. Yi and Z. Rong, *Chinese Phys. B*, 2013.
19. F. Schwierz, *Nat. Nanotechnol.*, 2010, 5, 487.
20. J. H. Chen, C. Jang, S. Xiao, M. Ishigami and M. S. Fuhrer, *Nat. Nanotechnol.*, 2008, 3, 206.
21. R. R. Nair, P. Blake, A. N. Grigorenko, K. S. Novoselov, T. J. Booth, T. Stauber, N. M. R. Peres and A. K. Geim, *Science*, 2008, **320**, 1308.
22. C. Lee, X. Wei, J. W. Kysar and J. Hone, *Science*, 2008, **321**, 385.
23. F. Wang, Y. Zhang, C. Tian, C. Girit, A. Zettl, M. Crommie and Y. R. Shen, *Science*, 2008, **320**, 206.
24. F. Bonaccorso, Z. Sun, T. Hasan and A. C. Ferrari, *Nat. Photonics*, 2010, **4**, 611.
25. F. Schedin, A. K. Geim, S. V. Morozov, E. W. Hill, P. Blake, M. I. Katsnelson and K. S. Novoselov, *Nat. Mater.*, 2007, **6**, 652.
26. Y. Zhu, S. Murali, M. D. Stoller, K. J. Ganesh, W. Cai, P. J. Ferreira, A. Pirkle, R. M. Wallace, K. A. Cychosz, M. Thommes, D. Su, E. A. Stach and R. S. Ruoff, *Science*, 2011, **332**, 1537.
27. X. Yang, C. Cheng, Y. Wang, L. Qiu and D. Li, *Science*, 2013, **341**, 534.

28. L. H. Hu, F. Y. Wu, C. T. Lin, A. N. Khlobystov and L. J. Li, *Nat. Commun.*, 2013, **4**, 1687.
29. S. M. Paek, E. J. Yoo and I. Honma., *Nano Lett.*, 2008, **9**, 72.
30. J. R. Williams, L. DiCarlo and C. M. Marcus., *Science*, 2007, **317**, 638.
31. Y. Zhang, T. T. Tang, C. Girit, Z. Hao, M. C. Martin, A. Zettl, M. F. Crommie, Y. R. Shen and F. Wang, *Nature*, 2009, **459**, 820.
32. X. Wang, S. M. Tabakman and H. Dai., *J. Am. Chem. Soc.*, 2008, **130**, 8152.
33. S. Kim, J. Nah, I. Jo, D. Shahrjerdi, L. Colombo, Z. Yao, E. Tutuc and S. K. Banerjee, *Appl. Phys. Lett.*, 2009, **94**, 062107.
34. L. Liao, Y. C. Lin, M. Bao, R. Cheng, J. Bai, Y. Liu, Y. Qu, K. L. Wang, Y. Huang and X. Duan, *Nature*, 2010, **467**, 305.
35. R. Cheng, J. Bai, L. Liao, H. Zhou, Y. Chen, L. Liu, Y. Lin, S. Jiang, Y. Huang and X. Duan, *Proc. Nat. Acad. Sci.*, 2012, **109**, 11588.
36. K. I. Bolotin, F. Ghahari, M. D. Shulman, H. L. Stormer and P. Kim, *Nature*, 2009, **462**, 196.
37. H. Zhou, W. J. Yu, L. Liu L, R. Cheng, Y. Chen, X. Huang, Y. Liu, Y. Wang, Y. Huang and X. Duan, *Nat. Commun.*, 2013, 4.
38. S. Bae, H. Kim, Y. Lee, X. Xu, J. Park, Y. Zheng, J. Balakrishnan, T. Lei, H. R. Kim, Y. I. Song, Y. Kim, K. S. Kim, B. Özyilmaz, J. Ahn, B. H. Hong and S. Iijima, *Nat. Nanotechnol.*, 2010, **5**, 574.
39. J. Y. Choi, *Nat. Nanotechnol.*, 2013, **8**, 311.
40. X. Li, Y. Zhu, W. Cai, M. Borysiak, B. Han, D. Chen, R. D. Piner, L. Colombo and R. S. Ruoff, *Nano Lett.*, 2009, **9**, 4359.
41. X. Liang, B. A. Sperling, I. Calizo, G. Cheng, C. A. Hacker, Q. Zhang, Y. Obeng, K. Yan, H. Peng, Q. Li, X. Zhu, H. Yuan, A. R. H. Walker, Z. Liu, L. Peng and C. A. Richter, *ACS Nano.*, 2011, **5**, 9144.
42. X. Li, W. Cai, I. H. Jung, G. Cheng, C. A. Hacker, Q. Zhang, Y. Obeng, K. Yan, H. Peng, Q. Li, X. Zhu, H. Yuan, A. R. H. Walker, Z. Liu, L. Peng and C. A. Richter, *ECS Trans.*, 2009, **19**, 41.
43. J. Song, F. Y. Kam, R. Q. Png, W. Seah, J. Zhuo, G. Lim, P. K. H. Ho and L. Chua, *Nat. Nanotechnol.*, 2013, **8**, 356.
44. A. K. Geim and I. V. Grigorieva, *Nature*, 2013, **499**, 419.
45. C. R. Dean, L. Wang, P. Maher, C. Forsythe, F. Ghahari, Y. Gao, J. Katoch, M. Ishigami, P. Moon, M. Koshino, T. Taniguchi, K. Watanabe, K. L. Shepard, J. Hone and P. Kim, *Nature*, 2013, **497**, 598.
46. A. F. Young, C. R. Dean, L. Wang, H. Ren, P. Cadden-Zimansky, K. Watanabe, T. Taniguchi, J. Hone, K. L. Shepard and P. Kim, *Nat. Phys.*, 2012, **8**, 550.
47. B. Hunt, J. D. Sanchez-Yamagishi, A. F. Young, M. Yankowitz, B. J. LeRoy, K. Watanabe, T. Taniguchi, P. Moon, M. Koshino, P. Jarillo-Herrero and R. C. Ashoori, *Science*, 2013, **340**, 1427.
48. L. A. Ponomarenko, R. V. Gorbachev, G. L. Yu, D. C. Elias, R. Jalil, A. A. Patel, A. Mishchenko, A. S. Mayorov, C. R. Woods, J. R. Wallbank, M. Mucha-Kruczynski, B. A. Piot, M. Potemski, I. V. Grigorieva, K. S. Novoselov, F. Guinea, V. I. Fal'ko and A. K. Geim, *Nature*, 2013, **497**, 594.

49. R. V. Gorbachev, A. K. Geim, M. I. Katsnelson, K. S. Novoselov, T. Tudorovskiy, I. V. Grigorieva, A. H. MacDonald, S. V. Morozov, K. Watanabe, T. Taniguchi and L. A. Ponomarenko, *Nat. Phys.*, 2012, **8**, 896.
50. S. Park and R. S. Ruoff, *Nat. Nanotechnol.*, 2009, **4**, 217.
51. K. Nakada, M. Fujita, G. Dresselhaus and M. S. Dresselhaus, *Phys. Rev. B*, 1996, **54**, 17954.
52. Y. W. Son, M. L. Cohen and S. G. Louie, *Phys. Rev. Lett.*, 2006, **97**, 216803.
53. V. Barone, O. Hod and G. E. Scuseria, *Nano Lett.*, 2006, **6**, 2748.
54. Y. W. Son, M. L. Cohen and S. G. Louie, *Nature*, 2006, **444**, 347.
55. X. Li, X. Wang, L. Zhang, S. Lee and H. Dai, *Science*, 2008, **319**, 1229.
56. J. Cai, P. Ruffieux, R. Jaafar, M. Bieri, T. Braun, S. Blankenburg, M. Muoth, A. P. Seitsonen, M. Saleh, X. L. Feng, K. Mullen and R. Fasel, *Nature*, 2010, **466**, 470.
57. M. Y. Han, B. Özyilmaz, Y. Zhang and P. Kim, *Phys. Rev. Lett.*, 2007, **98**, 206805.
58. Z. Chen, Y. M. Lin, M. J. Rooks and P. Avouris, *Physica E Low Dimens. Syst. Nanostruct.*, 2007, **40**, 228.
59. X. Wang and H. Dai, *Nat. Chem.*, 2010, **2**, 661.
60. J. Bai, X. Duan and Y. Huang, *Nano Lett.*, 2009, **9**, 2083.
61. L. Jiao, L. Xie and H. Dai, *Nano Res.*, 2012, **5**, 292.
62. L. A. Ponomarenko, F. Schedin, M. I. Katsnelson, R. Yang, E. Hill, K. Novoselov and A. Geim, *Science*, 2008, **320**, 356.
63. C. A. Palma and P. Samorì, *Nat. Chem.*, 2011, **3**, 431.
64. Z. Shi, R. Yang, L. Zhang, Y. Wang, D. Liu, D. Shi, E. Wang and G. Zhang, *Adv. Mater.*, 2011, **23**, 3061.
65. J. Bai, X. Zhong, S. Jiang, Y. Huang and X. Duan, *Nat. Nanotechnol.*, 2010, **5**, 190.
66. Z. Zeng, X. Huang, Z. Yin, H. Li, Y. Chen, H. Li, Q. Zhang, J. Ma, F. Boey and H. Zhang, *Adv. Mater.*, 2012, **24**, 4138.
67. M. Y. Han, J. C. Brant and P. Kim, *Phys. Rev. Lett.*, 2010, **104**, 056801.
68. M. Sprinkle, M. Ruan, Y. Hu, J. Hankinson, M. Rubio-Roy, B. Zhang, X. Wu, C. Berger and W. A. de Heer, *Nat. Nanotechnol.*, 2010, **5**, 727.
69. R. C. Haddon, J. Sippel, A. G. Rinzler and F. Papadimitrakopoulos, *MRS Bull.*, 2004, **29**, 252.
70. H. Zhang, B. Wu, W. Hu and Y. Liu, *Chem. Soc. Rev.*, 2011, **40**, 1324.
71. D. V. Kosynkin, A. L. Higginbotham, A. Sinitskii, J. R. Lomeda, A. Dimiev, B. K. Price and J. M. Tour, *Nature*, 2009, **458**, 872.
72. L. Jiao, L. Zhang, X. Wang, G. Diankov and H. Dai, *Nature*, 2009, **458**, 877.
73. L. Jiao, X. Wang, G. Diankov, H. Wang and H. Dai, *Nat. Nanotechnol.*, 2010, **5**, 321.
74. X. Jia, J. Campos-Delgado, M. Terrones, V. Meunier and M. S. Dresselhaus, *Nanoscale*, 2011, **3**, 86.
75. C. Tao, L. Jiao, O. V. Yazyev, Y. Chen, J. Feng, X. Zhang, R. B. Capaz, J. M. Tour, A. Zettl, S. G. Louie, H. Dai and M. F. Crommie, *Nat. Phys.*, 2011, **7**, 616.

76. A. L. Elías, A. R. Botello-Méndez, D. Meneses-Rodríguez, V. J. González, D. Ramírez-González, L. Ci, E. Muñoz-Sandoval, P. M. Ajayan, H. Terrones and M. Terrones, *Nano Lett.*, 2009, **10**, 366.
77. L. C. Campos, V. R. Manfrinato, J. D. Sanchez-Yamagishi, J. Kong and P. Jarillo-Herrero, *Nano Lett.*, 2009, **9**, 2600.
78. L. Ci, Z. Xu, L. Wang, W. Gao, F. Ding, K. F. Kelly, B. I. Yakobson and P. M. Ajayan, *Nano Res.*, 2008, **1**, 116.
79. S. S. Datta, D. R. Strachan, S. M. Khamis and A. T. C. Johnson, *Nano Lett.*, 2008, **8**, 1912.
80. C. W. Keep, S. Terry and M. Wells, *J. Catal.*, 1980, **66**, 451.
81. X. Wang, Y. Ouyang, X. Li, H. Wang, J. Guo and H. Dai, *Phys. Rev. Lett.*, 2008, **100**, 206803.
82. P. Gallagher, K. Todd and D. Goldhaber-Gordon, *Phys. Rev. B*, 2010, **81**, 115409.
83. C. Stampfer, J. Güttinger, S. Hellmüller, F. Molitor, K. Ensslin and T. Ihn, *Phys. Rev. Lett.*, 2009, **102**, 056403.
84. X. Wang, Y. Ouyang, L. Jiao, H. Wang, L. Xie, J. Wu, J. Guo and H. Dai, *Nat. Nanotechnol.*, 2011, **6**, 563.
85. T. Mueller, F. Xia and P. Avouris, *Nat. Photonics*, 2010, **4**, 297.
86. F. Xia, T. Mueller, Y. Lin, A. Valdes-Garcia and P. Avouris, *Nat. Nanotechnol.*, 2009, **4**, 839.
87. L. Ju, B. Geng, J. Horng, C. Girit, M. Martin, Z. Hao, H. A. Bechtel, X. Liang, A. Zettl, Y. R. Shen and F. Wang, *Nat. Nanotechnol.*, 2011, **6**, 630.
88. H. Yan, X. Li, B. Chandra, G. Tulevski, Y. Wu, M. Freitag, W. Zhu, P. Avouris and F. Xia, *Nat. Nanotechnol.*, 2012, **7**, 330.
89. http://www.graphene-flagship.eu/GF/index.php.
90. N. Weiss, H. Zhou, L. Liao, Y. Liu, S. Jiang, Y. Huang and X. Duan, *Adv. Mat.*, 2012, **24**, 5782.
91. M. Wilson, *Phys. Today*, 2006, **59**, 21.
92. P. Huang, C. Ruiz-Vargas, A. M. van der Zande, W. S. Whitney, M. P. Levendorf, J. W. Kevek, S. Garg, J. S. Alden, C. J. Hustedt, Y. Zhu, J. Park, P. L. McEuen and D. A. Muller, *Nature*, 2011, **469**, 389.
93. A. Liucan, G. Li and E. Andrei, *Solid State Comm.*, 2009, **149**, 1151.
94. L. Ma, J. Wang and F. Ding, *Chem. Phys. Chem.*, 2012, **14**, 47.
95. L. Xie, H. Wang, C. Jin, X. Wang, L. Jiao, K. Suenaga and H. Dai, *J. Am. Chem. Soc.*, 2011, **133**, 10394.

CHAPTER 2

Nanophotonic Light Trapping Theory for Photovoltaics

ZONGFU YU,*[a,b] AASWATH RAMAN[a] AND SHANHUI FAN[a]

[a] Department of Electrical Engineering and Ginzton Lab, Stanford University, Stanford, CA 94305; [b] Department of Electrical and Computer Engineering, University of Wisconsin, Madison WI 53706
*Email: zyu54@wisc.edu

2.1 Introduction

The extraordinary growth in photovoltaic panel deployment globally in the past decade has in turn renewed interest in fundamental research aimed at improving cell performance and lowering costs. Light trapping is one promising direction in which both these goals could be achieved simultaneously. Light trapping allows photovoltaic (PV) cells to absorb sunlight using an active material layer that is much thinner than the material's intrinsic absorption length. This then reduces the amount of materials used in PV cells, which cuts cell cost in general, and moreover facilitates mass production of PV cells that are based on less abundant materials. In addition, light trapping can improve cell efficiency, since thinner cells provide better collection of photo-generated charge carriers, and potentially a higher open circuit voltage.[1]

The theory of light trapping was initially developed for conventional cells where the light-absorbing film is typically many wavelengths thick.[2–4] From a ray-optics perspective, conventional light trapping exploits the effect of total internal reflection between the semiconductor material (such as silicon, with a refractive index $n \approx 3.5$) and the surrounding medium (usually

RSC Nanoscience & Nanotechnology No. 32
Nanofabrication and its Application in Renewable Energy
Edited by Gang Zhang and Navin Manjooran
© The Royal Society of Chemistry 2014
Published by the Royal Society of Chemistry, www.rsc.org

assumed to be air). By roughening the semiconductor-air interface one randomizes the light propagation directions inside the material. The effect of total internal reflection then results in a much longer propagation distance inside the material and hence a substantial absorption enhancement. For such light trapping schemes, the standard theory shows that the absorption enhancement factor has an upper limit of $4n^2/\sin^2\theta$,[2-4] where θ is the angle of the emission cone in the medium surrounding the cell. This limit of $4n^2/\sin^2\theta$ will be referred to in this paper as the *conventional limit*. This is in contrast to the $4n^2$ limit, which strictly speaking is only applicable to cells with isotropic angular response.

For nanoscale films with thicknesses comparable or even smaller than wavelength scale, some of the basic assumptions of the conventional theory are no longer applicable. Whether the conventional limit still holds thus becomes an important open question that has been pursued both numerically[5-15] and experimentally.[16-23] In this chapter, we review our recent work on nanophotonic light trapping theory where we use a statistical coupled mode theory that describes light trapping in general from a rigorous electromagnetic perspective.[24-26] Applying this theory, we show that the limit of $4n^2/\sin^2\theta$ is only correct in bulk structures. In the nanophotonic regime, the absorption enhancement factor can go far beyond this limit with proper design.

One particular category of photovoltaic systems that motivates our interest in nanophotonic light trapping is organic solar cells. Organic photovoltaic (OPV) research has grown in the past decade with the promise of low material costs and fast, scalable manufacturing.[27-29] In contrast to most inorganic cells, the active layers in organic cells have thicknesses that are far smaller than the wavelength, thereby placing them in the nanophotonic regime. As a result, exploring nanophotonic concepts is of particular relevance in organic solar cell design.

This chapter is organized as follows: in section 2.2 we consider an organic photovoltaic example system and numerically optimize a light trapping scheme. In section 2.3, we introduce the framework of the nanophotonic light trapping theory. In section 2.4, we apply the theory to periodic structures. In section 2.5, we show that the conventional light trapping limit can be significantly exceeded with the use of nanoscale modal confinement. Section 2.6 concludes the chapter.

2.2   Numerical Design Study of Organic Photovoltaic Light Trapping

2.2.1   Introduction

An important factor limiting the efficiency of current organic photovoltaic cells is the length-scale mismatch between the electronic carrier extraction, and optical absorption of the organic semiconductors used. While organic semiconductors are typically strong optical absorbers, it is difficult to efficiently extract photo-generated charge carriers from them. For example, in solar cells using recently developed organic bulk heterojunctions, the active layer needs to have a thickness of less than 100 nm for efficient carrier extraction.[30] These thin layers leave many photons unharvested, and have thus motivated much recent interest in optical design and light trapping[31-37] for organic solar cells.

Recent work on light trapping for organic solar cells has used metallic nanostructures,[38-40] or structured the active layer itself as a grating or photonic crystal.[36,37] Both of these approaches can present implementation challenges. The use of metallic nanostructures has the tendency to increase parasitic optical absorption loss due to the metal, which competes with absorption in the active layer. Structuring the active layer can be challenging since the active layer is solution-processed and must retain a short path to electrical contacts for efficient charge carrier extraction.

Here we propose and numerically study an alternative approach to light trapping in organic solar cells. Our approach leaves the active layer itself planar while altering the layers around it, and only uses low-loss dielectric components. We demonstrate that all-dielectric, top-surface one- and two-dimensional gratings can provide substantial broad-band absorption enhancement in organic solar cells. In particular, we show that by using ITO gratings on the top surface we can scatter into ITO modes that provide absorption enhancement since the active layer is close to the ITO layer.

In section 2.2.2, we introduce our model system where PCDTBT:PC$_{71}$BM is used as the bulk-heterojunction semiconductor. In section 2.2.3, we theoretically motivate the choice of ITO gratings. We demonstrate that an optimized 1D ITO-air grating can produce broad absorption peaks that enhance polarization-averaged photocurrent by 7.3% relative to a reference planar cell with an active layer thickness of 35 nm in section 2.2.4. To overcome the issue of polarization dependence, we consider 2D ITO-air gratings in section 2.2.5, which demonstrate 9.8% photocurrent enhancement. This enhancement can be increased up to 15.1% by reducing the thickness of the PEDOT:PSS layer. In section 2.2.6, we briefly examine more sophisticated grating schemes that achieve light trapping through strong broad-band scattering.

2.2.2 Materials and Methods

As a model system, we consider the cell structure shown in Figure 2.1(a). The active layer is a 1:4 blend of PCDTBT:PC$_{71}$BM spun from dichlorobenzene. PCDTBT:PC$_{71}$BM is a high-internal quantum efficiency organic semiconductor, which was used to set a record efficiency of 6.1% for organic solar cells in 2009.[30] Currently, to achieve superior absorption it must be thicker (80–100 nm) than is ideal for current extraction. There is thus significant desire to achieve a similar absorption for thinner layers of such organic semiconductors in general, in order to achieve higher internal quantum efficiency and overall power conversion efficiency. Materials above the active

(a)

(b)

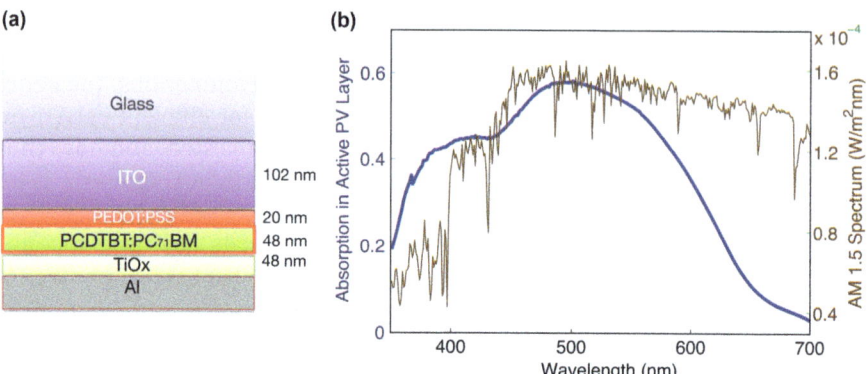

Figure 2.1   (a) Schematic of the optimized planar reference cell, and (b) the
absorption spectrum of the cell, with AM 1.5G spectrum also plotted
for reference.

layer include PEDOT:PSS, and ITO, which together form the cell's top contact, and function as a relatively transparent conductor. Below the active
layer is an optical spacer of TiOx (that also functionally acts a hole blocker)
and finally a reflecting substrate of Al. For all layers, refractive indices are
either taken from tabulated values when available, or by our own analysis of
experimentally available ellipsometry and absorption data. Our simulations
include both refractive indices and extinction coefficients, accounting for
absorption in all layers.

Even though we consider this specific model system, we emphasize that
the approach highlighted here is meant to be generally applicable to new
bulk heterojunctions being developed and synthesized. Similar to
PCDTBT:PC$_{71}$BM, these new organic semiconductors can achieve very high-
internal quantum efficiencies only in thinner layers (40–70 nm) than is ideal
for absorption. The light trapping techniques we discuss in this paper can
allow designers to keep active layers thin, which is beneficial for electronic
reasons, while still retaining the overall absorption of a thicker layer.

The rigorous coupled wave analysis (RCWA) method is used to simulate
and analyze the performance of candidate nanostructures. A scattering
matrix implementation was used,[41] where the electromagnetic field in the
spatial Fourier space for each layer is solved, and the scattering matrix of the
whole structure is determined by matching boundary conditions between
layers. For all structures considered in this section, a square lattice is used.
Ten Fourier orders sufficed to achieve convergence for both polarizations
considered.

2.2.3   Optimized Planar Structure

To assess the performance of various structures we compare their performance against an optimized reference planar cell where the ITO layer is
unstructured. The metric used to assess and optimize both the grating and

planar cells' performance is the maximum achievable photocurrent density J_{max}, defined as:

$$J_{\max} = \int d\lambda \left[e \frac{\lambda}{hc} \frac{dI}{d\lambda} \alpha(\lambda) \right] \qquad (2.1)$$

In this expression $dI/d\lambda$ is the intensity of light incident on the solar cell per unit area and wavelength. In this section, we use the AM 1.5G spectrum (ASTM G173-03). $\alpha(\lambda)$ is the fraction of light absorbed by the active layer of the cell. We specifically integrate this over the spectrum relevant to this organic semiconductor, from 375 to 750 nm.

For most parts of this analysis, the active PCDTBT:PC$_{71}$BM layer thickness is fixed to be 35 nm. This is deliberately thinner than active layers normally used in such OPV cells. Our aim is to highlight that, with light trapping, these thinner layers can achieve photocurrent generation on a par with thicker cells.

For the planar structure shown in Figure 2.1(a), an optimization is done for its ITO, and TiOx spacer layer thicknesses, resulting in heights of 102 nm and 48 nm, respectively. This optimized structure's absorption in the active layer for the relevant solar wavelength range of 375–750 nm is shown in Figure 2.1(b). It has a photocurrent density of 10.73 mA/cm^2 based on this absorption. All gratings and nanostructures examined in subsequent sections are judged relative to this baseline photocurrent density.

Comparing the absorption and the AM 1.5G spectra in Figure 2.1, we see that there will be substantial benefit to enhancing this cell's performance over the entire wavelength range. The benefit is particularly prominent in the 600–700 nm range where the active layer is weakly absorbing.

Based on the planar structure shown in section 2.2.2, we seek to enhance the absorption in the active layer through light trapping. In conventional cells, such as crystalline Si cells, the active layer, which has the highest refractive index in the layered structure, naturally forms a waveguide. Light trapping then involves scattering into the guided mode in the waveguide formed by the active layer. In the structure shown in Figure 2.1(a), however, the active layer has an index that is, at best, comparable to that of the ITO, and is also very thin. Thus the active layer alone does not form a waveguide. Instead, the guiding layer is the ITO layer. Therefore, in this section, we seek to achieve light trapping in the active layer by exploiting the guided modes in the ITO layer.

2.2.4   1D ITO Grating

We first consider a top-surface periodic dielectric grating (Figure 2.2(a)) that rests on top of the usual stack of layers present in an organic solar cell: a layer of the transparent conductor PEDOT:PSS, the bulk heterojunction, a TiOx optical spacer, and a reflecting Al back contact. The grating is introduced in the ITO layer. ITO is typically needed as a transparent front-surface

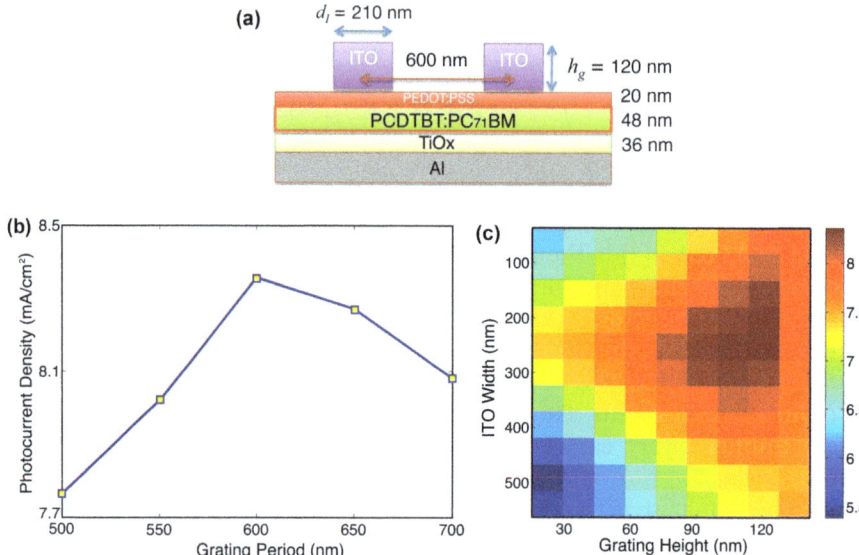

Figure 2.2 (a) Diagram of the optimized 1D periodic ITO grating structure. (b) Photocurrent density (mA cm^{-2}) for various grating periods, for the optimal device on other parameters, show that a 630 nm period grating is optimal for this device. (c) Photocurrent density for normally incident light as a function of ITO height and width for a grating with the optimal 630 nm period.

contact in organic cells, and is used here as the high-index component of the top-surface grating. Structuring only the top surface allows the thin solution-processed layers of the organic solar cell to remain planar, which is experimentally desirable to prevent shorting and roughness, encountered when the active layers themselves are structured.

2.2.4.1   Grating Optimization

We use RCWA to investigate the parameter space relevant to a 1D top-surface air-ITO grating. We optimize for the grating period a, the grating height h_g, the width of the ITO portion of the grating d_I and the height of the spacer layer h_s. The PEDOT:PSS layer is chosen to have a thickness of 20 nm[42] in order to optimize mode coupling to the active layer which is 35 nm in thickness.

For a grating period of 630 nm, grating height $h_g = 135$ nm, ITO width $d_I = 126$ nm and spacer height $h_s = 30$ nm, we find an optimal polarization-averaged photocurrent density of 11.53 mA/cm^2. This represents a 7.3% photocurrent enhancement over the planar reference cell. The optimized structure and these parameters are illustrated in Figure 2.2(a). In Figure 2.2(c), we show the effect on photocurrent density of varying the grating height and ITO width while fixing the period and spacer height at their optimal points. The performance of the optimized structure is fairly robust against small fluctuations in grating period, height, and ITO width.

2.2.4.2 Light Absorption in the Active Layer

We now examine the absorption spectrum in the active layer of this optimized structure to understand the source of the observed photocurrent enhancement. Note here again that by absorption we mean absorption in the active PCDTBT:PC$_{71}$BM layer only; that is to say, useful absorption. In Figure 2.3(a) we can see the absorption spectrum for *s*- and *p*-polarizations for the optimal grating structure, and that of the reference planar cell. For *s*-polarized light the grating structure has several broad peaks including a prominent peak centered at $\lambda_0 = 641$ nm with a width of $\Delta\lambda \approx 40$ nm, and a secondary peak near 488 nm.

The greatest contribution to the enhancement in photocurrent comes from the 641 nm peak, since this peak is both broad and located in a spectral region where the planar cell's absorption is weak. To better understand the physical origin of this peak we show the electric field intensity $|\mathbf{E}|^2$ at λ_0 in Figure 2.3(b). The field is strongly concentrated in the ITO layers with its maximum lying in the air space between the higher-index ITO part of the air–ITO grating. The field penetrates into the PEDOT:PSS and active layers, thereby yielding the absorption enhancement.

The *p*-polarization spectrum shows no significant deviation from that of the planar reference cell. Thus, while the *s*-polarization has a photocurrent density of 12.08 mA/cm^2, the *p*-polarization has a photocurrent density of 10.97 mA/cm^2.

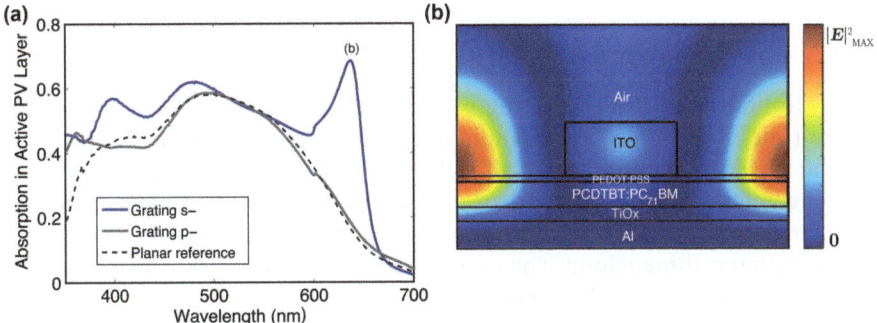

Figure 2.3 (a) Absorption in the active layer for the optimal 1D ITO grating organic cell for both polarizations (solid lines) compared to the optimal planar reference cell (dashed line). The broad peak at $\lambda_0 = 641$ nm for the grating in the *s*-polarization results in significant integrated photocurrent enhancement relative to the planar cell. (b) The electric field intensity $|\mathbf{E}|^2$ at the absorption peak of $\lambda_0 = 641$ nm. The field is concentrated in the ITO layer but penetrates down into the active layer, thereby yielding useful absorption.

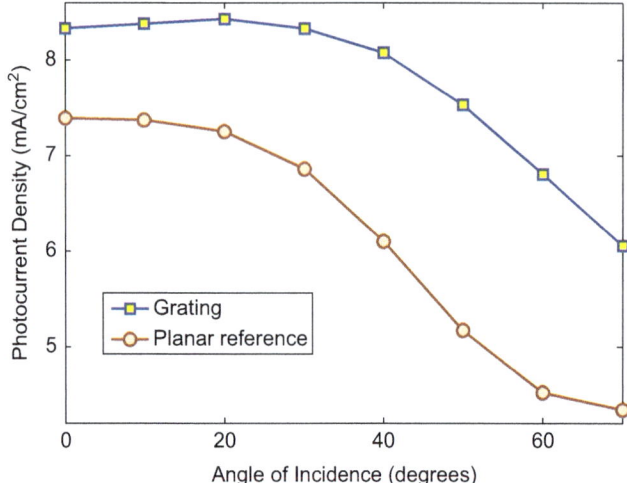

Figure 2.4   Polarization-averaged photocurrent density *vs.* (polar) angle of incidence for planar and optimized 1D ITO grating cells. The grating
structure outperforms the planar cell to large angles of incidence.

2.2.4.3   *Angular Response*

Finally, we calculate the angular response of the optimal structure identified
earlier. In Figure 2.4, we plot the photocurrent density as a function of incident (polar) angle. The polarization-averaged photocurrent density for the
1D ITO grating cell decreases with increasing angle of incidence but remains
above that of the planar reference cell up to 60°, indicating substantial
robustness in enhancement for a wide range of angles.

2.2.5   2D ITO Nanostructure

A key limitation of the 1D grating is its polarization dependence.[43,13] At the
same time, it is indeed remarkable that the simple 1D grating considered
in the previous section can itself deliver worthwhile polarization-averaged
enhancement. In this section, to overcome the lower performance in the
p-polarization, we examine a two-dimensional (2D) grating that is periodic
in both planar dimensions. The grating structure consists of a square lattice
of circular air holes in the ITO layer.

By using circular air holes in the ITO layer the absorption spectrum
becomes independent of *s*- and *p*-polarization. We optimize all tunable
parameters, including the air-hole diameter and height of the ITO layer, as
shown in Figure 2.5(a). For a grating period of 640 nm, an air-hole diameter
of $d_{air} = 598$ nm, an ITO layer height of $h_g = 118$ nm and a TiO$_x$ spacer layer
thickness of $h_s = 33$ nm, we find an optimal photocurrent density of
11.79 mA/cm². This is polarization-independent and represents a 9.8%
photocurrent enhancement over the planar reference cell. The photocurrent

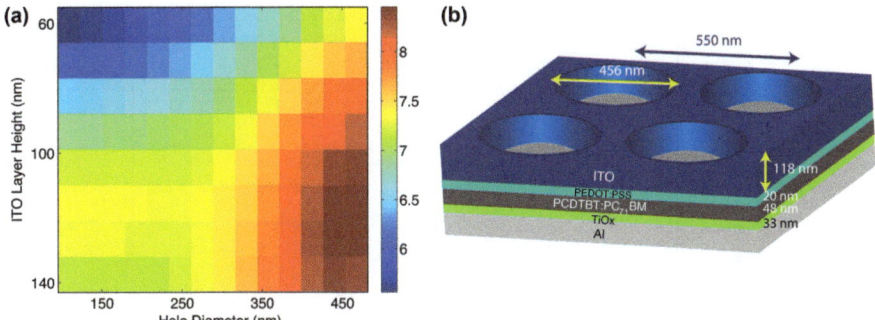

Figure 2.5 (a) Photocurrent density (mA cm^{-2}) for normally incident light as a function of ITO nanostructure height and air hole diameter for a 640 nm period. (b) The optimized 2D ITO-air hole nanostructure on top of the organic solar cell stack.

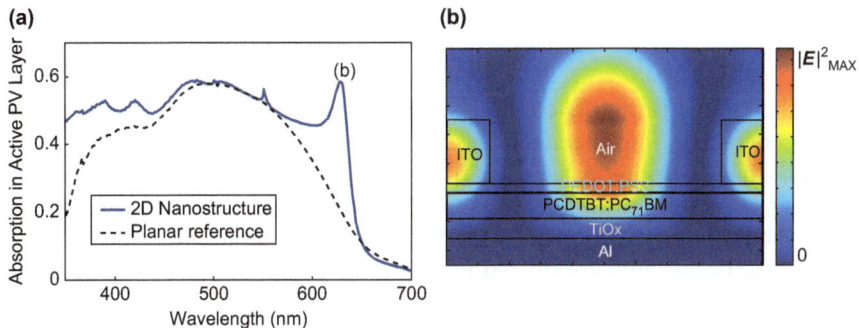

Figure 2.6 (a) Absorption in the active layer comparing the 2D grating structure (solid line) against the planar reference cell (dashed line). There is a prominent peak $\lambda_0 = 644$ nm that produces notable photocurrent enhancement for both polarizations. (b) Field plot of the electric field intensity $|\mathbf{E}|^2$ at peak $\lambda_0 = 644$ nm. As with the 1D grating, the field is concentrated in the air part of the ITO layer but penetrates down into the active layer, thereby yielding useful absorption.

achieved using this grating structure with 35 nm active layer thickness is equivalent to what can be achieved using a 45 nm thick active layer of PCDTBT:PC$_{71}$BM in an optimized planar cell. The optimized nanostructure is shown in Figure 2.5(b).

2.2.5.1 Absorption Spectrum

The absorption spectrum for this optimal structure is shown in Figure 2.6(a). As noted earlier, for a symmetric 2D grating the absorption spectrum is polarization-independent. There is a prominent absorption peak at $\lambda_0 = 644$ nm, whose electric field intensity we show in Figure 2.6(b). As can be

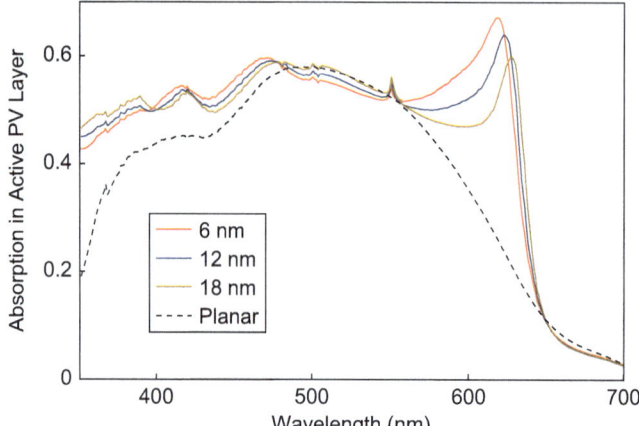

Figure 2.7   Absorption spectra for optimal 2D ITO grating structures while varying
the PEDOT layer height. Thinner PEDOT layers show greater absorption
at the peaks.

seen the field is primarily concentrated in the air hole, with a portion of the
field penetrating into the active layer thereby enhancing useful absorption.

Since the absorption peak arises from a mode whose field is concentrated in
the ITO layer above the organic layers, one expects that using a thinner PEDOT
layer should enable more of the field to penetrate into the active layer, thereby
increasing overall absorption. To test this, we fix all parameters in the optimized cell shown in Figure 2.7(b), vary only the thickness of the PEDOT layer,
and calculate the absorption spectrum for each case. As shown in Figure 2.7,
the height of the peak seen at λ_0, in the previous section, increases notably with
thinner PEDOT layers. The 6 nm thin layer in particular results in a 15.1%
photocurrent enhancement relative to the planar reference cell, while the 20
nm layer results in the 9.8% photocurrent enhancement noted earlier. Thus,
the closer the active layer is to the grating, the stronger the enhancement.

As a practical matter, it is generally deemed necessary for the PEDOT layer
to be at least 20 nm in thickness.[42] As an alternative to PEDOT, however,
transition metal oxides such as vanadium oxide and molybdenum oxide can
be reliably deposited and used with sub-10 nm thicknesses.[44,45] While their
dielectric constants are slightly higher than that of PEDOT, this does not alter
the observed optical effect of the ITO mode penetrating down into the active
layer. Thus, combined with the structures presented in this paper, these al-
ternatives should enable the stronger absorption enhancement noted above.

2.2.5.2   Planar ITO Layer

A practical concern with the above design might be that it reduces the surface area of the ITO contact, which can reduce the conductivity of the anode
and increase the charge collection distance for carriers in the PEDOT layer.

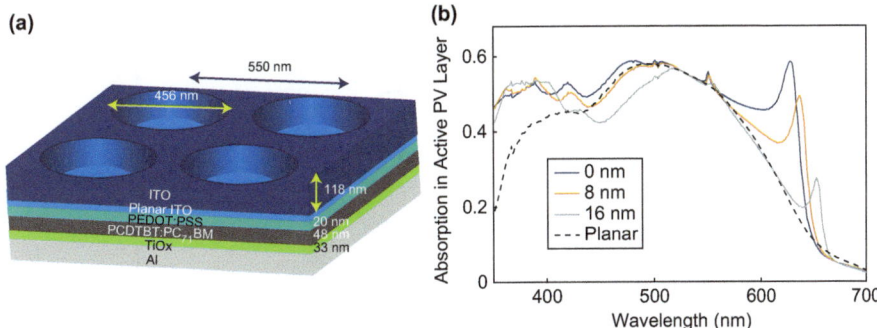

Figure 2.8 (a) A thin planar ITO layer introduced below the ITO-air grating layer, as shown in this diagram, may be electrically desirable for charge carrier extraction. (b) The effect of the thickness of this planar ITO layer on light absorption in the active layer is shown for increasing thicknesses. The resonance peak's prominence decreases with increasing thickness.

To alleviate this we include an additional planar layer of ITO in between the 2D ITO-air grating and the PEDOT layer, as shown in Figure 2.8(a). We examine the effect of the thickness of this planar ITO layer on light absorption in the active layer in Figure 2.8(b). An 8 nm planar layer of ITO reduces the photocurrent of the 2D grating device from 8.49 mA/cm^2 to 8.11 mA/cm^2, which still represents a 9.5% photocurrent enhancement over the planar reference cell. Moreover, combining a thin planar ITO layer with the thin metal oxide interlayer mentioned earlier, may be a practical compromise to maintaining a close distance between the active and grating layers (which produces strong absorption and photocurrent enhancement), while also maintaining desirable electrical properties.

2.2.5.3 Angular Response

We also calculate the angular response of the optimal 2D structure shown in Figure 2.5(b). In Figure 2.9, we plot the photocurrent density as a function of incident (polar) angle. The polarization-averaged photocurrent density for the ITO grating cell decreases with increasing angle of incidence but remains well above that of the planar reference cell up to 60°.

2.2.6 Multi-level Grating

Finally, we consider more sophisticated top-surface grating structures that can deliver stronger scattering than the ordered gratings considered in previous sections. One example is a multi-level grating structure, schematically shown in Figure 2.10(a), consisting of a 2D array of ITO blocks on the bottom, and a 1D ITO grating on top, whose axis is at a 45° angle relative to the bottom layer's grating.

This grating, deployed on our model system with an active layer thickness of 35 nm, results in a photocurrent of 11.79 mA/cm^2, an enhancement of

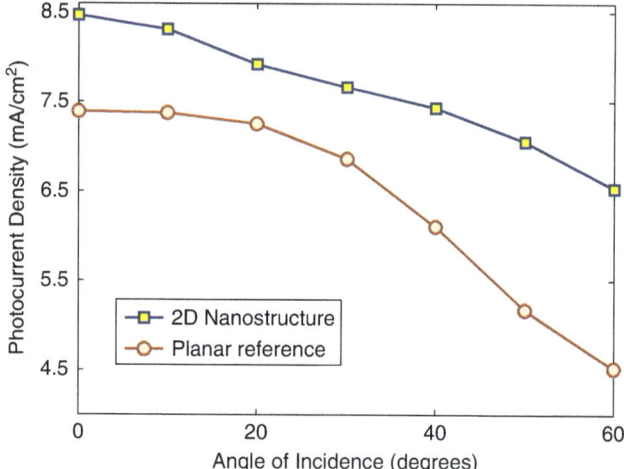

Figure 2.9   Polarization-averaged photocurrent density *vs.* (polar) angle of incidence for planar and optimized ITO grating cells.

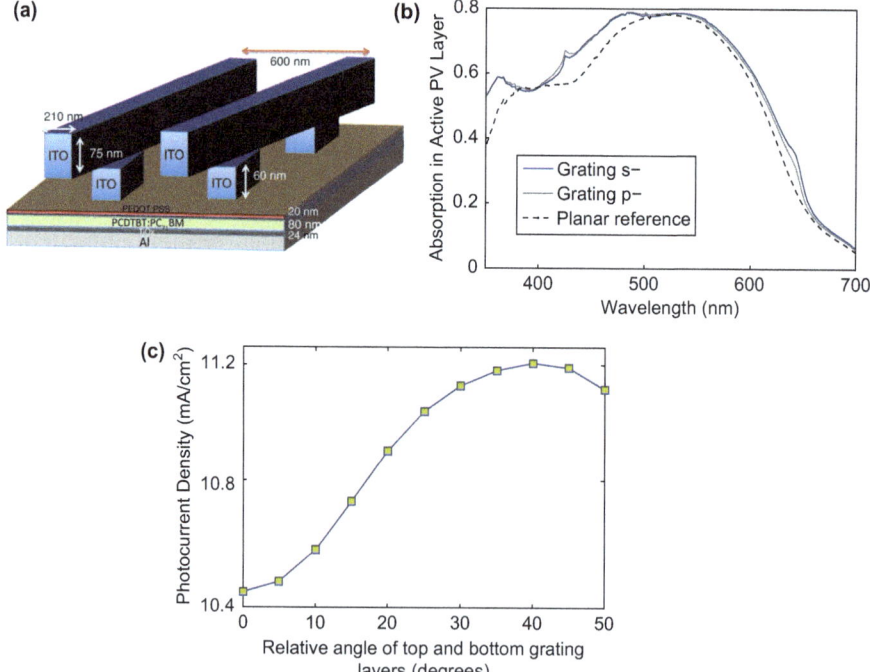

Figure 2.10   (a) A multi-level grating structure consisting of a 2D air-ITO grating on
bottom, and a 1D ITO grating on top whose axis is at a 40° angle
relative to the bottom layer. (b) Absorption in the active layer for the
multi-level grating for both polarizations compared to a planar reference cell, both with 80 nm-thick active layers. (c) Photocurrent density
vs. the top grating layer's angle relative to the lower layer, showing an
optimum at 40°.

10.1% compared with the optimized planar reference cell. The photocurrent achieved using this grating structure with 35 nm active layer thickness is equivalent to what can be achieved with a 44 nm thick active layer in an optimized planar cell.

We also apply this grating structure in a cell with an active layer thickness of 70 nm. We do this to compare our nanostructure's performance against the best PCDTBT:PC$_{71}$BM cell reported in the literature,[30] where the active layer is 70–80 nm thick. When the grating structure is applied, a relatively small, yet broad-band enhancement of absorption can be seen over large segments of the relevant wavelength range (Figure 2.10(b)), indicating the presence of strong scattering that is not strongly wavelength dependent. The 45° angle between the two grating layers is arrived at by an optimization procedure (Figure 2.10(c)). The use of this grating structure results in a photocurrent density of 14.67 mA cm^{-2}, a 3.1% photocurrent enhancement over the optimized 70 nm thick planar cell's photocurrent of 14.22 mA cm^{-2}.

We have thus shown that the use of such complex dielectric nano-structures can lead to enhancement in active layers with different thicknesses. This result indicates that our approach can be tailored to suit the different requirements of future bulk heterojunctions used in the active layer. Moreover, our results highlight the potential of complex, or randomly distributed, dielectric and ITO nanostructures (in particular, nanowires) to enable substantial light absorption enhancement in the active layers of the best organic solar cells in existence.

2.2.7 Conclusion

In this section, we have numerically studied a light trapping scheme for organic photovoltaic cells. We showed that nanostructured top-surface contacts made with ITO can significantly enhance light absorption and photocurrent generated in realistic bulk-heterojunction organic cells. Structures similar to the ITO-air gratings shown here can potentially be fabricated cheaply using electrospun ITO nanowires[46] or thermal nanoimprint lithography.[47] Thus, we believe that light absorption enhancement enabled by nanostructuring the transparent electrical contact is compatible with the low-cost manufacturing vision of organic solar cells.

By enabling partial decoupling between the issues of carrier extraction and light absorption, the designs presented here offer a new degree of freedom to organic photovoltaic researchers as new semiconductors are developed. In the next section, we introduce a physical theory that describes the benefits of nanophotonic light trapping, which has been observed numerically in this section.

2.3 Nanophotonic Light Trapping Theory

The organic photovoltaic system studied in the previous section is one example of a broad class of new nanoscale solar cells whose electromagnetic interactions cannot be captured by conventional ray optics. In this section, we introduce a rigorous wave electromagnetic theory of light trapping that fully elucidates light management challenges and opportunities in nanophotonic solar cells.

To illustrate our theory, we consider a high-index thin-film active layer with a high reflectivity mirror at the bottom, and air on top. Such a film supports guided optical modes. At the limit where the absorption of the active layer is weak, these guided modes typically have a propagation distance along the film that is much longer than the thickness of the film. Light trapping is accomplished by coupling incident plane waves into these guided modes, with either a grating with periodicity L (Figure 2.11(b)) or random Lambertian roughness (Figure 2.11(a)). It is well known that a

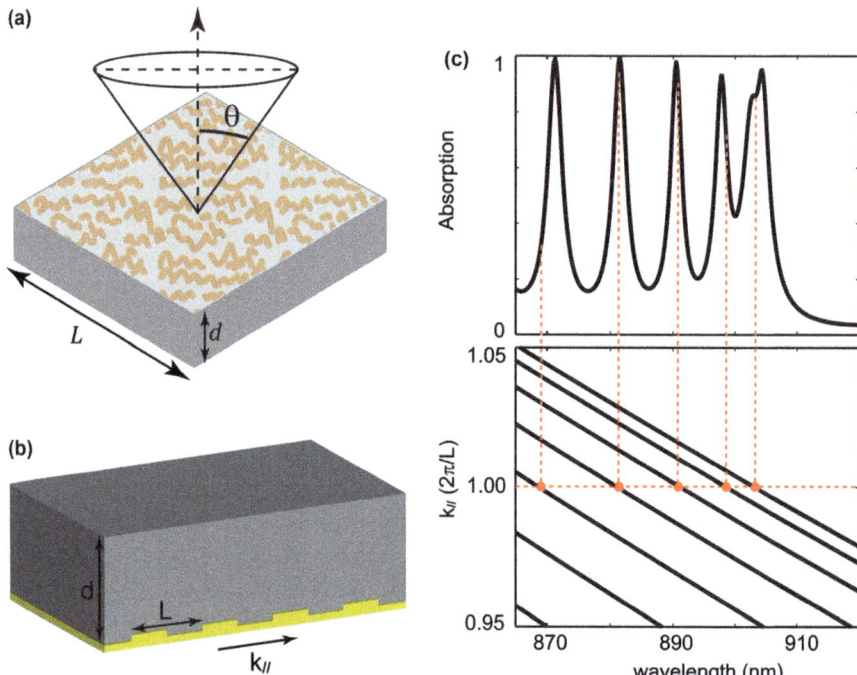

Figure 2.11   (a) Schematic of light trapping with random texture. (b) Light-trapping with the use of a periodic grating on a back-reflector (yellow). $d = 2$ mm. $L = 250$ nm. The depth and width of the dielectric groove in the grating are 50 nm and 175 nm respectively. The dielectric material is crystalline silicon. (c) Absorption spectrum (TM mode, normal incidence) and dispersion relation of waveguide modes for the structure in (b). The dispersion relation is approximated as $\omega = \frac{c}{n}\left[\left(\frac{m\pi}{d}\right)^2 + k_{//}^2\right]$, or equivalently in terms of free-space wavelength $\lambda = \frac{2\pi n}{\left(\frac{m\pi}{d}\right)^2 + k_{//}^2}$, where $m = 1, 2, 3, \dots$ is the band index indicating the field variation in the transverse direction. Resonances occur when $k_{//} = 2\pi/L$ (red dots).

system with random roughness can be understood by taking the $L \to \infty$ limit of the periodic system.[10,48] Thus, we will focus on periodic systems. As long as L is chosen to be sufficiently large, *i.e.* at least comparable to the free-space wavelength of the incident light, each incident plane wave can couple into at least one guided mode. By the same argument, such a guided mode can couple to external plane waves, creating a guided resonance.[49]

A typical absorption spectrum for such a film[6] is reproduced in Figure 2.11(c). The absorption spectrum consists of multiple peaks, each corresponding to a guided resonance. The absorption is strongly enhanced in the vicinity of each resonance. However, compared to the broad solar spectrum, each individual resonance has very narrow spectral width. Consequently, to enhance absorption over a substantial portion of the solar spectrum, one must rely upon a collection of these peaks. Motivated by this observation, we develop a statistical temporal coupled mode theory that describes the aggregate contributions from all resonances.

We start by identifying the contribution of a single resonance to the total absorption over a broad spectrum. The behavior of an individual guided resonance, when excited by an incident plane wave, is described by the temporal coupled mode theory equation:[50,51]

$$\frac{d}{dt}a = \left(j\omega_0 - \frac{N\gamma_e + \gamma_i}{2} \right) a + j\sqrt{\gamma_e}S \qquad (2.2)$$

Here a is the resonance amplitude, normalized such that $|a|^2$ is the energy per unit area in the film, ω_0 is the resonance frequency, and γ_i is the intrinsic loss rate of the resonance due to material absorption. S is the amplitude of the incident plane wave, with $|S|^2$ corresponding to its intensity. We refer to a plane wave that couples to the resonance as a *channel*. γ_e is the leakage rate of the resonance to the channel that carries the incident wave. In general, the grating may phase-match the resonance to other plane wave channels as well. We assume a total of N such channels. Equivalent to the assumption of a Lambertian emission profile.[2] We further assume that the resonance leaks to each of the N channels with the same rate γ_e. Under these assumptions, the absorption spectrum of the resonance is:[50]

$$A(\omega) = \frac{\gamma_i \gamma_e}{(\omega - \omega_0)^2 + (\gamma_i + N\gamma_e)^2 / 4} \qquad (2.3)$$

For light trapping purposes, the incident light spectrum is typically much wider than the linewidth of the resonance. When this is the case, we characterize the contribution of a single resonance to the total absorption by a *spectral cross-section*:

$$\sigma = \int_{-\infty}^{\infty} A(\omega)d\omega \qquad (2.4)$$

Notice that the spectral cross-section has units of frequency, and has the following physical interpretation: For an incident spectrum with bandwidth

$\Delta\omega \gg \sigma$, a resonance contributes an additional $\sigma/\Delta\omega$ to the spectrally-averaged absorption coefficient.

For a single resonance, from eqn (2.3) and (2.4), its spectral cross-section is:

$$\sigma = 2\pi\gamma_i \frac{1}{N + \gamma_i/\gamma_e} \tag{2.5}$$

which reaches a maximum value of

$$\sigma_{\max} = \frac{2\pi\gamma_i}{N} \tag{2.6}$$

in the *over-coupling* regime when $\gamma_e \gg \gamma_i$. We emphasize that the requirement to operate in the strongly over-coupling regime arises from the need to accomplish broad-band absorption enhancement. In the opposite narrow-band limit, when the incident radiation is far narrower than the resonance bandwidth, one would instead prefer to operate in the critical coupling condition by choosing $\gamma_i = N\gamma_e$, which results in $(100/N)\%$ absorption at the resonant frequency of ω_0. The use of critical coupling, however, has a lower spectral cross-section and is not optimal for the purpose of broad-band enhancement. The intrinsic decay rate γ_i differentiates between the two cases of broad-band and narrow-band. For light trapping in solar cells, we are almost always in the broad-band case where the incident radiation has bandwidth $\Delta\omega \gg \gamma_i$.

We can now calculate the upper limit for absorption by a given medium, by summing over the maximal spectral cross-section of all resonances:

$$A_T = \frac{\sum \sigma_{\max}}{\Delta\omega} = \frac{1}{\Delta\omega} \sum_m \frac{2\pi\gamma_{i,m}}{N} \tag{2.7}$$

where the summation takes place over all resonances (labeled by m) in the frequency range of $[\omega, \omega + \Delta\omega]$. In the over-coupling regime, the peak absorption from each resonance is in fact relatively small; therefore the total cross-section can be obtained by summing over the contributions from individual resonances. In addition, we assume that the medium is weakly absorptive such that single-pass light absorption is negligible.

Eqn (2.7) is the main result of this paper. In the following discussion, we will first use eqn (2.7) to reproduce the well-known $4n^2$ conventional limit, and then consider a few important scenarios where the effect of strong light confinement becomes important.

We first consider a structure with period L and thickness d that are both much larger than the wavelength. In this case, the resonance can be approximated as propagating plane waves inside the bulk structure. Thus, the intrinsic decay rate for each resonance is related to a material's absorption coefficient α_0 by $\gamma_{i,m} = \gamma_i = \alpha_0 \frac{c}{n}$. Eqn (2.7) can thus be simplified as:

$$A_T = \frac{2\pi\gamma_i}{\Delta\omega} \cdot \frac{M}{N} \tag{2.8}$$

where M is the number of resonances in the frequency range $[\omega, \omega + \Delta\omega]$ given by[52]

$$M = \frac{8\pi n^3 \omega^2}{c^3} \left(\frac{L}{2\pi}\right)^2 \left(\frac{d}{2\pi}\right) \delta\omega. \tag{2.9}$$

Each resonance in such a frequency range can couple to channels that are equally spaced by $\frac{2\pi}{L}$ in the parallel wavevector $\boldsymbol{k}_{//}$ space (Figure 2.12(a)). Moreover, since each channel is a propagating plane wave in air, its parallel wavevector needs to satisfy $|\boldsymbol{k}_{//}| \leq \omega/c$. Thus, the number of channels is:

$$N = \frac{2\pi \omega^2}{c^2} \left(\frac{L}{2\pi}\right)^2. \tag{2.10}$$

From eqn (2.8), the upper limit for the absorption coefficient for this system is then

$$A_T = \frac{2\pi\gamma_i}{\Delta\omega} \cdot \frac{M}{N} = 4n^2\alpha_0 d \tag{2.11}$$

resulting in the upper limit for the absorption enhancement factor F,

$$F \equiv \frac{A_T}{\alpha_0 d} = 4n^2 \tag{2.12}$$

which reproduces the $4n^2$ conventional limit, appropriate for the Lambertian emission case with $\sin\theta = 1$. The theory can be generalized to the case

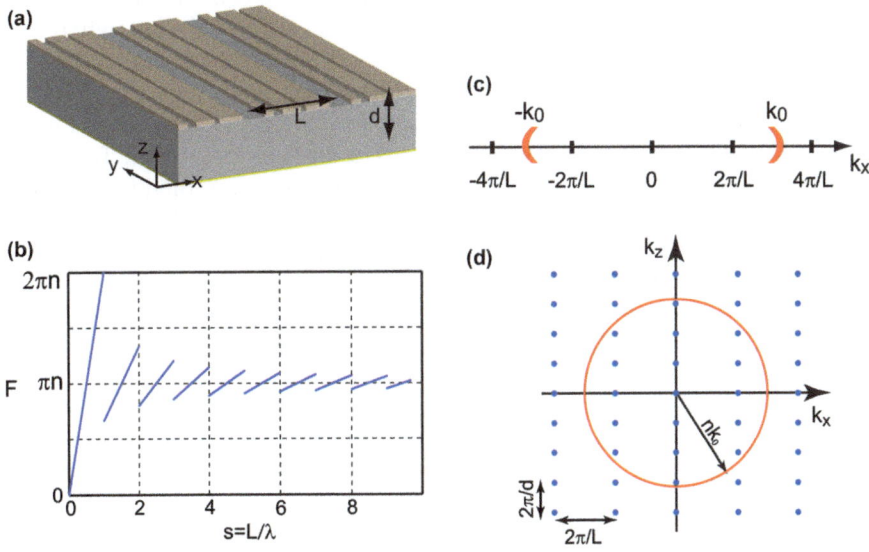

Figure 2.12 (a) Schematic of a grating structure. Brown ribbons are non-absorptive dielectric medium. The whole structure is placed on a perfect mirror (yellow). (b) Upper limit of absorption enhancement in 1D grating films without mirror symmetry. (c) Channels in 1D k-space. (d) Resonances in a film with 1D grating. Dots represent resonances.

of a restricted emission cone, and reproduces the standard result of $4n^2/\sin^2\theta$.

The analysis here also points to scenarios where the conventional limit is no longer applicable. Eqn (2.10) is not applicable when the periodicity is comparable to the wavelength, while eqn (2.9) is not valid when the film thickness is much smaller than the wavelength. In the next two sections, we consider both of these cases.

2.4   Light Trapping in Periodic Grating Structures

When the periodicity L is comparable to the wavelength λ, the discrete nature of the channels becomes important . To illustrate this effect, we assume that the film has a high refractive index (for example, silicon), such that the wavelength in the material is small compared with the periodicity. We also assume that the film has a thickness of a few wavelengths. In this case, all modes have approximately the same decay rate $\gamma_i = \alpha_0 \frac{c}{n}$, eqn (2.9) can still be used to count the number of resonances.

2.4.1   Theoretical Upper Limit of Enhancement Factor for 1D Grating Structures

We first consider a 1D grating defined by structures that are uniform in one dimension, *e.g.* y-direction, and are periodic in the other dimension (x-direction in Figure 2.12(a)) with a periodicity L. We consider incident light propagating in the xz-plane. We also assume a general case where the grating profile has no mirror symmetry along the x-direction (Figure 2.12(a)). The effect of symmetry will be described later.

For light incident from the normal direction, the periodicity results in the excitation of other plane waves with $k_x = 0, \pm 2\pi/L, \pm 4\pi/L$, in the free space above. Moreover, since these plane waves are propagating modes in air, one needs to have $k_x \leq k_0$, where k_0 is the wavevector of the incident light. These two requirements completely specify the number of channels available in k-space (Figure 2.12(a)).

We first consider the case $L \gg \lambda$. The spacing of the channel is $\frac{2\pi}{L} \ll \frac{2\pi}{\lambda} = k_0$. The discreteness of the channels is therefore not important. The total number of channels (Figure 2.12(c)) at wavelength λ thus becomes:

$$N = \frac{2k_0}{2\pi/L} = \frac{2L}{\lambda} \tag{2.13}$$

Notice that we consider only a single polarization. In the frequency range $[\omega, \omega + \Delta\omega]$, the total number of guided resonances supported by the film is (Figure 2.12(d)):

$$M = \frac{2n^2\pi\omega}{c^2}\left(\frac{L}{2\pi}\right)\left(\frac{d}{2\pi}\right)\Delta\omega \tag{2.14}$$

Combining eqn (2.8, 2.13, 2.14), we obtain the upper limit for absorption enhancement

$$F = \frac{A}{d\alpha} = \pi n \qquad (2.15)$$

In contrast to the bulk limit of $4n^2$, the enhancement factor is greatly reduced in structures that are uniform in one of the dimensions. Since the structure considered here is essentially a two-dimensional structure, we refer to this limit as the *2D bulk limit*.

In the case when the periodicity is close to the wavelength, the discreteness of the channels becomes important and eqn (2.13) is no longer valid. Instead, as shown in Figure 2.12(c), the number of channels is:

$$N = 2 \left\lfloor \frac{k_0}{2\pi / L} \right\rfloor + 1 = 2 \left\lfloor \frac{L}{\lambda} \right\rfloor + 1 \qquad (2.16)$$

where $\lfloor x \rfloor$ represents the largest integer that is smaller than x. We further assume the medium has a high refractive index such that the following conditions are satisfied:

$$L \gg \lambda / n$$
$$d \gg \lambda / n \qquad (2.17)$$

Under these conditions, the resonance in the film can still be approximated as forming a continuum of states, and eqn (2.14) is still applicable. The upper limit for the enhancement factor is thus calculated from eqns (2.8), (2.14) and (2.16). In Figure 2.12(b), we plot such an upper limit as a function of normalized frequency $s \equiv L/\lambda$. At low frequency, when $L/\lambda < 1$, there is only one channel, *i.e.* $N = 1$, while the number of resonances increases linearly with frequency. Hence the enhancement factor increases linearly with frequency, reaching its maximum value of $2\pi n$ at $L = \lambda$ At the frequency immediately above $s = 1$, the number of channels increases to $N = 3$, leading to a step-function drop of the enhancement factor. In general, such a sharp drop in enhancement factor occurs whenever new channels appear, *i.e.* whenever $L = m\lambda$, where m is an integer. Also, in between such sharp drops, the enhancement factor always increases as a function of frequency. In the limit of $L \gg \lambda$, the enhancement factor converges to the 2D bulk limit of πn.

We now analyze the effects of the symmetry of the grating profile on the light trapping limit. In contrast to the asymmetrical grating profile shown in Figure 2.12(a), a symmetrical grating has mirror symmetry in the x-direction (Figure 2.13). It is proposed to use asymmetrical gratings to reduce the reflection in the normal direction, and thus increase the absorption.[53] On the other hand, a semi-analytical method[10] finds no difference for gratings with different symmetries. Here, we analyze the effect of structure symmetry on the fundamental limit of light absorption enhancement with our rigorous electromagnetic approach.

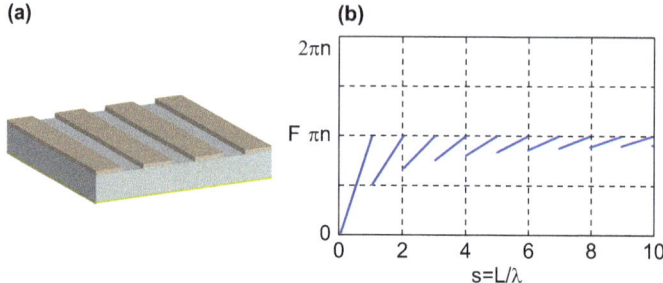

Figure 2.13   (a) 1D grating structure with mirror symmetry. (b) Limit of absorption enhancement in grating structures with mirror symmetry.

We first discuss the case of normally incident light. Due to mirror symmetry of the film, resonant modes either have an odd or even modal amplitude profile. The normally incident plane wave, which has even modal amplitude profile, cannot couple to modes with odd profiles. Therefore, for the symmetric case, the number of resonances that can contribute to the absorption is reduced by half when compared to the asymmetric case, *i.e.*:

$$M_{sym} = M/2 \qquad (2.18)$$

where M is given by eqn (2.14).

In the case where the period is smaller than the wavelength, there is only one channel $N_{sym} = N = 1$. Thus, when the period of the grating is less than the free-space wavelength, the symmetric case has a lower enhancement limit $F_{sym} = F/2$ (Figure 2.13(b)).

When the period of the grating is larger than the wavelength, there are more channels, and the effect of symmetry on channels also needs to be considered. Due to mirror symmetry, channels can be arranged as even and odd according to:

$$S_{even} = \frac{1}{\sqrt{2}} \left(S_{k_{//}} + S_{-k_{//}} \right)$$

$$\qquad (2.19)$$

$$S_{odd} = \frac{1}{\sqrt{2}} \left(S_{k_{//}} - S_{-k_{//}} \right)$$

Since the incident plane wave from the normal incidence has an even modal amplitude profile, only the even resonances can be excited, which only leak into even channels. Therefore, the number of channels available is also reduced by half $N_{sym} = N/2$ when $L \gg \lambda$. In this case, we have the same enhancement limit as that of asymmetrical gratings $F_{sym} = F$ (Figure 2.13(b)). Thus, the symmetry of the grating is important only when the periodicity of the grating is smaller or comparable to the wavelength of incident light. In structures with a period much larger compared to the wavelength, the symmetry of the grating does not play a role in determining the upper limit of absorption enhancement.

The upper limit in a 1D grating structure falls far below the conventional bulk limit of $4n^2$ due to a reduced number of resonances. To achieve a higher enhancement factor, it is always better to use 2D gratings that are periodic in both x and y directions, since a 2D grating allows access to all resonances supported by a film. In the next section, we present a detailed discussion on the upper limit of light trapping with a 2D grating based on eqn (2.8).

2.4.2 Theoretical Upper Limit of Enhancement Factor for 2D Grating Structures

We first consider the case where the grating has a square lattice, with a periodicity of L in both dimensions. In the frequency range $[\omega, \omega + \Delta\omega]$, assuming eqn (2.17) is satisfied, the total number of guided resonances supported by the film is given by eqn (2.9). Notice that the number of resonances increases quadratically as a function of frequency.

When a plane wave is normally incident upon the film, the grating can excite plane waves in other propagating directions in free space. The parallel wavevectors $G_{m,n}$ of these excited plane waves in free space are:

$$G_{m,n} = m\frac{2\pi}{L}\hat{x} + n\frac{2\pi}{L}\hat{y} \tag{2.20}$$

with m and n being integers. Hence these parallel wavevectors form a square lattice in the wavevector space (Figure 2.14(a), blue dots). Moreover, since these are propagating plane waves, their wavevectors need to lie within a circle as defined by $|G| < k_0 \equiv \frac{\omega}{c}$ (Figure 2.14(a), dashed line). The total number of different wavevector points, multiplied by two in order to take into account both polarizations, defines the number of channels N that are required for the calculation using eqn (2.8).

At low frequency when $s = L/\lambda < 1$, there are only two channels ($N = 2$), accounting for two polarizations, in the normal direction, while the number of resonances increases with frequency quadratically. Hence the enhancement factor increases quadratically with frequency and reaches its maximum value of $4\pi n^2$ at $s = 1$ (Figure 2.14(b)). Immediately above $s = 1$, the wavelength becomes shorter than the period L. The number of channels increases to $N = 10$, leading to a step-function drop of the enhancement factor. In the limit of $L \gg \lambda$, the calculated upper limit reproduces the 3D bulk limit of $4n^2$.[24]

Next, we analyze the angular response of the upper limit of the enhancement factor by considering a plane wave incident from a direction specified by an incidence angle θ and an azimuthal angle φ. Such an incident plane wave has a parallel wavevector $k_{//} = k_0 \sin(\theta)\cos(\varphi)\hat{x} + k_0 \sin(\theta)\sin(\varphi)\hat{y}$. In the presence of the grating, such a plane wave can excite other plane waves in free space with parallel wavevectors:

$$k = k_{//} + G_{m,n}, \tag{2.21}$$

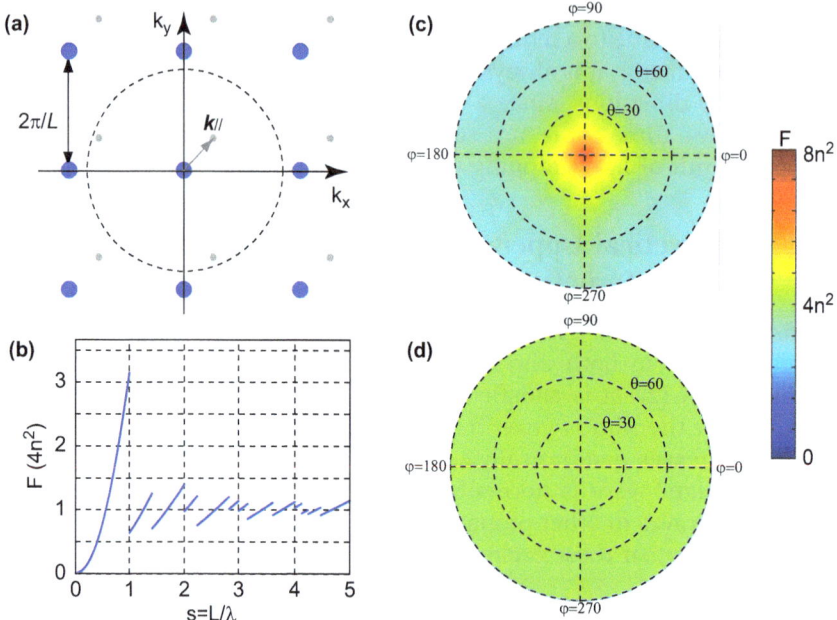

Figure 2.14 (a) Channels in 2D k-space. Blue and grey dots represent channels when the incident light comes in from normal and off-normal directions, respectively. The radius of the circle is k_0. (b) Limit of absorption enhancement in 2D grating structures. (c) Angular response of the average enhancement factor integrated over wavelength from $\lambda = p$ to $\lambda = 2p$ for a grating with a periodicity of $L = p$. (d) the same as (c) except the periodicity is $L = 5p$.

where $G_{m,n}$ is defined in eqn (2.20), provided that these parallel wavevectors are located in a circle in k-space $|k| < k_0$, as is required for propagating plane waves.

In comparison to the normal incident case described by eqn (2.20), eqn (2.21) defines a similar square lattice in the wavevector space, except that the positions of the lattice points in k-space are now shifted by the parallel wavevector $k_{//}$ of the incident plane wave (Figure 2.14(a)). As a result of such a shift, one can see that the number of channels N, which is proportional to the number of k-space lattice points that fall within the circle of $|k| < k_0$, is dependent on the incidence angle. Such angular dependency is particularly strong for the case where there are only small numbers of channels, which occurs when the periodicity is about wavelength scale.

As an example, Figure 2.14(a) illustrates the case where the normalized frequency $s = L/\lambda = 0.87 < 1$. For normal incidence, there is only one k-space lattice point (as shown by the blue dot in Figure 2.14(a)) within the circle of $|k| < k_0$ (dashed line in Figure 2.14(a)), corresponding to two channels. At the same frequency, for off-normal incident light with $k_{//} = 0.3k_0\hat{x} + 0.3k_0\hat{y}$, corresponding to a plane wave incident from a direction as defined by

$\theta = 25°$, $\varphi = 45°$, there are three k-space lattice points within the circle of $|k| < k_0$ (grey dots in Figure 2.14(a)), and hence six available channels. Thus, for a plane wave incident from such an off-normal incident direction, the upper limit for absorption enhancement should be only 1/3 of the upper limit for normal incident light. This example illustrates that there can be substantial angular dependency in absorption enhancement when the periodicity is comparable to the wavelength. In contrast, when the periodicity is much larger than the wavelength, the k-space lattice points are densely distributed and the number of channels $N \gg 1$. Therefore, the shift of the lattice points in k-space due to different incidence angles has negligible effect on the total number of channels. In the case where $L \gg \lambda$, the upper limit of enhancement is not sensitive to the incidence angles.

In Figure 2.14(b), we provide detailed analytic results regarding the angular dependency of the upper limit of absorption enhancement. We consider a grating with a period $L = p$. For each direction of incidence as specified by an angle of incidence θ and an azimuthal angle of φ, we calculate the upper limit F at each frequency using eqn (2.8). We then average the upper limit F calculated over the wavelength range between $\lambda = p$ to $\lambda = 2p$, and plot the spectral average $\bar{F}$ as a function of θ and φ in Figure 2.14(c). In this case, the grating period is smaller than the wavelength of interest. In the vicinity of the normal direction, $\bar{F}$ is approximately $8n^2$. Thus, it is possible to use a grating structure to obtain broad-band enhancement of absorption higher than $4n^2$ (Figure 2.14(c)). This result is consistent with Figure 2.14(b): the wavelength range corresponds to a range of $0.5 \leq s \leq 1$, where the enhancement factor is above $4n^2$. However, when the incident light deviates from normal direction, $\bar{F}$ starts to drop (Figure 2.14(c)). For incidence angles larger than $60°$, $\bar{F}$ drops well below $4n^2$, showing a strong angular dependency.

As a second example, we consider the case where the period $L = 5p$ nm, and the same wavelength range, which for this grating periodicity corresponds to the normalized frequency range $2.5 \leq s \leq 5$. In this case, the period is considerably larger than the wavelength, and the number of channels is much larger than 1. The spectrally-averaged upper-limit of enhancement has much weaker angular dependency. As shown in Figure 2.14(d), $\bar{F}$ is around $4n^2$ for all incidence angles, showing a near-isotropic response.

Finally, we briefly comment on the influence of the lattice periodicity of the grating structure. The analysis above has focused on square lattices. In practice, a triangular lattice is often found in closely packed nanoparticles and nanowires.[54] For a grating with a triangular lattice with period L, the channels form a triangular lattice in k-space (Figure 2.15(a)). The distance between the channel at the origin of k-space and its nearest neighbors is $k_T = (2/\sqrt{3})2\pi/L$, which is larger than that of the square lattice $k_S = 2\pi/L$. As a result, for normally incident light, the frequency range where the grating operates with only two channels is larger $(0 < s < 2/\sqrt{3})$ compared to the case of the square lattice $(0 < s < 1)$. This leads to a higher maximum

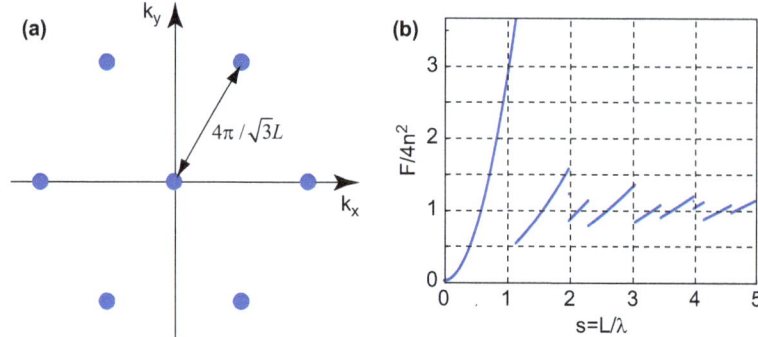

Figure 2.15   (a) Channels in 2D *k*-space for a grating with triangular lattice periodicity. The lattice constant is *L*. (b) Upper limit of absorption enhancement for gratings with triangular lattice periodicity.

enhancement factor $8n^2\pi/\sqrt{3}$ (Figure 2.15(b)). On the other hand, as the period becomes larger, the maximum enhancement ratios in gratings with different lattices all converge to the same bulk limit of $4n^2$.

To conclude this section on the theoretical analysis of 2D grating structures, we note that a grating with wavelength scale periodicity can achieve an absorption enhancement factor that is higher than $4n^2$ over a broad-range of wavelengths. Such enhancement, however, comes at the expense of substantial angular dependency. As a result, a grating structure in general, when the film thickness is a few wavelength thick, cannot overcome the conventional bulk limit of $4n^2/\sin^2\theta$. On the other hand, a grating structure with periodicity much larger than the wavelength has an enhancement factor approaching $4n^2$ with near-isotropic response.

2.5   Nanophotonic Light Trapping Beyond Conventional Limit

2.5.1   Light Trapping in Thin Films

We have described the light trapping properties for films with a thickness d much larger than the wavelength. In these situations, we use eqn (2.9) to count the number of resonances in the films. When the thickness d of the film is comparable to half the wavelength in the material, one can reach the single-mode regime where the film supports a single waveguide mode band for each of the two polarizations. In such a case, eqn (2.9) is no longer applicable. Instead, the number of resonances in the frequency range of $[\omega, \omega + \Delta\omega]$ can be calculated as:

$$M = 2 \times \frac{2\pi n_{wg}^2 \omega}{c^2} \left(\frac{L}{2\pi}\right)^2 \delta\omega \qquad (2.22)$$

where the first factor of two arises from counting both polarizations. (Here, to facilitate the comparison to the standard conventional limit, for simplicity, we have assumed that the two polarizations have the same group index n_{wg}). Notice that in this case the number of modes no longer explicitly depends upon the thickness d of the film.

In order to highlight the effect of such strong light confinement, we choose the periodicity to be a few wavelengths, in which case the number of channels can still be calculated using eqn (2.10). As a result we obtain the upper limit for the absorption enhancement factor:

$$F = 2 \times 4n_{wg}^2 \frac{\lambda}{4n_{wg}d} V \qquad (2.23)$$

where the factor $V = \frac{\alpha_{wg}}{\alpha_0}$ characterizes the overlapping between the profile of the guided mode and the absorptive active layer α_{wg} and n_{wg} are the absorption coefficient and group index of the waveguide mode respectively.

Eqn (2.23) in fact becomes $4n^2$ in a dielectric waveguide of $d \approx \lambda/2n$. Therefore, reaching a single-mode regime is not sufficient to exceed the conventional limit. Instead, to achieve the full benefit of nanophotonics, one must either ensure that the modes exhibit deep-subwavelength-scale electric-field confinement, or enhance the group index to be substantially larger than the refractive index of the active material, over a substantial wavelength range. Below, using both exact numerical simulations and analytic theory, we will design geometries that simultaneously satisfy both these requirements.

2.5.2 Light Trapping Enhancement Beyond the Classical Limit Using Nanoscale Modal Confinement

Guided by the theory above, we now numerically demonstrate a nanophotonic scheme with an absorption enhancement factor significantly exceeding the conventional limit. Motivated by the case of organic photovoltaic materials studied in the previous section, we consider a thin absorbing film with a thickness of 5 nm (Figure 2.16(a)), consisting of a material with a refractive index $n_L = \sqrt{2.5}$ and a wavelength-independent absorption length of 25 μm. The film is placed on a mirror that is approximated to be a perfect electric conductor (PEC). A PEC mirror is used for simulation convenience. In practice, it can be replaced by a dielectric cladding layer, which produces similar results.[24] Our aim here is to highlight the essential physics of nanophotonic absorption enhancement. The choice of material parameters therefore represents a simplification of actual material response. Nevertheless, we note that both the index and the absorption strength here are characteristic of typical organic photovoltaic absorbers in the weakly absorptive regime.[27] Furthermore, there is general interest in using thinner absorbers in organic solar cells given their short exciton diffusion lengths of about 3–10 nm.[55–57]

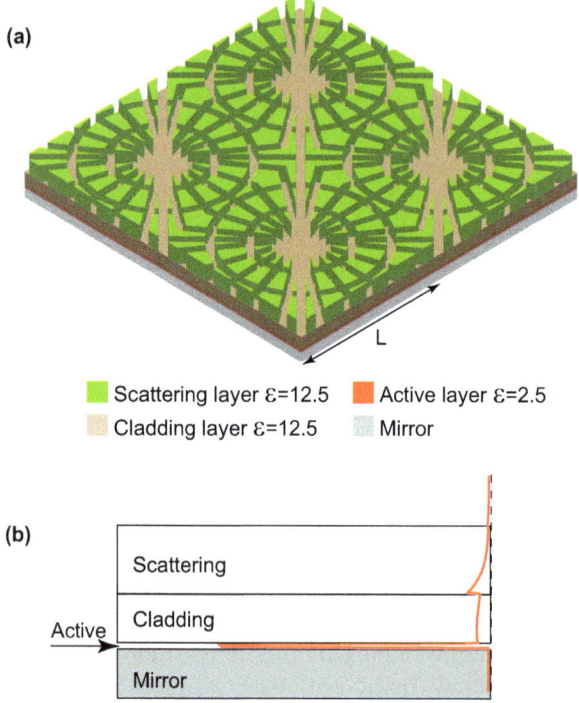

(a)

Scattering layer ε=12.5   Active layer ε=2.5
Cladding layer ε=12.5   Mirror

(b)

Scattering

Active →   Cladding

Mirror

Figure 2.16   Structure for overcoming the conventional light trapping limit. (a) A nanophotonic light-trapping structure. The scattering layer consists of a square lattice of air groove patterns with periodicity $L = 1200$ nm. The thicknesses of the scattering, cladding, and active layers are 80 nm, 60 nm, and 5 nm respectively. The mirror layer is a perfect electric conductor. (b) The profile of electric-field intensity for the fundamental waveguide mode. Fields are strongly confined in the active layer. To obtain the waveguide mode profile, the scattering layer is modeled by a uniform slab with an averaged dielectric constant.

In order to enhance the absorption in the active layer, we place a transparent cladding layer ($n_H = \sqrt{12.5}$) on top of the active layer. Such a cladding layer serves two purposes. First, it enhances density of state. The overall structure supports a fundamental mode with group index n_{wg} close to n_H, which is much higher than that of the absorbing material. Second, the index contrast between active and cladding layer provides nanoscale field confinement. Figure 2.16(b) shows the fundamental waveguide mode. The field is highly concentrated in the low-index active layer, due to the well-known slot-waveguide effect.[58] Thus, the geometry here allows the creation of a broad-band high-index guided mode, with its energy highly concentrated in the active layer, satisfying the requirement for high absorption enhancement.

In order to couple incident light into such nanoscale guided modes, we introduce a scattering layer with a periodic pattern on top of the cladding layer, with a periodicity L much larger than our wavelength ranges of interest. Each unit cell consists of a number of air grooves. These grooves are oriented along different directions to ensure that scattering strength does not strongly depend on the angles and polarizations of the incident light. We emphasize that there is no stringent requirement on these grooves as long as the scattering strength dominates over resonance absorption rates.

We simulate the proposed structure by numerically solving Maxwell's equations (Figure 2.17(a)). The device has a spectrally-averaged absorption enhancement factor of $F = 119$ (red line) for normally incident light. (All the

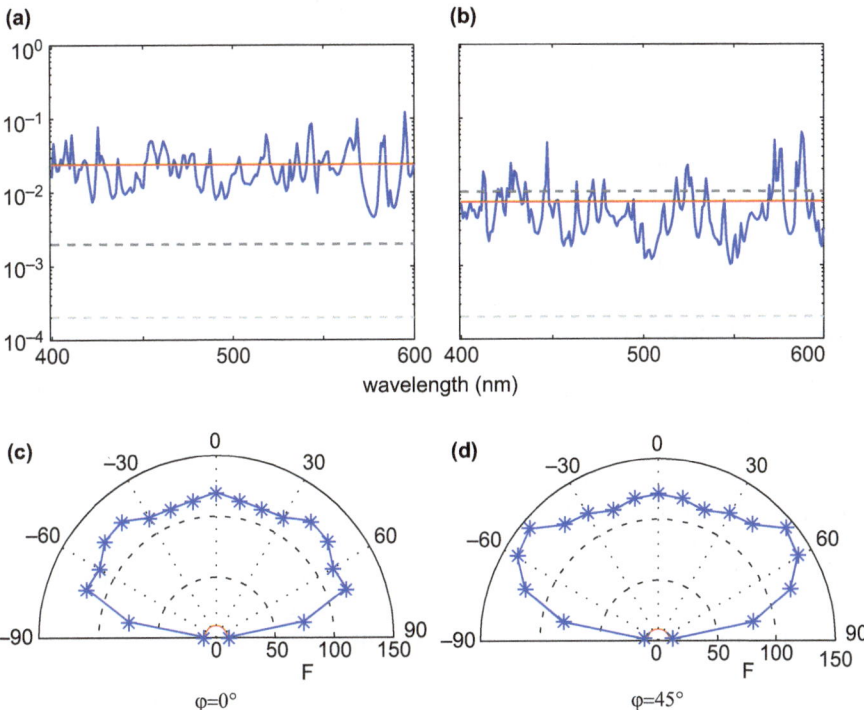

Figure 2.17 Absorption of light trapping structures. (a) Absorption spectrum for normally incident light for the structure shown in Figure 2.16(a). The spectrally-averaged absorption (red solid line) is much higher than both the single pass absorption (light gray dashed line) and the absorption as predicted by the limit of $4n_L^2$ (dark gray dashed line). The vertical axis is the absorption coefficient. (b) Absorption spectrum without nanoscale light confinement. The structure is the same as that of (a) except that the dielectric constant of the active layer is now the same as the cladding layer. The dark gray dashed line represents the absorption as predicted by the limit of $4n_H^2$ (c, d) Angular dependence of the spectrally-averaged absorption enhancement factor for the structure in Figure 2.16(a). Incident angles are labeled on top of the semi-circles. Incident planes are oriented at 0 (c) and 45 (d) degrees (azimuthal angles) with respect to the [10] direction of the lattice. The red circles represent the $4n_L^2$ limit.

absorption spectra and enhancement factors are obtained by averaging *s* and *p* polarized incident light.) This is well above the conventional limit for both the active material $(4n_L^2 = 10)$ and the cladding material $(4n_H^2 = 50)$. Moreover, the angular response is nearly isotropic (Figure 2.17(c,d)). Thus such enhancement cannot be attributed to the narrowing of angular range in the emission cone, and instead is due entirely to the nanoscale field confinement effect.

Using our theory, we calculate the theoretical upper limit of light trapping enhancement in this structure.[24] For wavelength $\lambda = 500$ nm, we obtain an upper limit of $F = 147$. The enhancement factor observed in the simulation is thus consistent with this predicted upper limit. The actual enhancement factor obtained for this structure falls below the calculated theoretical upper limit because some of the resonances are not in the strong over-coupling regime.

To illustrate the importance of nanoscale field confinement enabled by the slot-waveguide effect, we change the index of the material in the absorptive layer to n_H. Such a structure does not exhibit the slot-waveguide effect. The average enhancement in this case is only 37, falling below the conventional limit of 50 (Figure 2.17(b)).

2.6   Conclusion

We began this chapter by motivating our interest in nanophotonic light trapping with a numerical design study of a practical light trapping scheme for organic photovoltaic cells. We then introduced and developed a statistical coupled mode theory for nanophotonic light trapping, and showed that properly designed nanophotonic structures can achieve enhancement factors that far exceed the conventional limit. Our results presented here indicate substantial opportunity for nanophotonic light trapping using only low-loss dielectric components. The basic theory moreover, is generally applicable to any photonic structure, including nanowire[59,60] and plasmonic structures.[61] In plasmonic structures in particular, the presence of nanoscale guided modes may also provide opportunities to overcome the conventional limit.

Next-generation photovoltaic systems will seek to exploit the benefits of nanophotonics to continue highly-efficient power generation at increasingly lower costs. The theory presented in this chapter thus provides practical physical intuition and design guidance that will allow solar cell designers to optimize light absorption and cell performance in these next-generation systems. The theory also outlines paths to exceeding conventional limits on light trapping that can allow extremely thin, nanoscale active layers to generate as much power as much thicker layers. Conventional ray-optics light trapping theory proved to be enormously influential in the design of solar cells over the past three decades. Similarly, we anticipate that the nanophotonic light trapping theory presented here will influence the design of next-generation thin-film photovoltaic systems.

References

1. K. Taretto and U. Rau, *Progress in Photovoltaics: Research and Applications*, 2004, **12**, 573–591.
2. E. Yablonovitch, *J. Opt. Soc. Am. A*, 1982, **72**, 899–907.
3. A. Goetzberger, in *Fifteenth IEEE Photovoltaic Specialists Conference – 1981*. IEEE, 1981, 867–870.
4. P. Campbell and M. A. Green, *IEEE Trans. Electron Devices*, 1986, **33**, 234–239.
5. P. Sheng, A. N. Bloch and R. S. Stepleman, *Appl. Phys. Lett.*, **43**, 579–581.
6. P. Bermel, C. Luo, L. Zeng, L. C. Kimerling and J. D. Joannopoulos, *Opt. Express*, 2007, **15**, 16986–17000.
7. L. Hu and G. Chen, *Nano Letters*, 2007, 7, 3249–3252, DOI: 10.1021/nl071018b.
8. A. Chutinan and S. John, *Phys. Rev. A (At., Mol., Opt. Phys.)*, 2008, **78**, 023825–023815.
9. H. R. Stuart and D. G. Hall, *J. Opt. Soc. Am. A*, 1997, **14**, 3001–3008.
10. I. Tobias, A. Luque and A. Marti, *J. Appl. Phys.*, 2008, **104**, 034502.
11. P. N. Saeta, V. E. Ferry, D. Pacifici, J. N. Munday and H. A. Atwater, *Opt. Express*, 2009, **17**, 20975–20990.
12. R. A. Pala, J. White, E. Barnard, J. Liu and M. L. Brongersma, *Adv. Mat.*, 2009, **21**, 3504–3509.
13. S. B. Mallick, M. Agrawal and P. Peumans, *Opt. Express*, 2010, **18**, 5691–5706.
14. C. Lin and M. L. Povinelli, *Opt. Express*, 2009, **17**, 19371–19381.
15. S. Mokkapati, F. J. Beck, A. Polman and K. R. Catchpole, *Appl. Phys. Lett.*, 2009, **95**, 053115.
16. J. Müller, B. Rech, J. Springer and M. Vanecek, *Sol. Energy*, 2004, 77, 917–930.
17. M. D. Kelzenberg, S. W. Boettcher, J. A. Petykiewicz, D. B. Turner-Evans, M. C. Putnam, E. L. Warren, J. M. Spurgeon, R. M. Briggs, N. S. Lewis and H. A. Atwater, *Nature Mater.*, 2010, **9**, 239–244, doi:http://www.nature.com/nmat/journal/v9/n3/suppinfo/nmat2635_S1.html.
18. E. Garnett and P. Yang, *Nano Lett.*, 2010, **10**, 1082–1087.
19. S. Pillai, K. R. Catchpole, T. Trupke and M. A. Green, *J. Appl. Phys.*, 2007, **101**, 093105–093108.
20. L. Zeng, Y. Yi, C. Hong, J. Liu, N. Feng, X. Duan, L. C. Kimmerling and B. A. Alamariu, *Appl. Phys. Lett.*, 2006, **89**, 111111–111113.
21. L. Tsakalakos, J. Balch, J. Fronheiser, B. A. Korevaar, O. Sulima and J. Rand, *Appl. Phys. Lett.*, 2007, **91**, 233117–233113.
22. C. Rockstuhl, F. Lederer, K. Bittkau and R. Carius, *Appl. Phys. Lett.*, 2007, **91**, 171104–171106.
23. J. Zhu, Z. Yu, G. F. Burkhard, C. M. Hsu, S. T. Connor, Y. Xu, Q. Wang, M. McGehee, S. Fan and Y. Cui, *Nano Lett.*, 2009, **9**, 279–282.
24. Z. Yu, A. Raman and S. Fan, *Proc. Nat. Acad. Sci. USA*, 2010, **107**, 17491–17496.
25. Z. Yu, A. Raman and S. Fan, *Opt. Express*, 2010, **18**, A366.

26. Z. Yu and S. Fan, *Appl. Phys. Lett.*, 2011, **98**, 011106.
27. H. Hoppe and N. S. Sariciftci, *J. Mater. Res.*, 2004, **19**, 1924–1945.
28. A. C. Mayer, S. R. Scully, B. E. Hardin, M. W. Rowell and M. D. McGehee, *Materials Today*, 2007, **10**, 28–33.
29. D. Muhlbacher, M. Scharber, M. Morana, Z. Zhu, D. Waller, R. Gaudiana and C. Brabec, *Adv. Mater.*, 2006, **18**, 2884–2889.
30. S. H. Park, A. Roy, S. Beaupr, S. Cho, N. Coates, J. S. Moon, D. Moses, M. Leclerc, K. Lee and A. J. Heeger, *Nat. Photonics*, 2009, **3**, 297–302.
31. J. Y. Kim, S. Kim, H.-H. Lee, K. Lee, W. Ma, X. Gong and A. Heeger, *Adv. Mater.*, 2006, **18**, 572–576.
32. A. Hayakawa, O. Yoshikawa, T. Fujieda, K. Uehara and S. Yoshikawa, *Appl. Phys. Lett.*, 2007, **90**, 163517.
33. S.-B. Rim, S. Zhao, S. R. Scully, M. D. McGehee and P. Peumans, *Appl. Phys. Lett.*, 2007, **91**, 243501.
34. M. Niggemann, M. Glatthaar, A. Gombert, A. Hinsch and V. Wittwer, *Thin Solid Films*, 2004, **451–452**, 619–623.
35. S.-I. Na, S.-S. Kim, J. Jo, S.-H. Oh, J. Kim and D.-Y. Kim, *Adv. Func. Mater.*, 2008, **18**, 3956–3963.
36. J. R. Tumbleston, D.-H. Ko, E. T. Samulski and R. Lopez, *Opt. Express*, 2009, **17**, 7670–7681.
37. D.-H. Ko, J. R. Tumbleston, L. Zhang, S. Williams, J. M. DeSimone, R. Lopez and E. T. Samulski, *Nano Lett.*, 2009, **9**, 2742–2746.
38. N. C. Lindquist, W. A. Luhman, S.-H. Oh and R. J. Holmes, *Appl. Phys. Lett.*, 2008, **93**, 123308.
39. H. Shen, P. Bienstman and B. Maes, *J. Appl. Phys.*, 2009, **106**, 073109.
40. C. Min, J. Li, G. Veronis, J.-Y. Lee, S. Fan and P. Peumans, *Appl. Phys. Lett.*, 2010, **96**, 133302.
41. S. G. Tikhodeev, A. L. Yablonskii, E. A. Muljarov, N. A. Gippius and T. Ishihara, *Phys. Rev. B*, 2002, **66**, 045102.
42. C. Brabec, V. Dyakonov and U. Scherf, *Organic Photovoltaics: Materials, Device Physics, and Manufacturing Technologies*, Wiley-VCH, 2008.
43. Z. Yu, A. Raman and S. Fan, *Opt. Express*, 2010, **18**, A366–A380.
44. V. Shrotriya, G. Li, Y. Yao, C.-W. Chu and Y. Yang, *Appl. Phys. Lett.*, 2006, **88**, 073508.
45. M. D. Irwin, D. B. Buchholz, A. W. Hains, R. P. H. Chang and T. J. Marks, *Proc. Nat. Acad. Sci.*, 2008, **105**, 2783–2787.
46. H. Wu, L. Hu, T. Carney, Z. Ruan, D. Kong, Z. Yu, Y. Yao, J. Cha, J. Zhu, S. Fan and Y. Cui, *J. Am. Chem. Soc.*, 2011, **133**, 27–29.
47. Z. Yu, N. Sergeant, T. Skauli, G. Zhang, H. Wang and S. Fan, *Nat. Comm.*, 2013, **4**, 1730.
48. P. W. Anderson, *Phys. Rev.*, 1958, **109**, 1492.
49. S. Fan and J. D. Joannopoulos, *Phys. Rev. B*, 2002, **65**, 235112.
50. H. A. Haus, *Waves and fields in optoelectronics*, Prentice-Hall, New Jersey, 1984.
51. S. Fan, W. Suh and J. D. Joannopoulos, *J. Opt. Soc. Am. A*, 2003, **20**, 569–572.

52. C. Kittel, *Introduction to Solid State Physics*, John Wiley & Sons, Inc, New York, 7th edn, 1995, pp. 120–121.
53. C. Heine and R. H. Morf, *Appl. Opt.*, 1995, **34**, 2476–2482.
54. J. Zhu, C.-M. Hsu, Z. Yu, S. Fan and Y. Cui, *Nano Lett.*, 2010, **10**, 1979–1984.
55. A. Mayer, S. Scully, B. Hardin, M. Rowell and M. McGehee, *Mater. Today*, 2007, **10**, 28–33.
56. W. U. Huynh, J. J. Dittmer and A. P. Alivisatos, *Science*, 2002, **295**, 2425–2427.
57. G. Yu, J. Gao, J. C. Hummelen, F. Wudl and A. J. Heeger, *Science*, 1995, **270**, 1789–1791.
58. V. R. Almeida, Q. Xu, C. A. Barrios and M. Lipson, *Opt. Lett.*, 2004, **29**, 1209–1211.
59. M. Law, L. E. Greene, J. C. Johnson, R. Saykally and P. Yang, *Nat. Mater.*, 2005, **4**, 455–459.
60. B. M. Kayes, H. A. Atwater and N. S. Lewis, *J. Appl. Phys.*, 2005, **97**, 114302.
61. H. A. Atwater and A. Polman, *Nat. Mater.*, 2010, **9**, 205–213.

CHAPTER 3

Micro/nano Fabrication Technologies for Vibration-Based Energy Harvester

BIN YANG* AND JINGQUAN LIU

National Key Laboratory of Science and Technology on Micro/Nano Fabrication, Institute of Micro-Nano Science and Technology, Shanghai Jiao Tong University, Shanghai, China
*Email: binyang@sjtu.edu.cn

3.1 Introduction

Energy harvesting technologies have emerged as a prominent research topic and have been quickly developed in recent years due to their wide application, such as wireless sensor nodes for structural health monitoring, biomedical applications for implanted sensors, automobile applications for tire pressure sensors. Traditionally, electronic devices have relied on batteries for power as they are reliable, easily accessible and convenient to use. However, batteries can only provide energy over a finite period of time, after which they will have to be changed. Since batteries have to be replaced periodically, their usage is limited to applications in which battery replacement is convenient. This suggests that for autonomously operating remote devices where battery replacement is difficult, batteries are not a perfect solution in the long run. In view of this, researchers have been motivated to harvest energy from the environment to power such remote sensors, with ambient vibrations being one such source of energy. By combining an energy

RSC Nanoscience & Nanotechnology No. 32
Nanofabrication and its Application in Renewable Energy
Edited by Gang Zhang and Navin Manjooran
© The Royal Society of Chemistry 2014
Published by the Royal Society of Chemistry, www.rsc.org

harvester with a rechargeable battery, we can create battery-based power source of infinite lifetime ideally. Typically, the vibration-based energy harvesting devices employ one of the following energy transduction mechanisms: electrostatic, piezoelectric and electromagnetic mechanisms; or a hybrid of two of them improves their output performance. Moreover, with the emerging development of nanotechnology, the nanoelectronic devices and systems have been developed. This chapter demonstrates the micro/nanomachining technologies for the fabrication of micro/nanoscale energy harvesters including electrostatic and piezoelectric mechanisms. Some novel materials, structural optimized design, and process integration of micro- to nanoscale energy harvesters will be presented in detail in order to improve the output power and conversion coefficient.

3.2 MEMS-based Electrostatic Energy Harvester

Electrostatic energy harvesters employ either comb finger electrodes or parallel-plate electrodes as variable capacitors that are biased with external voltage sources or charged by an electret. During operation, the movable electrode of the capacitor electrodes will engage and disengage due to external vibrations such that the created capacitor is varied as a function of ambient vibrations. Electrostatic harvesters can be classified into three types: in-plane overlap varying, in-plane gap closing and out-of-plane gap closing, as shown in Figure 3.1. Roundy[1,2] demonstrates that the harvester of in-plane gap closing generates the highest output power density of 100 $\mu W\ cm^{-3}$ after an optimized design; out-of-plane gap closing is the next highest followed by in-plane overlap varying. Maximum power generation occurs for very small dielectric gaps.

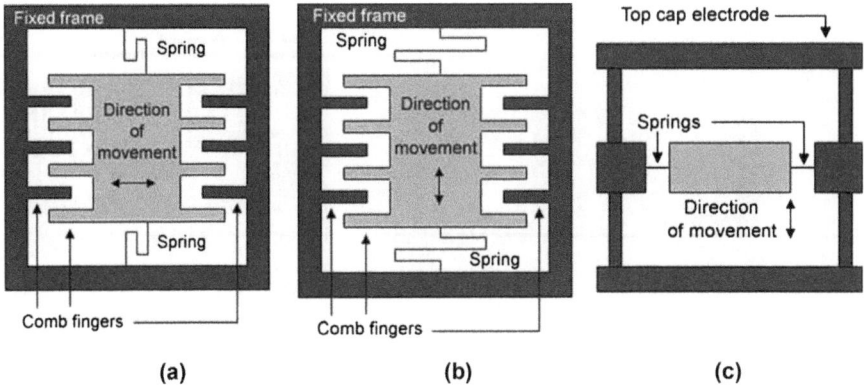

Figure 3.1 Three types of electrostatic energy harvester: (a) in-plane overlap varying; (b) in-plane gap closing; (c) out-of-plane gap closing.

3.2.1   Electrostatic Harvester with Bias Voltage

Hoffmann *et al.* have reported that the energy transduction coefficient of the electromagnetic mechanism decreases ten times faster than the one of the electrostatic mechanism when the volume of the MEMS energy harvester shrinks to 1% of the original size.[3] Therefore, the electrostatic MEMS energy harvester at microscale is the favored approach. The completely packaged electrostatic harvester with a package volume of about 0.2 cm[3] is proposed, as shown in Figure 3.2(a).[3] The test board contains two load resistors R_1 and R_2 (560 kΩ each) and a pair of multilayer ceramic capacitors with a total capacity of 1 μF. Figure 3.2(b) shows the corresponding comb electrodes interconnected on the chip *via* conductor paths to form two variable capacitors. Figure 3.3 shows the fabrication process of the harvester. The dry etching process is performed to form a cavity of 50 μm depth for free movement of the proof mass (Figure 3.3(b)). A highly p-doped device wafer is bonded onto the substrate wafer and thinned to the required thickness by chemical mechanical polishing (Figure 3.3(c)). The trench refill technology allows the fabrication of track crossing (Figure 3.3(d)) and conductor tracks are formed by wet etching of a 500 nm thick aluminium layer (Figure 3.3(e)). The device layer is etched by the dry process to form the proof mass, suspensions and comb electrodes (Figure 3.3(f)). The cap wafer (Figure 3.3(g)) with cavities is bonded together using a glass frit bonding technology (Figures 3.3(h, i)). The output power of this fabricated harvester is 3.5 μW when the matched loading resistor is 2$\times$560 kΩ. Because the bias voltage affects the output power, the bias voltage will become a significant design parameter with respect to the excitation conditions of the corresponding applications.

Currently, most of the electrostatic mechanisms are designed to harvest kinetic energy of a linear motion because the reported electrostatic energy harvesters comprise a set of springs, which is softer regarding to the moving direction of a mass along the first axis, *i.e.*, a linear motion, and is stiffer in

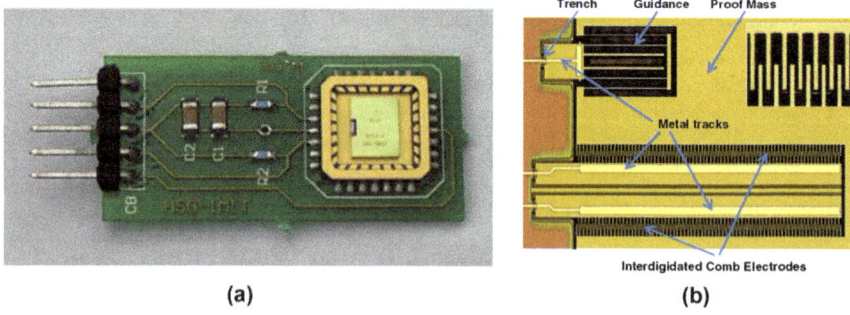

| (a) | (b) |

Figure 3.2   Packaged electrostatic energy harvester prototype: (a) Packaged energy harvester integrated with a PCB board; (b) Microscopic close-up view of the electrostatic structure.[3]

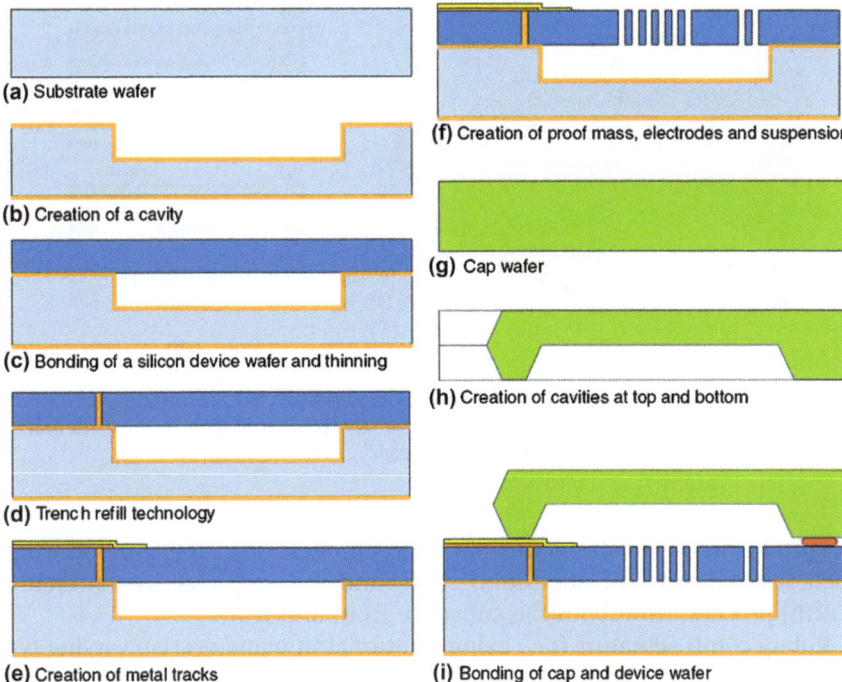

(a) Substrate wafer

(b) Creation of a cavity

(c) Bonding of a silicon device wafer and thinning

(d) Trench refill technology

(e) Creation of metal tracks

(f) Creation of proof mass, electrodes and suspension

(g) Cap wafer

(h) Creation of cavities at top and bottom

(i) Bonding of cap and device wafer

Figure 3.3 The fabrication of packaged harvester prototype.[3]

the two axes perpendicular to the first moving axis. However, the vibrations from the nature resources may not always be along one direction. It is common to see a pendulum-like vibration swing trace, *i.e.*, planar motion; it points out that planar vibrations comprising two components of kinetic energy along two axes are common in the natural environment. Besides, various ambient vibration sources have been reported.[4] For instance, the vibration frequency of a laptop during normal operation is 90.2 Hz, but that of the running compact disc read-only memory is 43.2 Hz.[5] Small microwave ovens have a peak acceleration of 2.5 m/s^2 at the resonant frequency of 121 Hz, washing machine have a peak acceleration of 0.5 m s^{-2} at 109 Hz, the casing on a kitchen blender has an acceleration of 6.4 m s^{-2} at 121 Hz, and a CD on a notebook computer has a peak acceleration of 0.6 m s^{-2} at 75 Hz.[4] A concept of a resonant rotating mass fixed on a spiral spring has been proposed as an inertial mechanism for harvesting kinetic energy from a rotational motion, *i.e.*, a kind of planar motion.[6,7] The design of the 2D resonator is shown schematically in Figure 3.4(a),[6] which consists of a seismic mass suspended by a circular spring system. The in-plane resonant frequency of this device is calculated, by finite element analysis, to be about 1230 Hz when the height of the springs and bridges and the spaced gap between the various concentric components are 230 μm and 150 μm, respectively. The resonant frequency varies with the square root of the spring

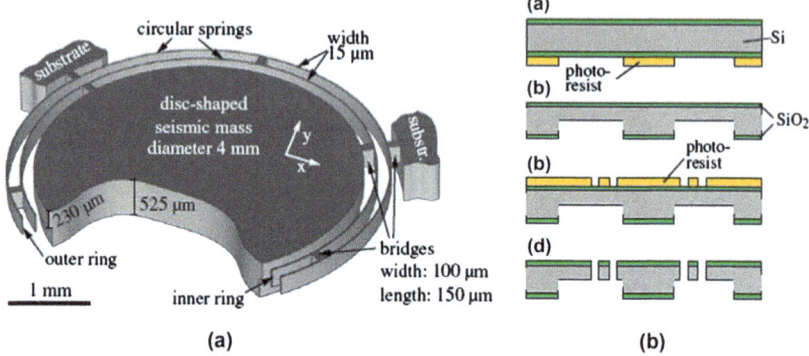

(a)

(b)

Figure 3.4   The design and fabrication process of 2D resonator for energy harvesting: (a) schematic view of 2D resonator with dimensions; (b) resonator fabrication process.[6]

height. Figure 3.4(b) shows the fabrication process of a 2D resonator, which includes two cycles of photolithography and the deep reactive-ion etching (DRIE) process, from both the substrate front and rear.

Rotary comb actuators (*i.e.*, using rotational in-plane overlap varying type electrodes), with a pivot or a virtual anchor at the center of the circle formed by the rotary comb, have been investigated as actuators for driving a vertical micromirror to move along a trace formed by radial line scanning against the pivot.[8,9] The rotary comb structure is typically formed by etching and releasing the vertical comb finger structure from the silicon device layer of a silicon-on-insulator (SOI) substrate. With the aid of integrated micromirrors, rotary comb actuators have been successfully applied to tunable external cavity diode lasers, variable optical attenuators and optical switches.[10–13]

Yang *et al.*[14] have proposed a MEMS energy harvester using rotary comb electrodes, which converts the kinetic energy of planar vibration from ambient sources and stores it as electrical energy. A schematic drawing of the one-sixth portion of the rotary comb based electrostatic energy harvester is shown in Figure 3.5(a). The suspended silicon structure consists of the 12 sets of fixed combs, a grid-like proof mass integrated with movable combs, six ladder springs and stoppers for avoiding pull-in phenomenon between movable and fixed combs. Figure 3.5(b) shows the lumped model for one set of movable and fixed combs. When the movable combs are applied with the excited acceleration, the initial overlap θ_0 is rotated by an angle θ, which leads to the total overlap angle $\theta_0 + \theta$ between the concentric interdigitated combs. All of these gaps are assumed to be parallel plates to simplify the derivation model.[15] Hence, the total capacitance can be calculated by Gauss's law:

$$C_{total} = C_{overlap} + C_{gap} \tag{3.1}$$

where $C_{overlap}$ is the overlap capacitance and C_{gap} is the gap capacitance.

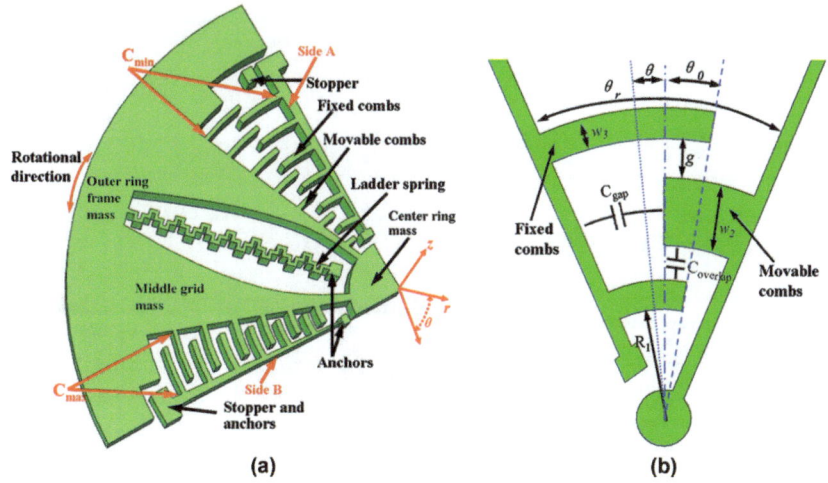

Figure 3.5 The structural design of rotary energy harvester: (a) schematic view of 2D rotary harvester; (b) the lumped models for one set of fixed and movable combs.[14]

The overlap capacitance for comb fingers is given by:

$$C_{overlap} = \varepsilon t(\theta + \theta_0) \times A(n) \tag{3.2}$$

where:

$$A(n) = \sum_{i=1}^{n} \left\{ \left[\ln\left(\frac{R_1 + 2(i-1)(W_f + g)}{R_1 + 2(i-1)(W_f + g) - g} \right) \right]^{-1} \right.$$
$$\left. + \left[\ln\left(\frac{R_1 + (2i-1)(W_f + g)}{R_1 + (2i-1)(W_f + g) - g} \right) \right]^{-1} \right\} - \left[\ln\left(\frac{R_1}{R_1 - g} \right) \right]^{-1}$$

and $W_f = (w_2 + w_3)/2$

The gap capacitance can be calculated by:

$$C_{gap} = \varepsilon t W_f B(n) \frac{1}{\dfrac{\theta_r - \theta_0}{2} - \theta} \tag{3.3}$$

where $B(n) = \sum_{i=1}^{n} \left\{ \frac{1}{R_1 + 2(i-1)(W_f + g)} + \frac{1}{R_1 + 2(i-1)(W_f + g)} \right\}$ and ε is the free-space permittivity with a value of 8.854×10^{-6} pF μm^{-1}, t is the thickness of combs, R_1 is the radius of the concave surface of the first movable combs, g is the gap between fixed and movable combs, w_2 is the width of movable combs, w_3 is the width of fixed combs. θ_r is the initial angle between fixed and movable arms.

Figure 3.6(a) shows the micromachining process of a rotary comb energy harvester based on a SOI wafer with a 30 μm thick heavily doped device layer.

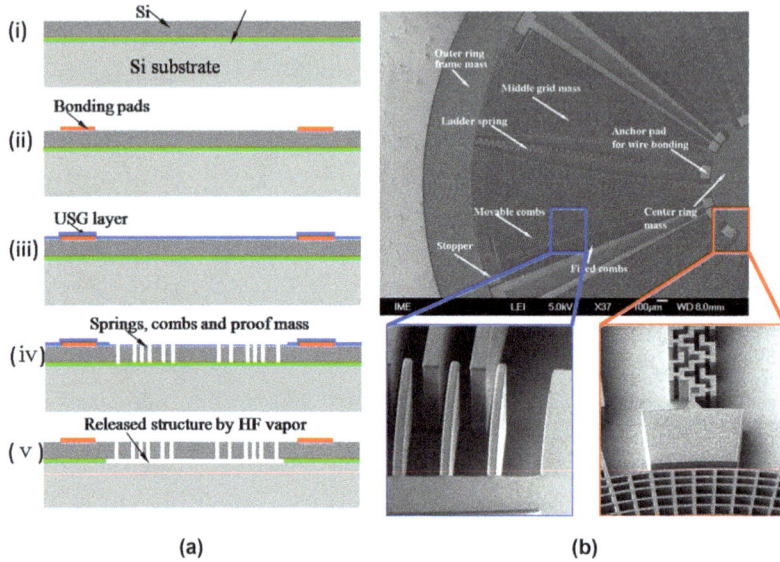

Figure 3.6 The energy harvester based on a rotary comb: (a) fabrication process; (b) fabricated rotary harvester.[14]

A 0.75 μm aluminium metal layer is sputtered on the SOI wafer and patterned to form wire bonding pads. A 5000 Å PECVD undoped silicon glass (USG) is deposited as an insulating layer. After a photolithography pattern transfer process, the hard mask layer of 5000 Å USG is prepared on the silicon device layer by reactive-ion etching (RIE). As shown in step (iii) of Figure 3.6(a), the rotary comb energy harvester is made from the 30 μm silicon device layer of the SOI wafer by DRIE technology. In this step, the mixture gas of C_4F_8, SF_6 and O_2 is provided in an inductively coupled plasma (ICP) system, where C_4F_8 is used as the passivation precursor, SF_6 and O_2 are served as the etching gases. In order to avoid a crack in the process of dicing after the release of whole structure, the dicing process for small chips is employed first before the release process. HF vapor is used to etch away silicon oxides layer in order to release the energy harvester structure from the substrate. Figure 3.6(b) shows the top view of the fabricated devices and an enlarged view of the ladder spring and combs.

Figure 3.7(a) and (b) shows the prototype after wire bonding to a dual-in-line package (DIP) with 40 footprints and a vacuum-packaged prototype, respectively. The measured output powers under different accelerations in air and at the pressure level of 3 Torr are also shown in Figure 3.7(a) and (b), respectively. The maximum measured output power for vibrations of 0.5 g, 1 g, 1.5 g, 2 g and 2.5 g is 0.11 μW, 0.17 μW, 0.24 μW, 0.3 μW and 0.35 μW, respectively, when the loading resistance matches the parasitic resistance of 80 MΩ at the resonant frequency. Compared to the one in air, the maximum electrical output power of 0.25 g acceleration at the pressure level of 3 Torr has obviously improved and arrives at 0.39 μW.

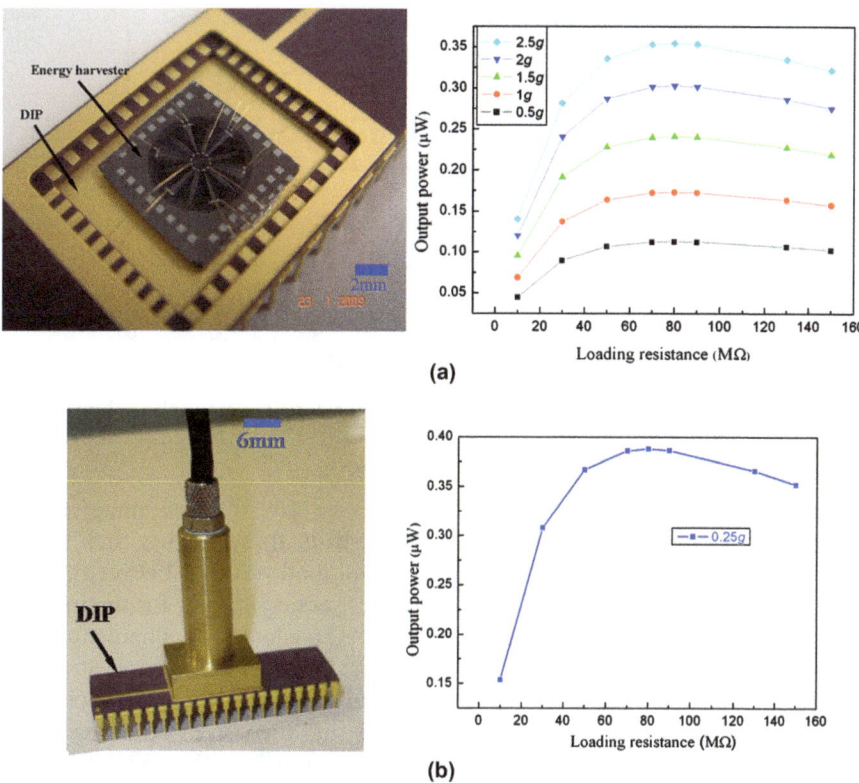

Figure 3.7 (a) The testing output power under air conditions; (b) The testing output power at the pressure level of 3 Torr.[14]

With consideration of the air damping effect, it is concluded that the out-of-plane gap closing mechanism faces significant air damping.[4,16] A nonresonant based electrostatic MEMS of the out-of-plane gap closing mechanism has been proposed by Miao *et al.*[17] Output voltages of up to 220 V were obtained in which it is referring to a net generated power of 120 nJ per cycle. On the other hand, Kuehne *et al.* has reported a resonant based electrostatic MEMS of the out-of-plane gap closing mechanism in 2008.[18] This device provides an output power of 4.28 µW under vibration with a frequency of 1 kHz and an amplitude of 1.96 m s^{-2}, *i.e.*, 0.2 g. Due to the air damping effect, the two plates of electrodes cannot contact so as to achieve the maximum capacitance state in the electrostatic energy harvesters. It is the main reason that leads to a low energy harvesting capability for the electrostatic approach. Lee *et al.* have studied how the device configurations can affect energy output.[19] The comparison between three major electrostatic mechanisms (*i.e.*, in-plane overlap, in-plane gap closing and out-of-plane gap closing mechanisms) is obtained by normalizing the output energy with the device volume. The peak value of the energy output density for the in-plane gap closing and the best value for the out-of-plane gap closing mechanism in various cases of load volumes can be

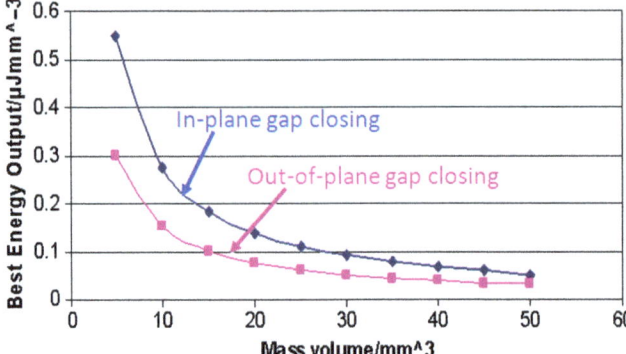

Figure 3.8   Curves of the optimum energy output density versus the volume of the movable mass.[19]

deduced and plotted in Figure 3.8. It is observed that the best energy output density for the in-plane gap closing mechanism is always higher than that of the out-of-plane gap closing mechanism for all load volumes between 5 mm^3 and 50 mm^3. Additionally, the ratio of the best energy output for the in-plane gap closing mechanism to the out-of-plane gap closing mechanism is approximately consistent at 1.8 for load volumes between 5 mm^3 and 50 mm^3.

  Since the operation of the converter depends on the mechanical oscillation of a proof mass, mechanical contact switches that turn on or off according to the position of the proof mass can work synchronously in the converter. Such switches were employed in a nonresonant converter[20] and the switches were modeled as ideal diodes.[21] Chiu *et al.*[22] have proposed a novel integrated mechanical switch for a resonant electrostatic vibration-to-electricity energy harvester. The switches SW1 and SW2 are realized as lateral contact mechanical switches, as shown in Figure 3.9(a) and (b), respectively. Compared to the traditional design with diodes or active electric circuit components, the mechanical switches have some advantages, such as zero leakage current, a very low capacitive coupling effect, low power consumption, synchronous operation and monolithic integration with the whole device structure. Recently Basset *et al.* have presented a novel silicon-based and batch-processed MEMS electrostatic harvester.[23,24] The output power of 61 nW on a 60 MΩ resistive load under a vibration level of 0.25 g at 250 Hz is reported.

3.2.2   Electret-based Electrostatic Harvester

To get rid of the requirement for external bias, an electret layer is introduced to form an electrical potential and then electricity is harvested from alternatively changed relative positions of electrodes due to vibrations. An electret is an insulating material that exhibits a net electrical charge or dipole moment. The net charge or dipole moment in the electret can be used to provide a biasing electric field such as for a MEMS electret microphone[25,26] and electrostatic energy harvester.[27–31]

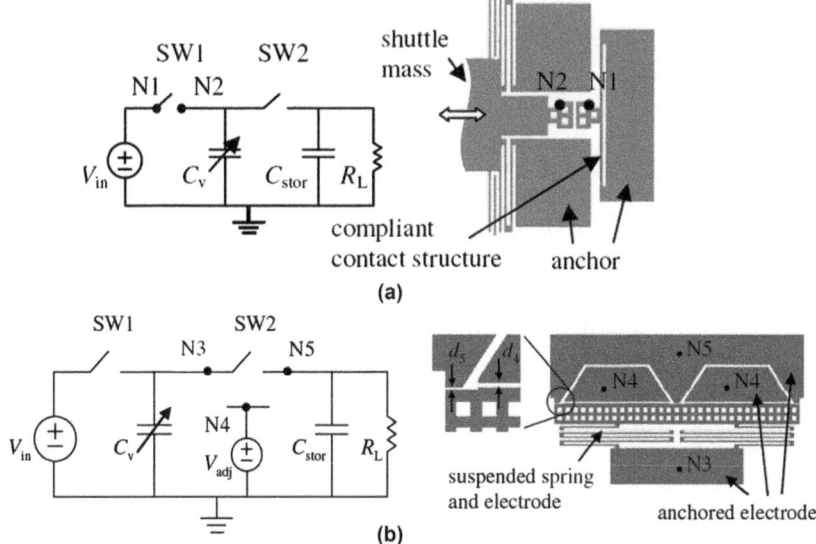

Figure 3.9 The layout design of mechanical switches: (a) mechanical switch SW1: schematic and layout; (b) mechanical switch SW2: schematic and layout.[22]

Tai *et al.* have reviewed some electret materials and compared their charge densities.[32] Tsutsumino *et al.* proposed CYTOP®-based energy harvesters.[33,34] Boland *et al.* developed an energy harvester with Teflon AF® as the electrets.[35–37] Sterken *et al.* demonstrated silicon oxide/silicon nitride electret micro-energy harvesters.[38,39] Among various electret materials, polymers such as Teflon and CYTOP® have a research interest because of their ease of processing. Parylene HT®, a new electrets material, has been reported as an electrostatic harvester, as shown in Figure 3.10,[32] which includes the PEEK rotor block, the stator electrode and the device assembly. 7.32 μm of parylene HT® is deposited onto the PEEK rotor and then charged *via* corona charging. The harvester with parylene HT®-coated PEEK rotors can collect 7.7 μW at 10 Hz and 8.23 μW at 20 Hz.

SiO_2/Si_3N_4 electrets are used for capacitive energy harvesting because of their long charge storage stability, their ability to work in a harsh environment[40] and the compatibility to CMOS processing. However, a relative low volume resistivity of 10^{16} Ω cm would result in a fast decay of the stored charges. Because ion implantation provides a well-controlled charge density and a well-defined energy distribution of the charge density far below the surface, Meschdder *et al.* have reported properties of SiO_2 electret films charged by ion implantation for MEMS-based energy harvesting systems.[41] Although an oxide/nitride electret could have a higher charge density,[42] high-temperature processes often render them inferior to polymer counterparts in certain applications.

In our environment, the natural frequency of most from vibration sources is very low. Low-frequency vibration is only sufficient to excite an extremely low

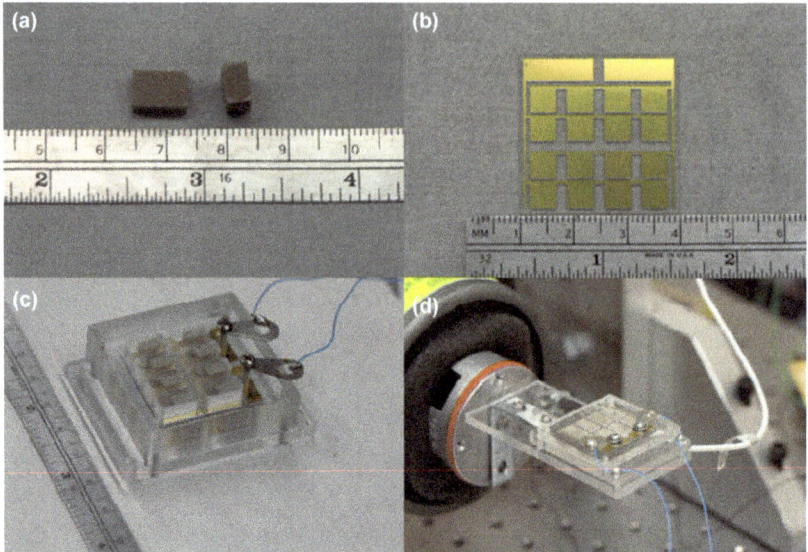

Figure 3.10 An electret generator with parylene HT R-coated PEEK rotors: (a) parylene HT R-coated PEEK rotors, (b) a bare stator, (c) the assembled generator and (d) the generator assembly mounted on a shaker.[32]

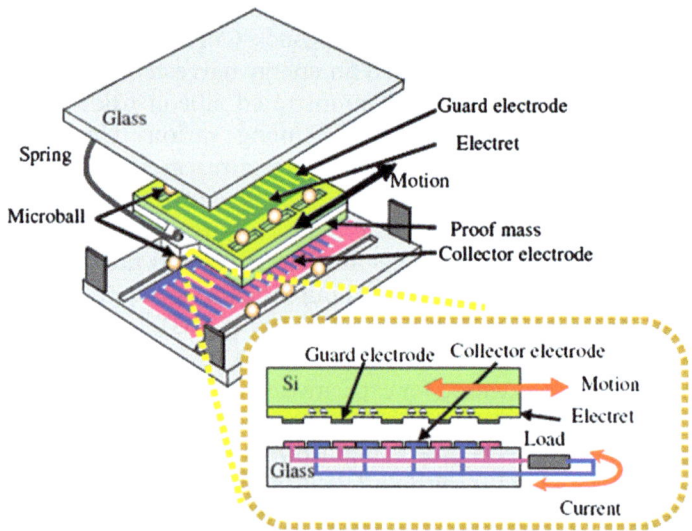

Figure 3.11 The new structure of an electrostatic micro energy harvester.[31]

output power. To overcome this difficult challenge, a new electrostatic micro power generator that (1) can enable both separation gap control and long-range movement at low frequency, and (2) can increase the number of times power generation (electrostatic induction) is proposed.[31] Figure 3.11 shows an electrostatic micro-energy harvester supported on microball bearings. This harvester generates 40 μW of power output at very low-frequency vibration (2 Hz). But energy loss occurred because the mover collided with the edges of the device, which can be improved by refining this design. Recently Bartsch *et al.* have reported a two-dimensional electret-based resonant micro-energy harvester to extract energy from ambient vibrations with arbitrary planar motion directions. It shows that the resonant frequencies and extracted harvester powers depend on the design of the circular spring system and capacitor dimensions within the 0.3–2 kHz and 1–100 pW ranges.[43] In order to optimize the electrets-based electrostatic harvester, the nonlinear dynamic model is proposed and analyzed, showing the importance of properly accounting for the nonlinearity in the optimization process.[44] The important step is the choice of a minimum set of optimization parameters.

3.3 MEMS-based Piezoelectric Energy Harvester

Piezoelectric energy harvesters employ the mechanical strain of piezoelectric elements under loaded force as a result of direct piezoelectric effect. The piezoelectric harvester will become a potential choice when compared with electromagnetic and electrostatic harvesters due to its high energy density.[45] Roundy *et al.*[46] reported a kind of prototype of tiny, piezoelectric cantilever (9–25 mm in length) with a relatively heavy mass on the free end, which can generate 375 mW from a vibration source of 2.5 m s^{-2} at 120 Hz. The scale of the device, however, is larger than that of most MEMS devices; furthermore the device fabrication is limited by manual assembly. The primary reason for choosing the cantilever structure for most reported harvesters is because it is the most compliant structure for a given input force compared to a doubly supported beam and diaphragm.

3.3.1 Sol-gel Process for PZT Film

For the first time, Fang *et al.*[47] successfully developed a PZT MEMS power-generating device based on the d_{31} mode of piezoelectric harvesters that uses top and bottom laminated electrodes. The proof mass on the free end (tip) of the cantilever is used to decrease the structure's natural frequency for low-frequency vibration application. As depicted in Figure 3.12(a), the composite cantilever is made up of an upper piezoelectric thick film, sandwiched between a pair of metal (Pt/Ti) electrodes, and with a lower non-piezoelectric element. The working principle of this device demonstrates as follows: when the base frame of the device is excited by environmental vibration sources, some movable parts of the device will move relatively to the base frame, the relative displacement causes the piezoelectric material in the system to be tensed or compressed. This in turn induces charge shift and accumulation due to the piezoelectric effect. The magnitude of this electric charge voltage is proportional to the stress induced by the relative displacement.

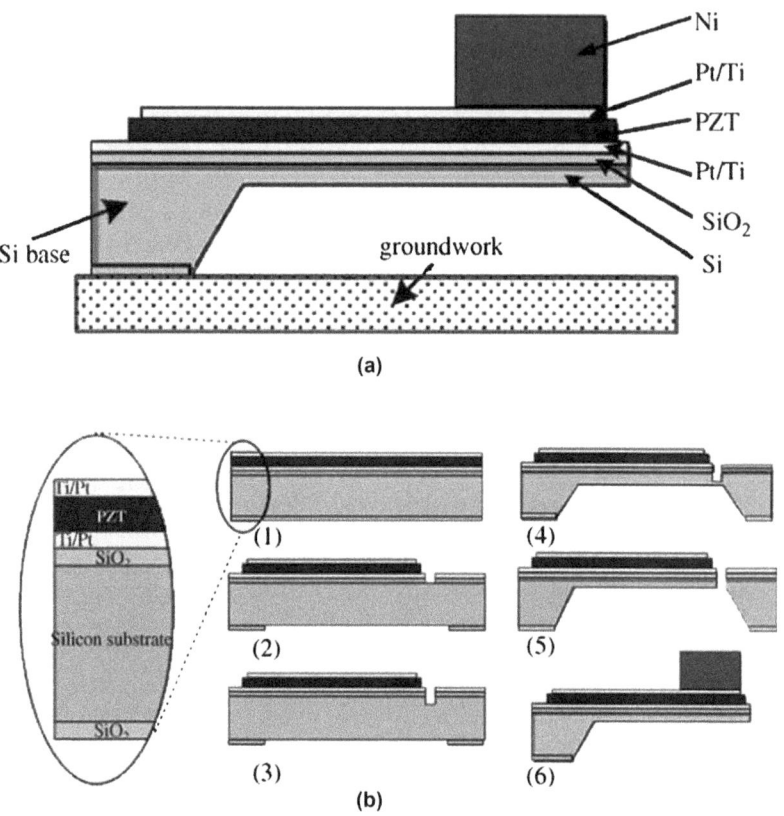

Figure 3.12   The design and fabrication of MEMS-based energy harvester: (a) the structural design of a MEMS-based piezoelectric harvester; (b) the fabricated process of MEMS-based piezoelectric harvester.[47]

Figure 3.12(b) shows the brief fabrication process. The layer of 2 μm thick silicon oxide serves to improve the adhesion of the bottom Ti/Pt electrode to the wafer surface and will act as a mask during the silicon wet etching in the later process. Following that, the bottom electrode, of 30 nm thick Ti and (111) oriented 300 nm thick Pt successively, was sputtered on the oxide layer. And then, PZT films were deposited by the sol-gel method:[48,49] a precursor solution was prepared from Pb acetate, Zr isopropoxide and Ti tetrobutoxide, in 2-methoxyethanol solvent, where the Zr/Ti ratio is 52/48. The 0.25M PZT precursor with 20 mol% excess lead content in solutions was prepared and spin-on coated on Pt/Ti/SiO$_2$/Si substrates at 3000 rpm for 20 s, followed by pyrolysis at 300 °C for 2 min. The process was repeated and PZT films with different coating layers were obtained, respectively. Finally, the rapid thermal annealing process at 650 °C for 30 min was carried out to obtain the perovskite phase PZT film. A crackless PZT film layer of 1.64 mm thick was achieved after depositing 15 coats. On the PZT film, top electrodes Ti/Pt were then sputtered. Figure 3.13(a) demonstrates the photo of the final prototype.

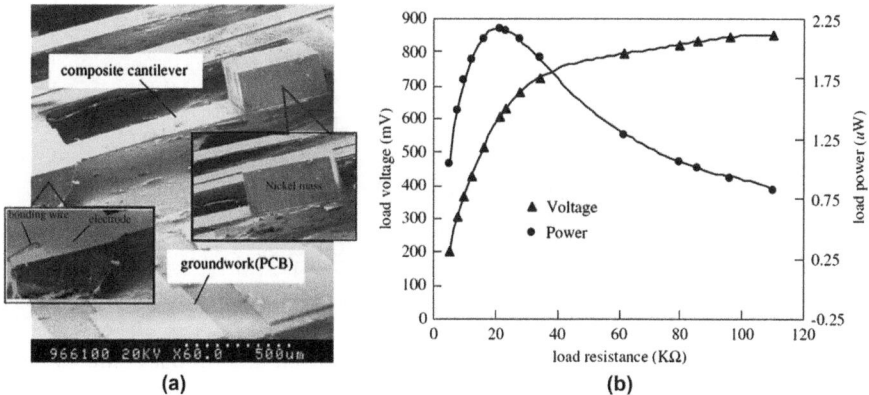

Figure 3.13   The testing of a MEMS-based piezoelectric harvester: (a) SEM photo of MEMS-based harvester; (b) load voltage and power delivered to the load versus resistance.[47]

The dimension of the device is of cantilever length×width: 2000×600 μm², with a silicon layer thickness of 12 μm, PZT layer thickness of 1.64 μm and the added Ni metal mass of length×height: 600×500μm². Figure 3.13(b) shows its voltage increases with increased load, up to 898 mV at 112 kΩ. The maximum AC voltage and power are 608 mV and 2.16 μW, respectively, when the matched resistance is 21.4 kΩ.

Because the Ni proof mass was assembled at the tip of the cantilever manually,[47] it results in the mismatch of the location center of the cantilever and proof mass. As a result, the cantilever will generate the twisted deformation. Therefore, the output performance and its conformance would be reduced. Renaud *et al.* fabricated a MEMS PZT cantilever with an integrated proof mass that can generate 40 μW at its resonant frequency of 1.8 kHz and amplitude of 180 nm.[50] In fact, the frequencies of environmental vibration sources are relatively low (normally less than 200 Hz) and vary in a certain frequency range. As a result, it is promising solutions to develop the energy harvesting mechanisms with low-frequency, wideband operation range or tunable resonant frequency. Shen *et al.*[51,52] have reported a unimorph PZT piezoelectric harvester with a micromachined Si proof mass for a low-frequency vibration energy harvesting application. Figure 3.14 shows the 45° view of the fabricated cantilever taken from (a) the front and (b) the back side by SEM (scanning electron microscope). It is observed that the clearly defined straight cantilever beam is slightly bending up due to residual tensile stress in the PZT film. It is hard to completely eliminate the internal stress in such a long cantilever, and the small curvature of the beam has demonstrated the practicable layer structure of the device and the fabrication process. The top of the proof mass is more precisely formed by the ICP RIE process, as shown in Figure 3.14(b). The average power and power density determined at 0.75 g and its resonant frequency of 183.8 Hz are 0.32 μW and 416 μW cm⁻³, respectively.

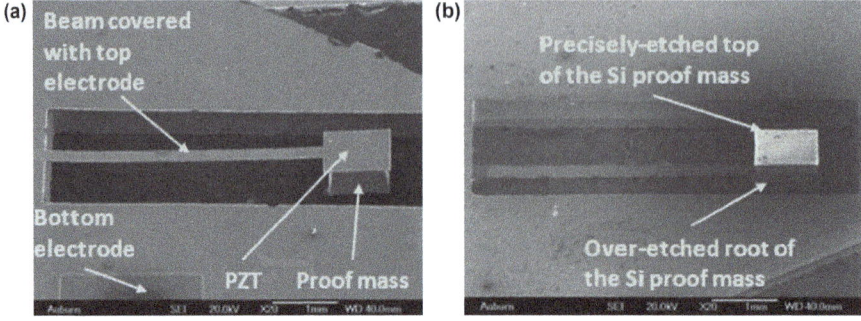

Figure 3.14 The SEM pictures of a cantilever (a) front and (b) back side 45° view fabricated on a SOI wafer.[52]

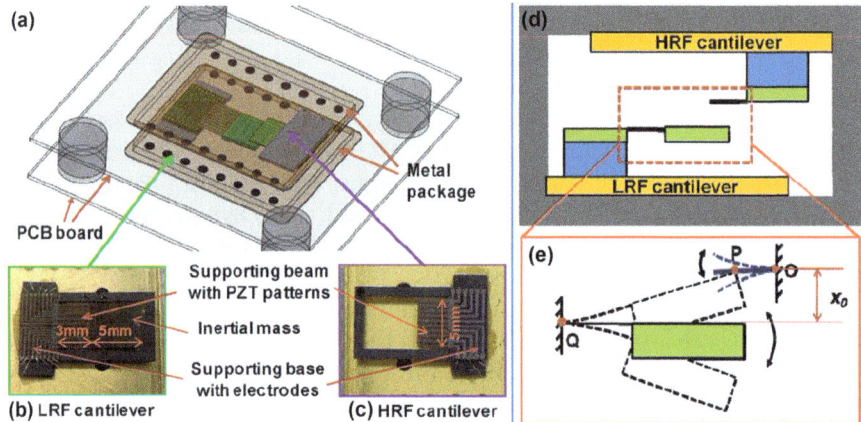

Figure 3.15 (a) Schematic drawing of the piezoelectric cantilever, with a large proof mass to gain, in low resonant frequency; (b) fabricated low-resonant-frequency (LRF) cantilever; (c) fabricated high-resonant-frequency (HRF) cantilever; (d) arrangement; (e) working principle of the piezoelectric cantilever for achieving wideband operation frequency.[53]

Lee *et al.*[53] have proposed a microfabricated piezoelectric energy harvester, which can realize a quite low resonant frequency as well as a wideband operation frequency range. Figure 3.15(a) shows a schematic illustration of the MEMS-based piezoelectric energy harvesting cantilever device operating in bending mode. Figure 3.15(b) shows the working principle of wideband mechanism. The input acceleration increases to a certain level, the amplitude of the half cycle δm for upward bending is increased. However, the amplitude of the other half cycle for downward bending is retarded by the carrier base to H. Therefore, the structure stiffness is changed at this stage, and the frequency bandwidth can be widened as the frequency sweeps up in the neighborhood of the resonant frequency. The fabrication process of this device is shown in Figure 3.15(c). A 3 μm thick (100)-oriented PZT thin film layer is deposited by the sol-gel process. The gold wires are bonded from bonding pads of the device to metal pins of the DIP, as shown in Fig. 3.15(d). The energy harvester has a wideband and steadily increased power generation from 19.4 nW to 51.3 nW within the operation frequency bandwidth ranging from 30 Hz to 47 Hz at 1.0 g.

The conventional piezoelectric cantilever normally contains a straight beam. The S-shaped PZT cantilever designed for achieving an extremely low resonance of 27.4 Hz is reported by Lee *et al.*[54] This device would be more applicable to ambient vibrations at low frequencies and low accelerations. The multilayer deposition process of Pt/Ti/PZT/Ti/Pt/SiO$_2$ on the SOI wafer is shown in Figure 3.16(a). The microfabricated device is assembled onto a dual in-line package (DIP) with a spacer chip of 2.1-mm thick in between as shown in Figure 3.16(b). In order to investigate the effect of stop distances on the frequency bandwidth, the top stopper is set at P1 (1.7 mm), P2 (1.2 mm) and P3 (0.7 mm), as shown in Figure 3.16(c). The output performance under the varied stopper is shown in Figure 3.16(d). For example, for a stop distance of P2, the operation bandwidth increases from 2.6 to 6.0 Hz, while the output voltage at resonance remains relatively constant at around 28 mV and the normalized power at resonance decreases from 78.4 to 9.3 nW g^{-2}.

3.3.2　Aerosol Deposition Process for PZT Film

The aerosol deposition method was originally invented by Akedo and co-workers.[55–57] Its larger deposition rate and lower annealing temperatures of PZT films is suitable for MEMS fabrication.[58,59] The aerosol deposition method for the PZT layer has been proven to be a quick, efficient and easy to-pattern MEMS process.[60,61] The deposition rate of the PZT film is up to 0.1 μm min^{-1}. The thickness of the films deposited at room temperature by an aerosol deposition method can be varied from 2 μm to over 15 μm. Figure 3.17 shows the schematic diagram of the aerosol deposition equipment.[60] The primary principle of aerosol deposition is to accelerate PZT particles to impact the substrates at high speed. As a result, the PZT film can be produced due to particle densification.[55]

A 5 μm PZT film for a d_{31} and d_{33} mode MEMS generator was prepared by an aerosol deposition process.[62] The piezoelectric MEMS harvester was a laminated cantilever structure, which was composed of a supporting silicon membrane, a piezoelectric layer, and laminated electrodes or interdigitated electrodes. Figure 3.18(a) and (b) show the fabrication process of d_{31} and d_{33} mode harvesters, respectively. The silicon proof mass is integrated at the tip of the piezoelectric cantilever by the DRIE process. The poling conditions of d_{31} and d_{33} mode harvesters are under 100 V and 180 V at 160 °C for 30 mins, respectively. The SEM of the fabricated d_{31} and d_{33} mode piezoelectric MEMS harvester are shown in Figure 3.18(c) and (d), respectively. The fabricated device under the d_{31} mode obtained the maximum output power of 2.765 μW at 255.9 Hz under the excited acceleration level of 2.5 g. The d_{33} mode piezoelectric harvester shows a maximum open-circuit output voltage of

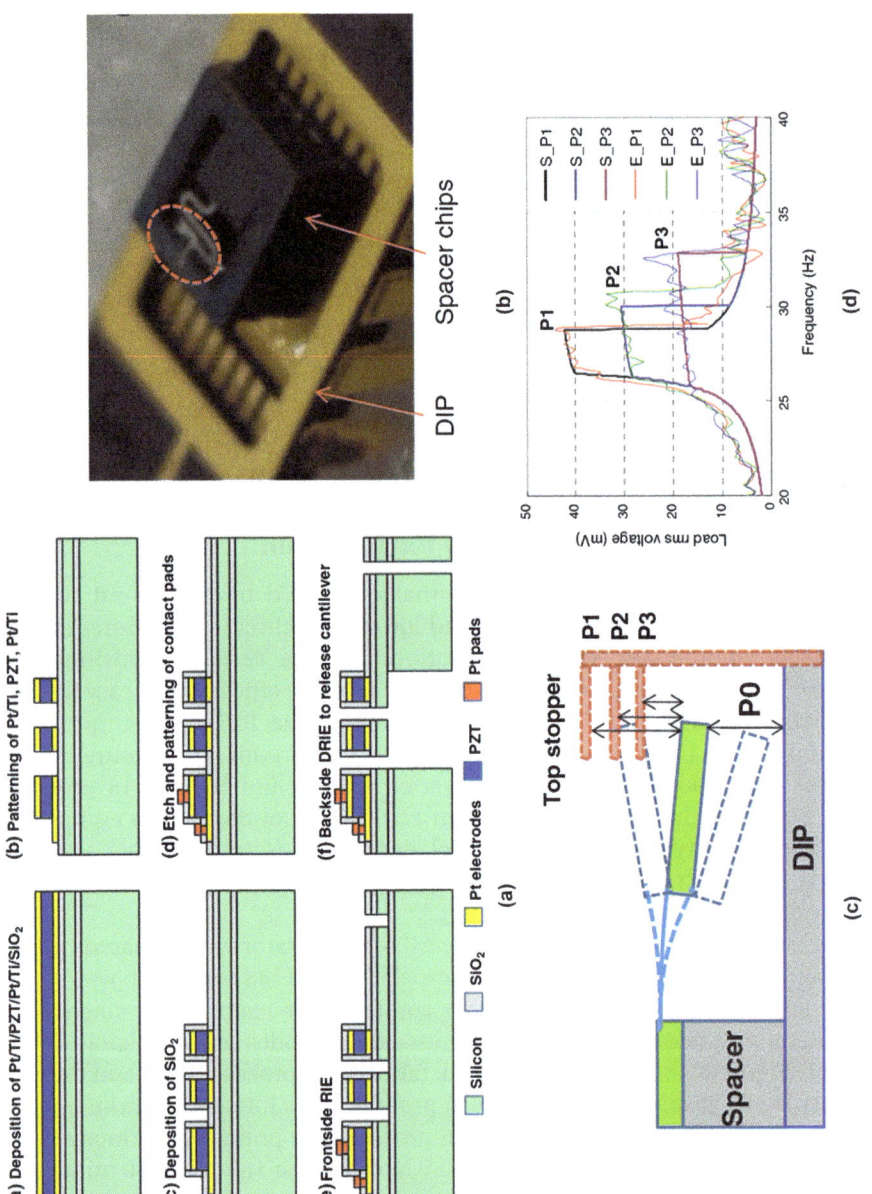

Figure 3.16 (a) Fabrication process of a low frequency harvester; (b) Microfabricated piezoelectric cantilever device assembled with DIP; (c) Configuration of the PZT cantilever device with an adjustable top stopper; (d) Simulated and experimental results of the output voltages with various stop distances (P1, P2 and P3) and 0.2 g.[54]

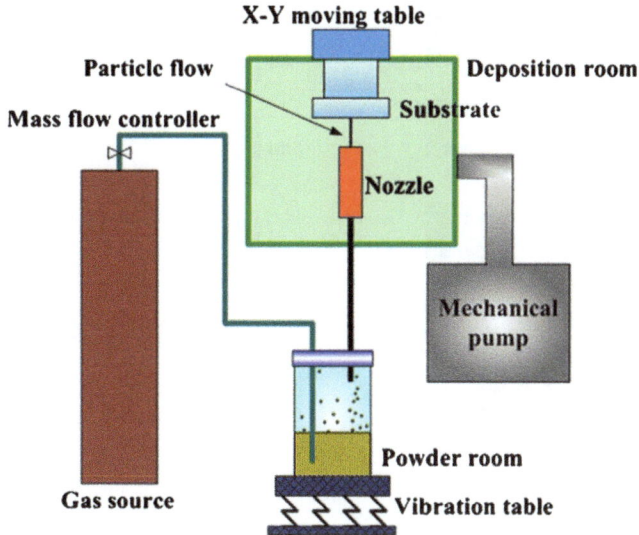

Figure 3.17   Schematic diagram of the aerosol deposition equipment.[60]

4.127 V_{P-P} and a maximum output power of 1.288 µW with a 2.292 V_{P-P} output voltage at a resonant frequency of 214 Hz at a 2 g acceleration level. The piezoelectric d_{33}-mode conversion can achieve a higher voltage output for a very low-pressure source, limited size and the simplicity in electrode arrangement. However, the d_{31}-mode conversion may generate a higher output power and have a greater advantage in MEMS application.

3.3.3   Bulk PZT Bonding Techniques for PZT Film

Besides the aforementioned sol-gel and aerosol process for deposited piezoelectric PZT films, other microfabrication processes of PZT thin or thick film for harvesters have been reported. A 1-µm-thick film was deposited by spin coating and the structure of PZT films was analyzed by X-ray diffraction.[63] In addition, a hydrothermal method[64] and screen printing[65,66] have been developed for thick films over 5 µm in thickness. A sintering temperature larger than 550 °C can provide high piezoelectricity,[67] which limits the application of the following MEMS fabrication process. Meanwhile, compared to bulk PZT ceramics, these films were with poor piezoelectric performance, which resulted in low output power for the application of the energy harvester. In order to improve the output performance of piezoelectric devices, bulk PZT bonding and thinning techniques have been developed.[68–70] PZT thick films were prepared with Au as the bonding intermediate layer between PZT and silicon and using a mechanical lapping process.[71] A tight bonding of more than 20 MPa was obtained at a bonding temperature of 450, 500 and 550 °C and a pressure of 0.8 MPa. The output power of an energy harvester with a 20 µm thickness PZT film was 205 µW at

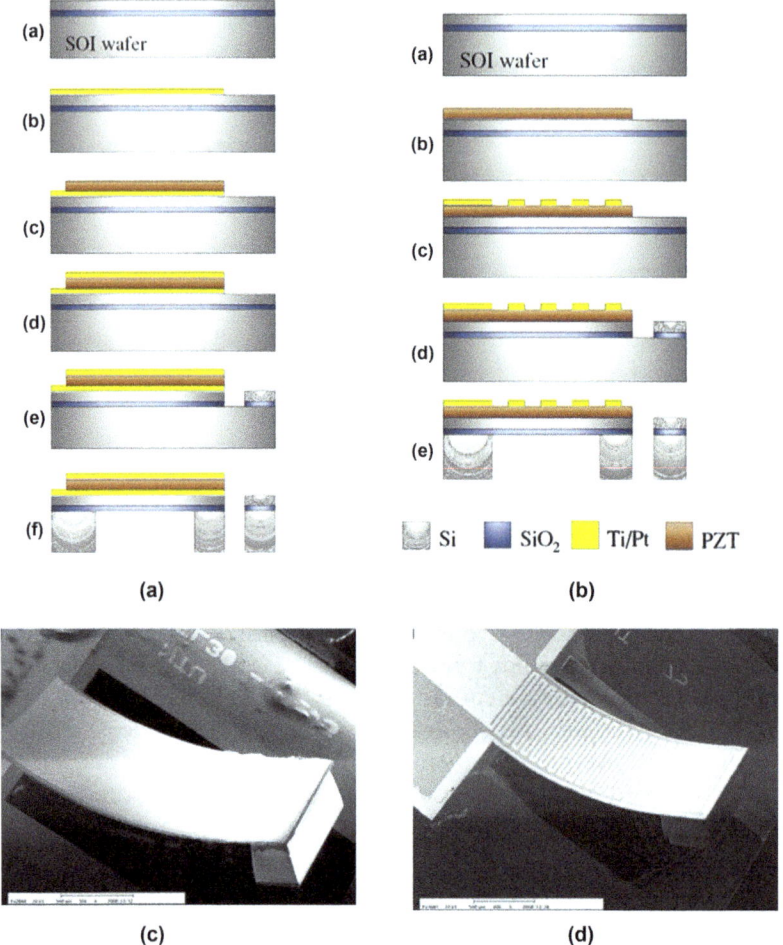

Figure 3.18   (a) Processing procedure of d31 mode piezoelectric harvester; (b) processing procedure of d33 mode piezoelectric harvester; (c) Photograph of d31 mode; (d) Photograph of d33 mode.[62]

1.5 g and its active device dimension was $7 \times 7 \times 0.55$ mm³.[69] In order to reduce the cost of the bonding layer and be free of special equipment, non-conductive epoxy resins were deployed as the intermediate layer and the PZT thickness was controlled by the wet-etching rate.[70,71] Wang *et al.*[72] developed the integration of bulk PZT and silicon by CYTOP, a patternable bond layer and the thinning of the bulk PZT by the chemical mechanical polishing method.

A novel process to fabricate a cantilever energy harvester is proposed based on bulk PZT bonding and thinning.[73] The bonding strength of the PZT-conductive epoxy-Si wafers can be measured by a conventional tensile test[74] using a micro tensile tester (Instron Model 5868). Figure 3.19 shows

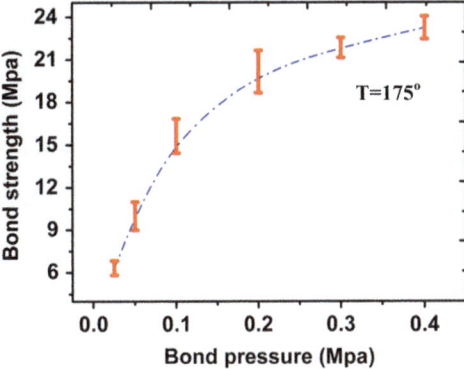

Figure 3.19 The relationship between bond strength pressure of the curing at the temperature of 175°C.[73]

the bonding strengths of the PZT-conductive epoxy-Si bonded wafers as a function of applied pressures at the curing temperature of 175 °C during bonding. It can be seen that the bonding strength was more than 10 MPa at an applied pressure of 0.1 MPa. The increasing of bond pressures contributed to improve the bond strength, which resulted from the decrease of the amount of voids. But the bonded wafers could be easy to be deformed or damaged under higher bond pressure. Therefore, the applied pressure of 0.1 MPa during bonding would be a relatively ideal value for its application. Figure 3.20(a) shows the fabrication process using an SOI micromachining technology. The thickness of the silicon device layer and after thinning PZT layer is 10 µm and 15 µm, respectively. A Au/Cr/PZT/conductive-epoxy/Cr/Au/SiO$_2$/Si/SiO$_2$ multilayer cantilever structure with a nickel mass is successfully fabricated. Figure 3.20(b) shows the multiple cantilevers with different lengths and widths and the assembled Ni proof mass with different dimensions. Figure 3.20(c) shows the cross-section view with the high quality bonding interface after thinning down the PZT. The well-developed grain structure of the bulk PZT is also clearly seen, which is dense and guarantees the excellent piezoelectricity and cannot be obtained by the conventional methods of PZT films.

3.3.4 Other Piezoelectric Materials for Energy Harvester

The coupling coefficient of piezoelectric material plays an important role in energy harvesters, since it directly affects the energy conversion effectiveness. Different piezoelectric materials have been deployed widely in MEMS energy harvesters. Besides PZT thin film or thick film introduced in the above two sections, AlN[75,76] and ZnO[77] thin film for the application of energy harvesters have been proposed. AlN and ZnO are both wurtzite-structured materials, which show an immediate piezoelectric response when grown with a c-axis orientation.[78] However, PZT needs to be poled after deposition.

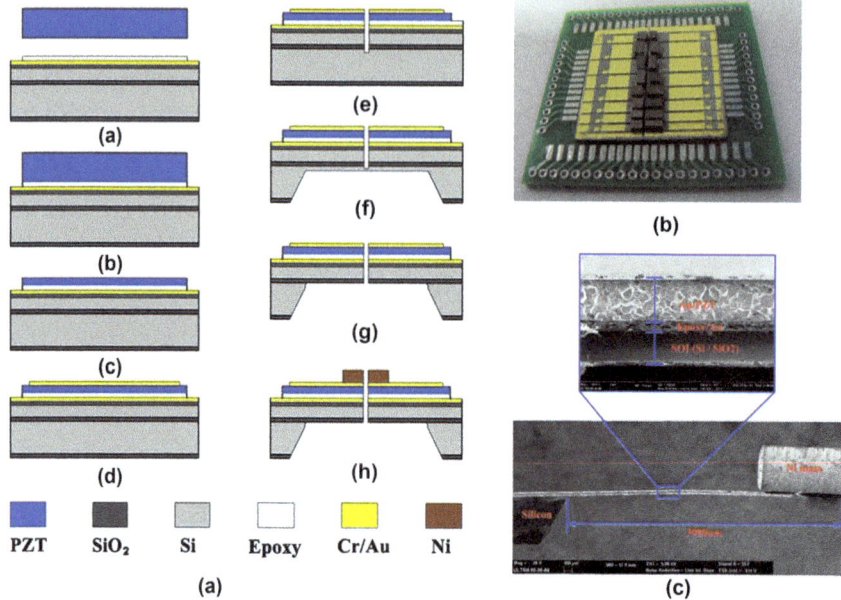

Figure 3.20   MEMS-based harvester based on PZT bonding and thinning techniques: (a) fabrication process; (b) photography of MEMS cantilever arrays harvester; (c) SEM of cross-section view.[73]

Several studies have shown that the stress is formed in AlN layers during deposition. The residual stress is related to and can be controlled by the partial pressure of the reactive gas (*i.e.*, nitrogen), the deposition rate, the kinetic energy of sputtered ions, the deposition temperature and the film thickness.[79–83] Meanwhile, the piezoelectric and the dielectric properties of AlN layers are affected by their residual stress level.[84–86] Since the output performance of piezoelectric energy harvesters is directly related to the properties of the piezoelectric stack,[87] the process parameters for AlN film are important ones to improve their performance. Karakakaya *et al.*[76] have investigated the effect of the built-in stress level of AlN layers on the properties of a piezoelectric vibration energy harvester. Figure 3.21(a) shows the fabrication process for the piezoelectric energy harvesters with AlN piezoelectric layers. Test cantilevers in various lengths from 50 μm to 1000 μm and widths from 30 μm to 200 μm are realized by steps 1 to 6 followed by a final release etching performed with tetramethylammonium hydroxide (TMAH) to form a cantilever stack of $SiO_2/Si_3N_4/Ta/Pt/AlN/Al$ layers. SEM images of the released cantilevers with 800 nm AlN layers that have different stress levels are given in Figure 3.21(b). It is observed that the initial curvature of the cantilevers is upwards, regardless of the residual stress in the AlN layers, which means that the net bending moment of the multilayered beams have the same sign. The larger compressive residual stress will reduce the electromechanical coupling and the quality factor of the testing structure.

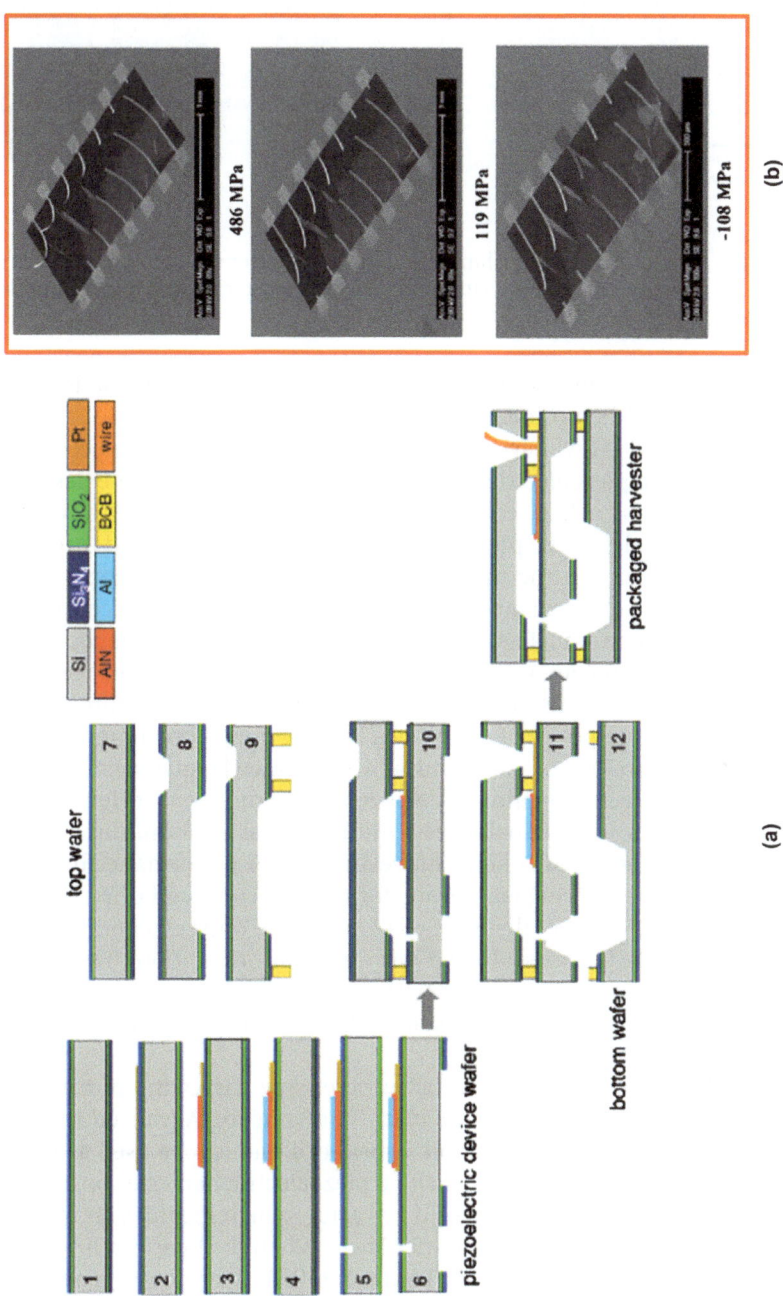

Figure 3.21 MEMS-based harvester based on AlN thin films: (a) fabrication process; (b) SEM image of the released test cantilevers (SiO$_2$/Si$_3$N$_4$/Ta/Pt/AlN/Al) with 800 nm AlN layers with different stress levels.[76]

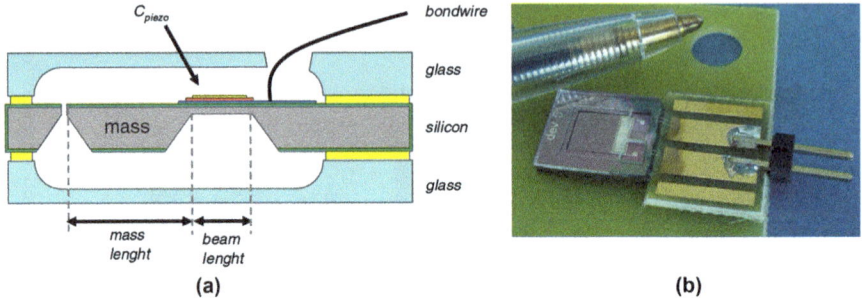

Figure 3.22 (a) Vibration energy harvester packaged in between glass substrates at the rest position; (b) vibration energy harvester mounted on a supportive board with electrical connectors.[88]

In order to optimize the output power of AlN-based harvester, the large residual stress should be avoided.

Elfrink *et al.*[88] in IMEC have presented a MEMS-based piezoelectric harvester with AlN as a piezoelectric material. Figure 3.22(a) shows the design of the harvester and its package configuration at the rest position. Both the top and bottom electrodes of the piezoelectric capacitor are connected through a small contact opening in the upper glass substrate. Electrical contacts are made with wire bonds to a small PCB board on which standard connectors are soldered, as shown in Figure 3.22(b). The AlN films are deposited by reactive sputtering from an Al target. The AlN layer typically consists of a thin amorphous layer at the AlN–Pt interface and a columnar structure on top. The maximum output power of 60 µW has been obtained on an unpackaged device at an acceleration of 2.0 g and at a resonance frequency of 572 Hz. Because the piezoelectric quality of AlN is related to the crystal orientation of the film apart from its residual stress, the influence of aluminium nitride crystal orientation on MEMS energy harvesting device performance has been researched.[89] The sputtering conditions for the AlN consisted of 20 sccm N_2 at 6 mT, 2 kW, and a target distance of 45 mm. The optimal conditions are obtained when AlN is *c*-axis aligned with a (002) orientation. The experimental results show that the wafer with (102) and (103) peaks had a significantly different piezoelectric constant, which caused the power generated to be significantly reduced.

ZnO thin film is used as the piezoelectric energy harvesting material because of its ease of fabrication. Choi *et al.*[90] at Samsung Advanced Institute of Technology have reported a flexible hybrid cell that can be used as both a solar energy harvester and touch sensitive piezoelectric harvester on a single platform. Figure 3.23 shows the hybrid cell array, which includes an inverted organic solar cell array with a ZnO buffer layer. When the hybrid cell is under light illumination ((i), (ii) and (iv) in Figure 3.23(b)), solar power generates a positive direct current (DC) signal ((i), (ii) and (iv) in Figure 3.23(c)). When the hybrid cell is touched, the hybrid cell below the finger is in the dark ((iii) in Figure 3.23(b)). A 50 nm thick ZnO thin film layer was first sputtered onto

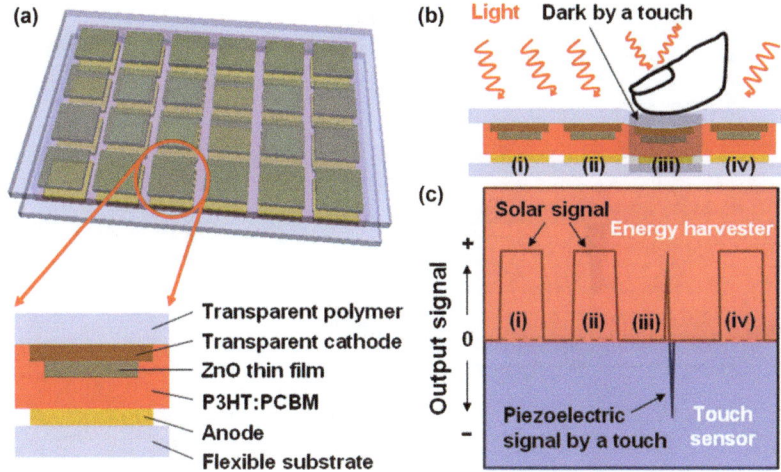

Figure 3.23   Piezoelectric touch-sensitive flexible hybrid nanoarchitecture. (a) Hybrid cell array on a flexible substrate and a unit cell configuration; (b) hybrid cells under light illumination ((i), (ii) and (iv)) and under a finger touch (iii). (c) Corresponding signal generation. The output signals can be separated: the positive pulse (red region) is recognized as a solar/piezoelectric energy harvester and the negative pulse (blue region) is used as a touch sensor.[90]

indium tioxide (ITO)-coated polyethersulfon (PES) substrates by a radio-frequency (RF) magnetron sputter at room temperature. Figures 3.24(a)–(d) show the operations of the signal controller with the hybrid cells under a range of conditions. When this device is touched by finger, the piezoelectric elements of the ZnO layer generate AC signals.

Because there is a large gap between the lower performance of many prototypes and the practical requirement of wireless microsystems, some attempts have been made to increase the efficiency of piezoelectric energy harvesting devices using a new piezoelectric material, PMN-PT piezoelectric single crystal.[91,92] It has been shown that PMN-PT material exhibits outstanding piezoelectric properties that considerably surpass the PZT ceramics by a factor of 4 to 5.[93] However, the current research on piezoelectric energy harvesting devices with PMN-PT material is focused on bulk prototypes, which have a large volume. Tang *et al.*[94] have proposed a PMN-PT based MEMS piezoelectric harvester. The cross-section view of the assembled prototype with the high quality bonding interface after thinning down the PMN-PT is shown in Figure 3.25(a). Figure 3.25(b) shows the maximum output power and output voltage against different applied acceleration levels when the prototype is at resonance. The piezoelectric MEMS generator was tested up a 2.0 g acceleration level. A maximum output power of 5.929 mW for the fabricated device was obtained and the output voltage was 3.08 V_{P-P} under the optimal resistance of 200 kΩ at a 2.0 g acceleration. Meanwhile, the open-circuit output voltage was found to be 5.84 V_{P-P} for this device.

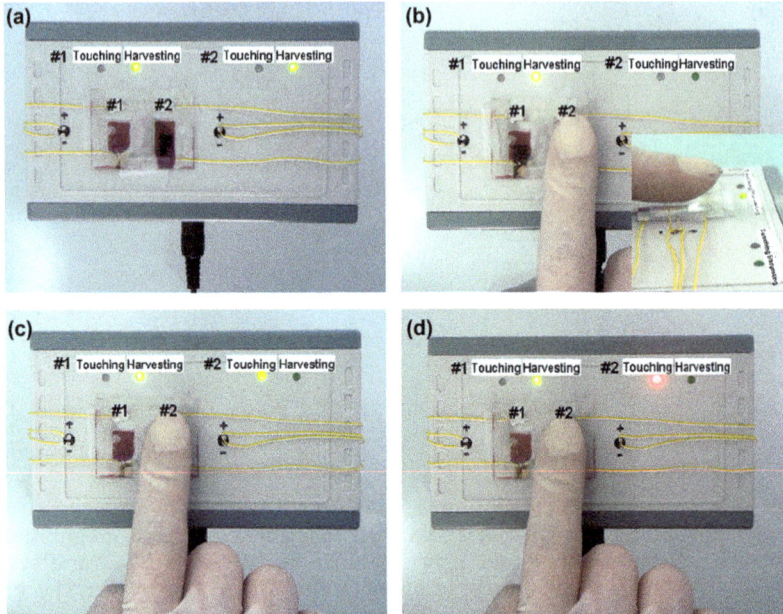

Figure 3.24   Demonstration of decoupled performance of a hybrid cell. (a) Solar cells from the hybrid cells turn on under room lighting. (b) A dark condition by the finger before touching the #2 hybrid cell. The inset presents a close-up photograph for the dark condition by a finger before touching. When the #2 hybrid cell is touched, the instant negative signal turns on the touching LEDs: (c) green for a weak touch and (d) red for a strong touch. While touching the #2 hybrid cell, the #1 hybrid cell harvests solar energy continuously, as shown by the harvesting LED of the #1 cell in (a)–(d).[90]

Figure 3.25   PMNT-based harvester after assembling proof mass.

3.4   Nano-based Energy Harvester

3.4.1   PZT Nanowire-based Energy Harvester

PZT is a good piezoelectric material due to its higher voltage for energy harvesting applications. As a ceramic bulk material, a thin film PZT is more fragile in comparison to organic PVDF, but has demonstrated very good mechanical strength in nanowire form. PZT nanofibers prepared by an electrospinning process exhibit an extremely high piezoelectric voltage constant (g_{33}, 0.079 Vm N^{-1}), high bending flexibility, and high mechanical strength.[95] Lin *et al.* have reviewed piezoelectric nanofibers for the application of energy harvesting, including PZT nanofibers and PVDF-based nanofibers.[96] Figure 3.26 shows the different fabrication processes of PZT nanowires, including epitaxially grown (Figure 3.26(a)),[97] transfer process (Figure 3.26(b))[98] and the electrospinning method (Figure 3.26(c)).[99] Figure 3.27(a) shows the free vibration test using the PZT nanogenerator as a damper. The output voltage from the nanogenerator was measured when a Teflon cantilever was placed on top of the nanogenerator, as shown in Figure 3.27(b). The measured result revealed that the amplitude of noise signal was only at about the 10 mV level. This confirmed that the power output from the PZT nanogenerator was in fact the energy harvested from mechanical vibration.

Gu *et al.*[100] have proposed a simple approach of fabrication with vertically ultralong PZT nanowire arrays from electrospinning fibers to make a maximum output peak voltage of 209 V. The electrospinning for the orientated nanofibers and suspending calcination techniques are used to prepare PZT nanofibers with an average crystal size of about 16 nm. The orientated PZT nanofiber film is cut and stacked layer-by-layer to form a multilayer film, as shown in Figure 3.28(a). Figure 3.28(f)–(h) shows that the synthesized VANA can be bent, stretched, or twisted to a large degree without breaking its structure, which shows its potential application in flexible energy harvesting and self-powered systems.

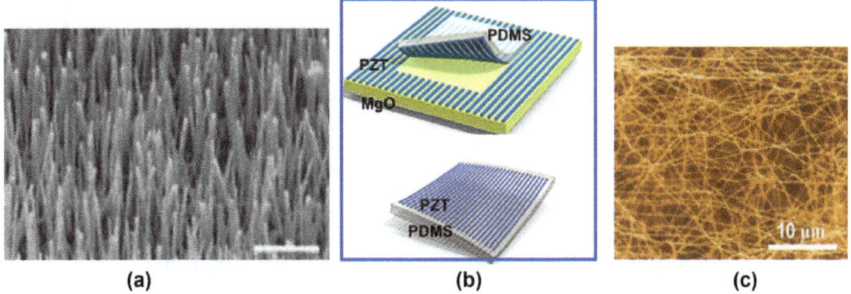

(a)   (b)   (c)

Figure 3.26   (a) SEM image of the epitaxially grown PZT nanowire arrays by hydrothermal decomposition;[97] (b) A transfer process using lithography to define PZT ribbons and PDMS substrate to transfer the PZT fibers;[98] (c) PZT nanofibers by electrospinning.[99]

(a)

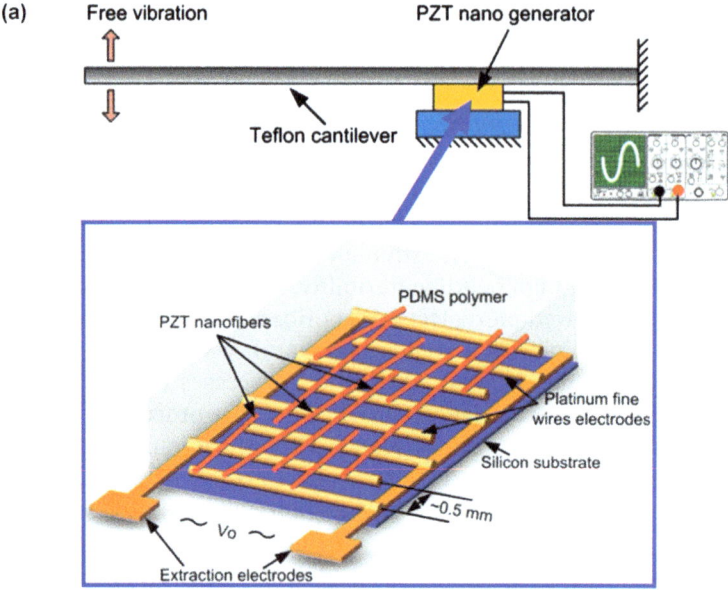

(b)

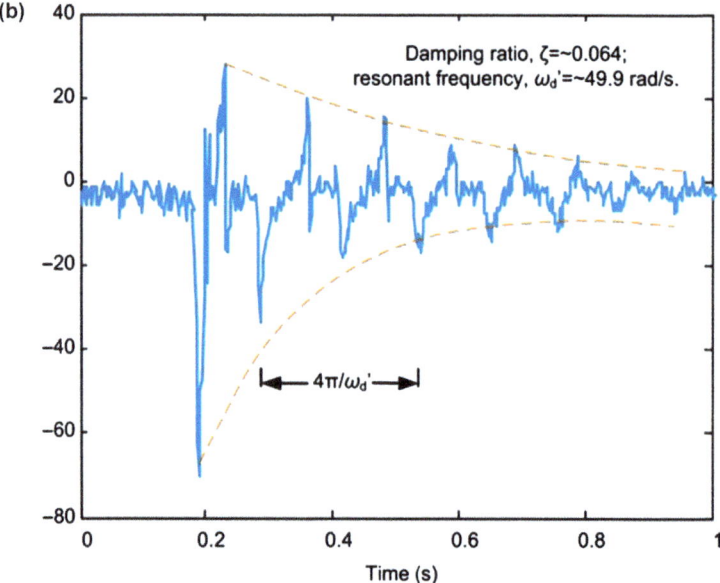

Figure 3.27   Energy harvested from the free vibration of a Teflon cantilever. (a) Schematic of the experimental setup. (b) The open circuit voltage output when the cantilever was under free vibration.[99]

3.4.2   ZnO Nanowire-based Energy Harvester

The known ZnO of one-dimensional (1D) nanomaterials has three key advantages.[101] First, it exhibits both semiconducting and piezoelectric (PZ)

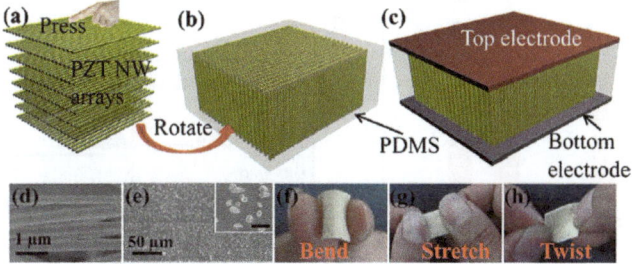

Figure 3.28 Fabrication process and structure characterization of the NG. (a–c) Experiment setup for fabricating the high-output NG using the regionally orientated electrospinning nanofibers. (d) Field emission scanning electron microscope (SEM) image of the regionally oriented electrospinning PZT nanofibers. (e) Top-view SEM images of the fabricated VANA. The inset is a magnified image with scale bar of 2 μm. The white spots are the exposed tops of PZT nanowires. (f–h) Optical photographs of VANA under different deformations that show its flexibility and robustness.[100]

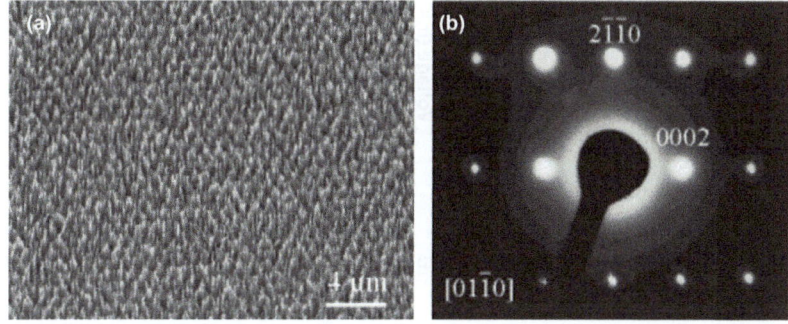

Figure 3.29 Experimental design for converting nanoscale mechanical energy into electrical energy by a vertical piezoelectric (PZ) ZnO NW. (a) Scanning electron microscopy images of aligned ZnO NWs grown on a Al_2O_3 substrate; (b) transmission electron microscopy images of ZnO NWs.[102]

properties that can form the basis for electromechanically coupled sensors and transducers. Second, ZnO is relatively biosafe and biocompatible, and it can be used for biomedical applications with little toxicity. Finally, ZnO exhibits the most diverse and abundant configurations of nanostructures known so far, such as nanowires (NWs), nanobelts (NBs), nanosprings, nanorings, nanobows, and nanohelices. Figure 3.29(a) shows the piezoelectric ZnO nanowires grown on c-plane-oriented α-Al_2O_3 substrate by a vapor-liquid-solid (VLS) process, using Au particles as a catalyst.[102,103] Figure 3.29(b) shows the NW grown along the (0001) direction and has side surfaces of (01$\bar{1}$0). To improve the power generation capabilities of the system, the ultrasonic waves driven by the motion of the NWs are used to

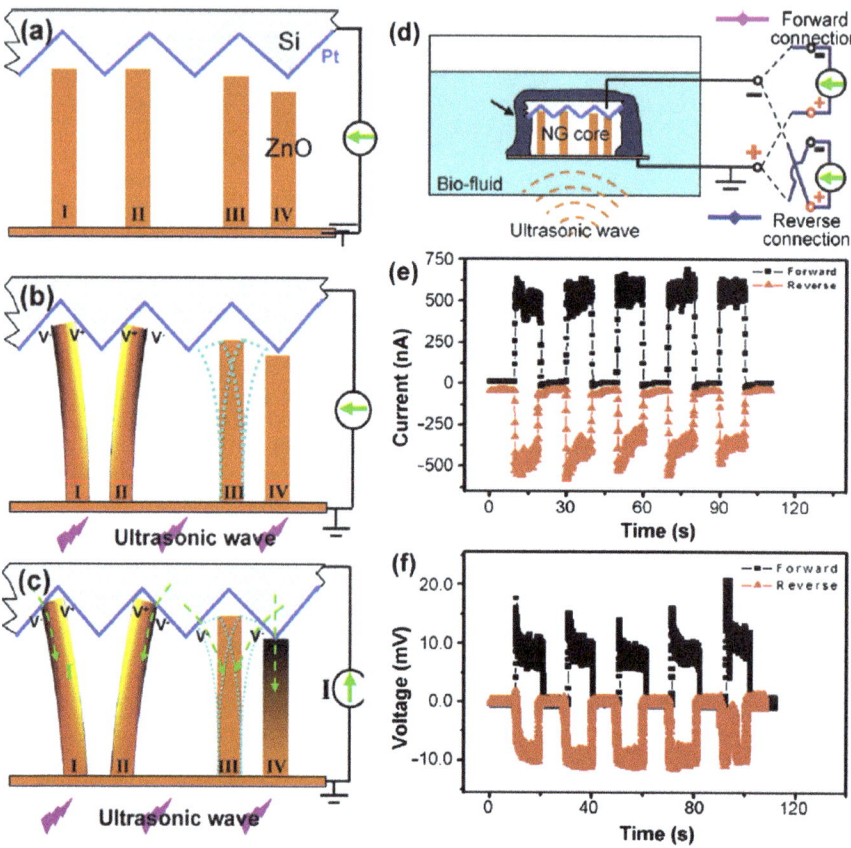

Figure 3.30 The mechanism of the nanogenerator driven by an ultrasonic wave.[104]

replace the AFM tip with a simpler source of mechanical energy, which results in the production of a continuous current.[104,105] The output voltage of such a nanowires generator has recently been improved to 10 mV driven by an ultrasonic wave.[104] Figure 3.30 shows the working principle of a nano-based harvester driven by an ultrasonic wave, which includes four possible configurations of contact between a NW and the zigzag electrode, the equivalent electric circuit, the current and voltage outputs. The output power of the nanogenerator fabricated with a substrate of area $= 2$ mm^2 is $W_{wave} = I_A V \approx 1$ pW.

Recently, p-doped ZnO nanowire arrays for generator application had also been grown on (001) silicon substrate using thermal vapor deposition.[106] The electric output of such a p-type generator is 10–15 mV in magnitude under a conductive atomic force of 5 nN. Besides using silicon substrate, well aligned ZnO nanowires with an average length of ≈ 5 µm had also been grown on GaN/AlN substrate through a vapor-solid process.[107] A theoretical model that predicts the electrical power generated by an array of vertically

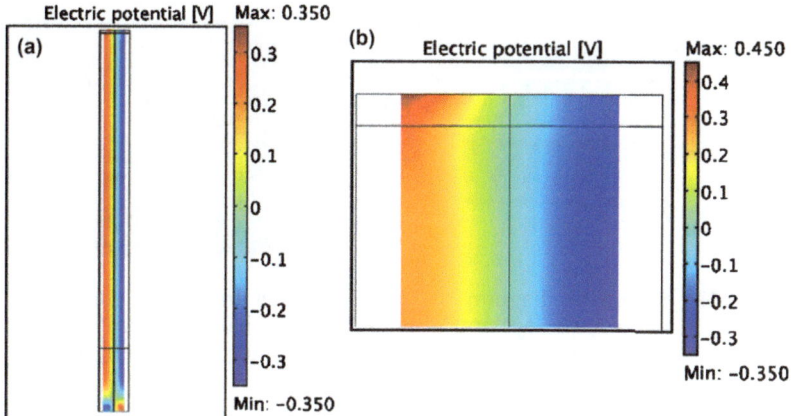

Figure 3.31 Color online FEM calculation of a ZnO NW bent by a lateral force of 80 nN. (a) The whole NW; (b) upper part of the NW.[112]

aligned ZnO nanoribbons is proposed.[108] The theory suggests that an array of vertically aligned ZnO nanoribbons can produce as much as 1 nW mm^{-2} and 1–100 nW mm^{-3} electricity, which can power the device such as a 30 pW processor.[109] These values are significantly less than the power density of MEMS-based systems of comparable size (0.1–1 μW mm^{-3}).[110,111] Finite element method (FEM) calculations for analyzing the bending of a semi-conducting piezoelectric ZnO nanowire for harvesters are presented.[112] Figure 3.31 shows the results of the electric potential for a bent ZnO nanowire using the finite element method of COMSOL multiphysics. The electric potential on the shell element with the applied force is slightly higher than in the rest of the NW. The generated electrical potential is 0.3 V based on theoretical calculations.

The vapor-phase method, as compared to the low-cost and simple aqueous methods, usually employs a vacuum, sophisticated equipment and a higher temperature, which restricts the type of substrate used. Among the aqueous grown methods, the hydrothermal method has emerged as a powerful method for the fabrication of one-dimensional nanostructures, offering significant advantages such as controllable structures and a cost-effective, low-temperature, substrate-independent technique that allows for easy integration. There are some reports of hydrothermal growth of ZnO nanostructures. Vayssieres *et al.*[113–115] have reported the preparation of highly oriented ZnO microrods and microtubes and the diameter of the ZnO rods from 1–2 μm to 100–200 nm. Govender *et al.*[116] synthesized ZnO nanocolumns on gold-coated tin-oxide substrates by a low-temperature solution method. Greene *et al.*[117] reported the wafer-scale growth of ZnO nanorods and studied the temperature dependent photoluminescence (PL) properties.

Vibration-based generators provide the maximum output power when such devices operate at their mechanical resonances. However, the vibration frequency of available ambient vibration sources is random and

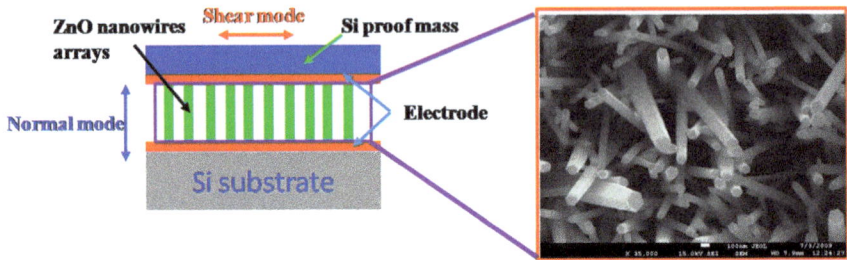

Figure 3.32   Schematic diagram of the power generator, which consists of ZnO
                   nanowire arrays and Si proof mass for low frequency vibration har-
                   vesting. The harvester is excited and energy is scavenged by normal
                   mode and shear mode.

varies from one case to another. For example, the vibration frequency of a
laptop during normal operation is 90.2 Hz, but the one of a running compact
disc read-only memory is 43.2 Hz.[118] Therefore, it is necessary to develop the
generators with multiple modes to improve conversion efficiency. Yang
et al.[119] discussed a low-frequency vibration-based energy harvester using
ZnO cylindrical nanowires arrays (NWA) grown by hydrothermal method for
process integration and low cost. Figure 3.32 shows the cross-section
drawing of the piezoelectric power generator, which includes the NWA deposited on a silicon substrate, Cr/Au layer as the bottom electrode, and a
silicon proof mass with the top electrode. The proof mass helps to scavenge
kinetic energy from low-frequency vibrations, hence allowing the generator
to achieve a better performance. The generator is excited at two vibration
resonant modes, *i.e.*, normal mode A and shear mode B, so as to maximize
the output voltage generated. In normal mode A, the proof mass moves
up and down along the longitudinal direction of the ZnO nanowires.
In shear mode B, the direction of the proof mass movement is perpendicular
to the length of the ZnO nanowires. The average length of ZnO NWA is
about 2.5 μm. The diameter of the ZnO nanowires varies from 53 nm
to 75 nm.

The friction will be limited by the coefficient of friction μ and the maximum axial load F_0, which is defined as the force that can be applied on the
cylindrical nanowire prior to buckling. This corresponds to an interfacial
force as:

$$V_f = \mu F_0 = \mu \frac{\pi^3 Y d^4}{256 T^2} \tag{3.4}$$

where μ is the coefficient of friction between the ZnO tip and the silicon
proof mass, $Y = C_{11} - C_{13}^2/C_{33}$ is the Young's modulus for plane stress, C is
the fourth-rank stiffness tensor, and d and T are the diameter and length of
the ZnO nanowire, respectively. For ZnO, the values of C_{11}, C_{13}, C_{33} are
210 GPa, 105 GPa and 210 GPa, respectively. For values of $\mu = 0.3$, $d = 70$ nm,
$T = 2.5$ μm, the interfacial force V_f is derived to be 15.39 nN.

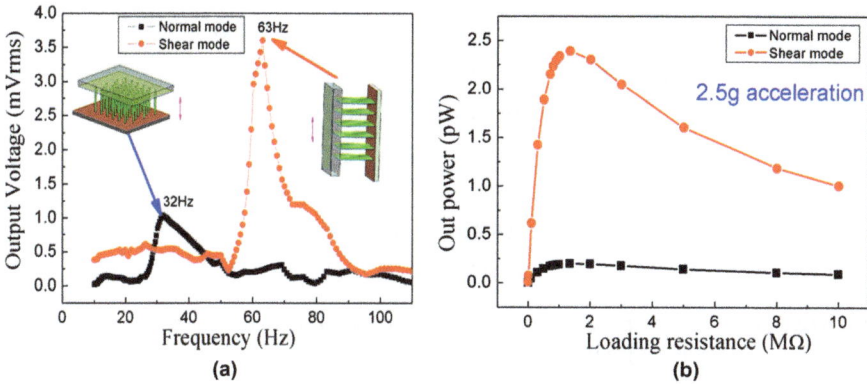

Figure 3.33   (a) The output voltages of the generator when it is excited under 2.5 g acceleration during normal mode A and shear mode B; (b) the output power of the generator at 2.5g acceleration during normal mode A and shear mode B.

The total electrostatic energy stored in a ZnO nanowire is:

$$E = \frac{\pi^5 \mu^2 \varepsilon_{11} Y^2 d^6}{32768 T^3} \left\{ e_{15} + \frac{\varepsilon_{11} C_{44}}{e_{15}} \right\}^{-2} = 0.0093 \frac{\mu^2 \varepsilon_{11} Y^2 d^6}{T^3} \left\{ e_{15} + \frac{\varepsilon_{11} C_{44}}{e_{15}} \right\}^{-2}$$

(3.5)

where e is the third-rank stress constant tensor, ε is the second-rank electric permittivity tensor. For ZnO, $C_{44} = 42.5$ GPa, $e_{31} = -0.61$ C m^{-2}, $e_{33} = 1.14$ C m^{-2}, $e_{15} = -0.59$ C m^{-2}, and $\varepsilon_{11} = \varepsilon_{33} = 7.38 \times 10^{-11}$ F m^{-1}.[120]

Figure 3.33 shows the output voltages and powers of the ZnO NWA from 10 Hz to 110 Hz at different accelerations under two modes. The open-circuit output voltage of the resonant generators at normal and shear mode are 1.1 mV and 3.59 mV, respectively. For an acceleration of 2.5 g, the maximum output power obtained from normal and shear resonant modes are 0.2 pW and 2.39 pW, respectively.

3.4.3   Organic PVDF Nanofiber-based Energy Harvester

PVDF has superior piezoelectric properties due to its polar crystalline structure. Figure 3.34 shows the crystalline structure of PVDF, including the α-phase as the most abundant form and β-phase as the polar structure. The β-phase has a piezoelectric response due to its oriented hydrogen and fluoride(CH2–CF2) unit cells along with the carbon backbone, which is realized by electrical poling and mechanical stretching processes.

PVDF nanofibers fabricated by the conventional electrospinning process are under a high bias voltage (>10 kV), which could transform some non-polar α-phase structures to polar β-phase structures for piezo-electricity.[121–124] The near field electrospinning (NFES) process, as shown in

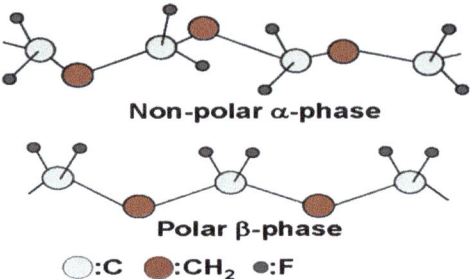

Non-polar α-phase

Polar β-phase

◯:C ⬤:CH₂ ●:F

Figure 3.34 Schematic diagrams of PVDF crystalline structures, including non-polar α-phase and β-phase.

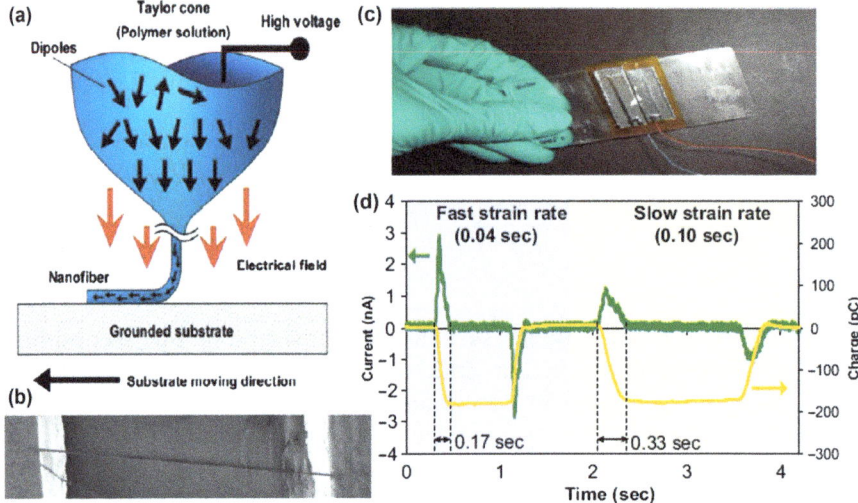

Figure 3.35 (a) Schematic diagram of the near-field electrospinning; (b) SEM microphotograph showing an electrospun PVDF nanofiber on two contact pads; (c) an optical image of a single nanofiber nanogenerator on top of a plastic substrate with two electrical wires; (d) testing results of a PVDF nanofiber nanogenerator.[125]

Figure 3.35(a), also possesses the inherent high electric field with in-situ mechanical stretching for possible alignment of dipoles along the longitudinal direction of the nanofiber.[125] The single PVDF nanofiber generates 0.5–3 nA of current and 5–30 mV of voltage under repeated long term reliability tests without noticeable performance degradation, as shown in Figure 3.35(d). Hansen *et al.* have demonstrated an energy harvesting system of a PVDF nanogenerator, which is fabricated by a conventional far-field electrospinning process.[126] Figure 3.36(a) shows the PVDF nanofiber network fabricated using a modified far-field electrospinning process. Figure 3.36(b) illustrates the integrated system with both a piezoelectric

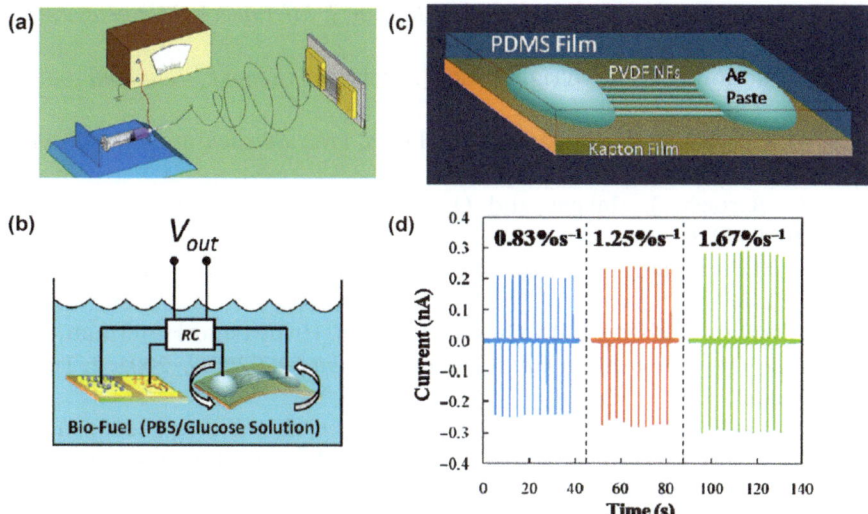

Figure 3.36   (a) Schematic diagram of the far-field electrospinning; (b) testing schematic; (c) with the PDMS and a post-electric poling process; (d) testing results under different strain rates.[126]

PVDF nanofiber nanogenerator and a biofuel cell. Figure 3.36(c) shows the silver paste fixed at the ends of the PVDF nanofibers. With a fixed strain rate of 1.67%/s, voltage and current outputs as high as 20 mV and 0.3 nA were recorded, respectively, as shown in Figure 3.36(d).

P(VDF-TrFE) generally exhibits good piezo- and ferroelectric behavior and a single all-trans polar crystalline phase (β-phase) that is stable at room temperature.[127] Persano *et al.* have introduced a large area, flexible piezoelectric material that consists of sheets of electrospun fibers of the polymer poly(vinylidenefluoride-co-trifluoroethylene).[128] Mandal *et al.* have demonstrated a 43 mm-thick pressure type nanogenerator by far-field electrospinning of P(VDF-TrFE) with an output voltage of about 400 mV under compression tests.[129]

Piezoelectric nanofibers, such as PVDF and PZT, are highly flexible and easy to fabricate for possible integration in implantable and/or flexible devices, as well as textile applications such as electric clothing.[96] This tremendous variety in possible fields for fiber-based nanogenerators makes them a hot topic of research. Several groups have demonstrated first prototypes of nanofiber nanogenerators. Further development of fiber-based nanogenerators focuses on the improvement of energy conversion efficiency.

References

1. S. Roundy, P. Wright and K. Pister, *Proc. IMECE*, 2002, 1.
2. S. Roundy, *PhD Thesis University of California*, Berkeley, 2003.

3. D. Hoffmann, B. Folkmer and Y. Manoi, *J. Micromech. Microeng.*, 2009, **19**, 094001.
4. S. Roundy, P. K. Wright and J. Rabaey, *Comput. Commun.*, 2003, **26**, 1131.
5. L. K. Swee, M. W. Neil and N. R. Harris, *Proc. Eurosensors XXII (Dresden, Germany)*, 2008, 395.
6. U. Bartsch, J. Gaspar and O. Paul, *J. Micromech. Microeng.*, 2010, **20**, 035016.
7. E. M. Yeatman, *Proc. Inst. Mech. Eng., Part C: J. Mech. Eng. Sci.*, 2008, **222**, 27.
8. J. D. Jill, Y. Zhang, J. D. Grade, H. Lee, S. Hrinya and H. Jerman, *IEEE Proc. Optical FiberCommun. Conf. and Exhibit (OFC2001)*, 2001, **2**, TuJ2-1-TuJ2-3.
9. J. A. Yeh, C. N. Chen and Y. S. Lui, *J. Micromech. Microeng.*, 2005, **15**, 201.
10. T. S. Lim, C. H. Ji, C. H. Oh, H. Kwon, Y. Yee and J. U. Bu, *IEEE J. Sel. Topic Quantum Electron.*, 2004, **10**, 558.
11. J. A Yeh, S. S. Jiang and C. Lee, *IEEE Photon. Technol. Lett.*, 2006, **18**, 1170.
12. C. C. Tu, K. Fanchiang and C. H. Liu, *Microsyst. Technol.*, 2006, **12**, 1099.
13. T. K. HouMax, J. Y. Huang, S. S. Jiang and J. A. Yeh, *J. Micro/Nanolith. MEMS MOEMS*, 2008, **7**, 043015.
14. B. Yang, C. Lee, R. K. Kotlanka, J. Xie and S. P. Lim, *J. Micromech. Microeng*, 2010, **20**, 065017.
15. C. C. Tu, K. Fanchiang and C. H. Liu, *Microsyst. Technol.*, 2006, **12**, 1099.
16. S. Roundy, *J. Intell. Mater. Syst. Struct.*, 2005, **16**, 809.
17. P. Miao, P. D. Mitcheson, A. S. Holmes, E. M. Yeatman, T. C. Green and B. H. Stark, *Microsyst. Technol.*, 2006, **12**, 1079.
18. I. Kuehne, A. Frey, D. Marinkovic, G. Eckstein and H. Seidel, *Sens. Actuators A*, 2008, **142**, 263.
19. C. Lee, Y. M. Lim, B. Yang, R. K. Kotlanka, C. H. Heng, J. H. He, M. Tang, J. Xie and H. Feng, *Sens. Actuators A*, 2009, **156**, 208.
20. P. D. Mitcheson, P. Miao, B. H. Stark and E. M. Yeatman, *Sens. Actuators A*, 2004, **115**, 523.
21. Y. Chiu, C. T. Kuo and Y. S. Chu, *Microsyst. Technol.*, 2007, **13**, 1663.
22. Y. Chiu and V. F. G. Tseng, *J. Micromech. Microeng.*, 2008, **18**, 104004.
23. P. Basset, D. Galayko, A. Mahmood Paracha, F. Marty, A. Dudka and T. Bourouina, *J. Micromech. Microeng.*, 2009, **19**, 115025.
24. A. Mahmood Paracha, P. Basset, D. Galayko, F. Marty and T. Bourouina, *IEEE Electro. Device Lett.*, 2009, **30**, 481.
25. W. Hsieh, T. Hsu and Y. Tai, *Solid-State Sensors and Actuators*, 1997, **1**, 425–428.
26. W. Hsieh, T. J. Yao and Y. C. Tai, *Solid-State Sensors and Actuators*, 1999, 7, 1064.
27. W. Ma, R. Zhu, L. Rufer, Y. Zohar and M. Wong, *J. Micromech. Syst.*, 2007, **16**, 29.
28. T. Sterken, P. Fiorini, K. Baert, R. Puers and G. Borghs, *Proc. Transducers'03*, 2003, **8**, 1291.

29. Y. Arakawa, Y. Suzuki and N. Kasagi, *Power MEMS Conference (Kyoto, Japan)*, 2004, 187–190.
30. Y. Sakane, Y. Suzuki and N. Kasagi, *J. Micromech Microeng.*, 2008, **18**, 104011.
31. Y. Naruse, N. Matsubara, K. Mabuchi, M. Izumi and S. Suzuki, *J. Micromech. Microeng.*, 2009, **19**, 094002.
32. H. W. Lo and Y. C. Tai, *J. Micromech. Microeng.*, 2008, **18**, 115025.
33. T. Tsutsumino, Y. Suzuk and N. Kasagi, *Transducers 2007, Int. Conf. Solid-State Sensors, Actuators and Microsystems*, 2007, **6**, 863.
34. T. Tsutsumino, Y. Suzuki, N. Kasagi and Y. Sakane, *19th IEEE Int. Conf. Micro Electro Mechanical Systems, 2006 (MEMS 2006) (Istanbul)* ed. Y. Suzuki, 2006, **19**, 98.
35. J. Boland, Y. H. Chao, Y. Suzuki and Y. C. Tai, *IEEE 16th Ann. Int. Conf. Micro Electro Mechanical Systems, 2003 (MEMS-03) (Kyoto)*, pp. 538–541.
36. J. S. Boland, J. D. Messenger, H. W. Lo and Y. C. Tai, *Micro Electro Mechanical Systems*, 2005, **18**, 618.
37. J. S. Boland and Y. C. Tai, *18th IEEE Int. Conf. Micro Electro Mechanical Systems, 2005 (MEMS 2005)*, pp. 618–621.
38. T. Sterken, P. Fiorini, G. Altena, C. A. V. Hoof and R. A. Puers, *Int. Conf. Solid-State Sensors, Actuators and Microsystems Conf.*, 2007, 129.
39. T. Sterken, P. Fiorini, K. Baert, R. A. Puers and G. A. Borghs, *Int. Conf. Solid-State Sensors, Actuators and Microsystems*, 2003, **12**, 1291.
40. S. J. Zhang, *Int. Conf. on Power MEMS* (Freiburg, Germany, Nov.) 2007, 105.
41. U. Mescheder, B. Muller and P. Urbanovic, *J. Micromech. Microeng.*, 2009, **19**, 094003.
42. H. Amjadi, *IEEE Trans. Dielectr. Electr. Insul.*, 1999, **6**, 852.
43. F. Peano and T. Tambosso, *J. Microelectromechanical Syst.*, 2005, **14**, 4294.
44. U. Bartsch, J. Gaspar and O. Paul, *Proc. IEEE MEMS Conf.*, Sorrento, Italy, 2009, 10436.
45. S. Roundy and P. K. Wright, *Smart Mater. Struct.*, 2004, **13**, 1131.
46. S. Roundy, E. S. Leland, J. Baker, E. Carleton, E. Reilly, E. Lai, B. Otis, J. M. Rabaey, P. K. Wright and V. Sundararajan, *IEEE Pervasive Ompt*, 2005, **4**, 28.
47. H. B. Fang, J. Q. Li, Z. Y. Xu, L. Dong, L. Wang, D. Chen, B. Cai and Y. Liu, *Microelectron. J.*, 2006, **37**, 1280.
48. J. Cheng and Z. Meng, *Thin Solid Films*, 2001, **385**, 5.
49. J.-R. Cheng, W. Zhu, N. Li and L. E. Cross., *J. Appl. Phys. Lett.*, 2002, **81**, 4085.
50. M. Renaud, T. Sterken, A. Schmitz, P. Fiorini, V. Van Hoof and R. Puers, *Proc. Int. Conf. on Solid-State Sensors, Actuators and Microsystems (Lyon, France)*, 2007, 891.
51. D. Shen, J.-H. Park, J. Ajitsaria, S.-Y. Choe, H. C. Wikle and D.-J. Kim, *J. Micromech. Microeng.*, 2008, **18**, 055017.

52. D. N. Shen, J. H. Park, J. H. Noh, S. Y. Choe, S. H. Kim, H. C. III Wikle and D. J. Kim, *Sens. Actuators A.*, 2009, **154**, 103.

53. H. Liu, C. J. Tay, C. G. Quan, T. Kobayashi and C. Lee, *J. Microelectromech. Syst.*, 2011, **20**, 1131.

54. H. Liu, C. Lee, T. Kobayashi, C. J. Tay and C. G. Quan, *Microsyst. Technol.*, 2012, **18**, 497.

55. J. Akedo, *Mater. Sci. Forum*, 2004, **449**, 43.

56. J. Akedo and M. S. Lebedev, *J. Appl. Phys.*, 2003, **42**, 5931.

57. M. Lebedev and J. Akedo, *Ferroelectrics*, 2002, **270**, 117.

58. M. Lebedev, J. Akedo and Y. Akiyama, *Jpn. J. Appl. Phys.*, 2000, **39**, 5600.

59. J. H. Park, J. Akedo and H. Sato, *Sens. Actuators A*, 2007, **135**, 86.

60. X. Y. Wang, C. Y. Lee, C. J. Peng, P. Y. Chen and P. Z. Chang, *Sens. Actuators A*, 2008, **143**, 469.

61. X. Y. Wang, C. Y. Lee, Y. C. Hu, W. P. Shih, C. C. Lee, J. T. Huang and P. Z. Chang, *J. Micromech. Microeng.*, 2008, **18**, 055034.

62. B. S. Lee, S. C. Lin, W. J. Wu, X. Y. Wang, P. Z. Chang and C. K. Lee, *J. Micromech. Microeng.*, 2009, **19**, 065014.

63. S. B. Kim, H. Park, S. H. Kim, H. C. Wickle, J. H. Park and D. J. Kim, *J. Microelectromech. Syst.*, 2013, **22**, 26.

64. T. Kanda, M. K. Kurosawab, H. Yasuis and T. Higuchi, *Sens. Actuators A*, 2001, **89**, 16.

65. S. B. Beeby, A. Blackburn and N. M. White, *J. Micromech. Microeng.*, 1999, **9**, 218.

66. R. A. Dorey, R. W. Whatmore, S. P. Beeby, R. N. Torah and N. M. White, *Integr. Ferroelectr.*, 2003, **54**, 651.

67. X. Y. Wang, C. Y. Lee, C. J. Peng, P. Y. Chen and P. Z. Chang, *Sens. Actuators A*, 2008, **143**, 469.

68. K. Tanaka, T. Konishi, M. Ide and S. Sugiyama, *J. Micromech. Microeng.*, 2006, **16**, 815.

69. E. E. Aktakka, R. L. Peterson and K. Najafi, *Transducer's 11(Beijing, China)*, 2011, 1649.

70. X. H. Xu and J. R. Chu, *J. Micromech. Microeng.*, 2008, **18**, 065001.

71. X. H. Xu, B. Q. Li, Y. Feng and J. R. Chu, *J. Micromech. Microeng.*, 2007, **17**, 2439.

72. Z. H. Wang, J. M. Miao and C. W. Tan, *Sens. Actuators A*, 2009, **149**, 277.

73. G. Tang, J. Q. Liu, B. Yang, J. B. Luo, H. S. Liu, Y. G. Li, C. S. Yang, D. N. He, V. D. Dao, K. Tanaka and S. Sugiyama, *J. Micromech. Microeng.*, 2012, **22**, 065017.

74. O. Vallin, K. Jonsson and U. Lindberg, *Mater. Sci. Eng.*, 2005, **50**, 109.

75. M. Marzencki, Y. Ammar and S. Basrour, *Sens. Actuators A*, 2008, **145**, 363.

76. K. Karakaya, M. Renaud, M. Goedbloed and R. van Schaijk, *J. Micromech. Microeng.*, 2008, **18**, 104012.

77. M. P. Lu, J. H. Song, M. Y. Lu, M. T. Chen, Y. F. Gao, L. J. Chen and Z. L. Wang, *Nano Lett.*, 2009, **9**, 1223.

78. S. Trolier-McKinstry and P. Muralt, *J. Electroceram.*, 2004, **12**, 7.
79. F. Martin, P. Muralt, M. A. Dubois and A. Pezous, *J. Vac. Sci. Technol. A*, 2004, **22**, 361.
80. G. Este and W. D. Westwood, *J. Vac. Sci. Technol. A*, 1987, **5**, 1892.
81. B. K. Gan, M. M. M. Bilek, D. R. McKenzie, M. B. Taylor and D. G. McCullogh, *J. App. Phys.*, 2004, **95**, 2130.
82. M. A. Dubois and P. Muralt, *J. App. Phys.*, 2001, **89**, 6389.
83. F. Engelmark, G. Fucntes, I. V. Katardjiev, A. Harsta, U. Smith and S. Berg, *J. Vac. Sci. Technol. A*, 2000, **18**, 1609.
84. E. Iborra, J. Olivares, M. Clement, L. Vergara, A. Sanz-Hervas and J. Sangrador, *Sens. Actuators A.*, 2004, **115**, 501.
85. R. Naik, J. J. Lutsky, R. Reif and C. G. Sodini, *IEEE Trans. Ultrason. Ferroelectr. Freq.Control*, 2000, **47**, 292.
86. T. Adam, J. Kolodzey, C. P. Swann, M. W. Tsao and J. F. Rabolt, *Appl. Surf. Sci.*, 2001, **175**, 428.
87. M. Renaud, K. Karakaya, T. Sterken, P. Fiorini, C. Van Hoof and R. Puers, *Sens. Actuators A.*, 2007, **145**, 380.
88. R. Elfrink, T. M. Kamel, M. Goedbloed, S. Matova, D. Hohlfeld, Y. V. Andel and R. V. Schaijk, *J. Micromech. Microeng.*, 2009, **19**, 094005.
89. N. Jackson, R. O'Keeffe, F. Waldron, M. O'Neill and A. Mathewson, *J. Micromech. Microeng.*, 2013, **23**, 075014.
90. D. Choi, K. Y. Lee, K. H. Lee, E. S. Kim, T. S. Kim, S. Y. Lee, S. W. Kim, J. Y. Choi and J. M. Kim, *Nanotechnology*, 2010, **21**, 405503.
91. C. Sun, L. Qin, F. Li and Q. M. Wang, *J. Intell. Mater. Syst. Struct.*, 2009, **20**, 559.
92. A. Mathers, K. S. Moon and J. G. Yi, *IEEE Sens. J.*, 2009, **9**, 731.
93. I. A. Ivan, M. Rakotondrabe, J. Agnus, R. Bourquin, N. Chaillet, P. Lutz, J. C. Poncot, R. Duffait and O. Bauer, *Rev. Adv. Mater. Sci.*, 2010, **24**, 1.
94. G. Tang, J. Q. Liu, B. Yang, J. B. Luo, H. S. Liu, Y. G. Li, C. S. Yang, V. D. Dao, K. Tanaka and S. Sugiyama, *IEE ElectronicsLett*, 2012, **13**, 784.
95. X. Chen, S. Xu, N. Yao, W. Xu and Y. Shi, *Appl. Phys. Lett.*, 2009, **94**, 253113.
96. J. Chang, M. Dommer, C. Chang and L. W. Lin, *Nano energy*, 2012, **1**, 356.
97. S. Xu, B. J. Hansen and Z. L. Wang, *Nat. Commun.*, 2010, **1**, 93.
98. Y. Qi, N. T. Jafferis, K. Lyons, C. M. Lee, H. Ahmad and M. C. McAlpine, *Nano Lett.*, 2010, **10**, 524.
99. X. Chen, S. Xu, N. Yao and Y. Shi, *Nano Lett.*, 2010, **10**, 2133.
100. L. Gu, N. Cui, L. Cheng, Q. Xu, S. Bai, M. Yuan, W. Wu, J. Liu, Y. Zhao, F. Ma, Y. Qin and Z. L. Wang, *Science*, 2006, **312**, 242.
101. Z. L. Wang and J. H. Song, *Science*, 2006, **312**, 242.
102. X. D. Wang, C. J. Summers and Z. L. Wang, *Nano Lett.*, 2004, **3**, 423.
103. X. D. Wang, J. H. Song, P. Li, J. H. Ryou, R. D. Dupuis, C. J. Summers and Z. L. Wang, *J. Am. Chem. Soc.*, 2005, **127**, 7920.
104. Z. L. Wang, *Nano Res.*, 2008, **1**, 1.

105. X. D. Wang, J. H. Song, J. Liu and Z. L. Wang, *Science*, 2007, **316**, 102.
106. M. P. Lu, J. H. Song, M. Y. Lu, M. T. Chen, Y. F. Gao, L. J. Chen and Z. L. Wang, *Nano Lett.*, 2009, **9**, 1223.
107. J. Liu, P. Fei, J. Zhou, R. Tummala and Z. L. Wang, *Appl. Phys. Lett.*, 2008, **92**, 173105.
108. C. Majidi, M. Haataja and D. J. Srolovitz, *Smart Mater. Struct.*, 2010, **19**, 055027.
109. M. Seok, S. Hanson, Y. S. Lin, Z. Foo, D. Kim, Y. Lee, N. Liu, D. Sylvester and D. Blaauw, *IEEE Symp. VLSI Circuits*, 2008, 188.
110. N. E. Dutoi and B. L. Wardl, *Integr. Ferroelectr.*, 2007, **83**, 13.
111. S. R. Anton and H. A. Sodano, *Smart Mater. Struct.*, 2007, **16**, 21.
112. M. A. Schubert, S. Senz, M. Alexe, D. Hesse and U. Gosele, *Appl. Phys. Lett.*, 2008, **92**, 122904.
113. L. Vayssieres, K. Keis, S. E. Lindquist and A. Hagfeldt, *J. Phys. Chem. B*, 2001, **105**, 3350.
114. L. Vayssieres, K. Keis, A. Hagfeldt and S. E. Lindquist, *Chem. Mater.*, 2001, 4395.
115. L. Vayssieres, *Adv. Mater.*, 2003, 464.
116. K. Govender, D. S. Boyle, P. O'Brien, D. Brinks, D. West and D. Coleman, *Adv. Mater.*, 2002, **14**, 1221.
117. L. E. Greene, M. Law, J. Goldberger, F. Kim, J. C. Johnson, Y. Zhang, R. J. Saykally and P. Yang, *Angew. Chem. Int. Ed.*, 2003, **42**, 3131.
118. L. K. Swee, M. W. Neil and N. R. Harris, *Proc. Eurosensors XXII (Dresden, Germany)*, 2008, 395.
119. B. Yang, C. Lee, G. W. Ho, W. L. Ong, J. Liu and C. S. Yang, *J. Microelectromech. Syst.*, 2012, **21**, 776.
120. H. J. Ding and W. Q. Chen, *NY: Nova Science*, 2001, 7.
121. D. Farrar, K. Ren, D. Cheng, S. Kim, W. Moon, W. L. Wilson, *et al.*, *Adv. Mater.*, 2011, **23**, 3954.
122. Y. R. Wang, J. M. Zheng, G. Y. Ren, P. H. Zhang and C. Xu, *Smart Mater. Struct.*, 2011, **20**, 045009.
123. J. Zheng, A. He, J. Li and C. Han, *Macromol. Rapid Commun.*, 2007, **28**, 2159.
124. C. Ribeiro, V. Sencadas, J. L. G. Ribelles and S. Lanceros-Mendez, *Soft Mater.*, 2010, **8**, 274.
125. C. Chang, V. H. Tran, J. Wang, Y. K. Fuh and L. Lin, *Nano Lett.*, 2010, **10**, 726.
126. B. J. Hansen, Y. Liu, R. Yang and Z. L. Wang, *ACS Nano*, 2010, **4**, 3647.
127. X. Y. Jin, K. J. Kim and H. S. Lee, *Polymer*, 2005, **46**, 12410.
128. L. Persano, C. Daddeviren, Y. Su, Y. Zhang, S. Girardo, D. Pisignano, Y Huang and J. A. Rogers, *Nat. Commun.*, 2013, **4**, 1633.
129. D. Mandal, S. Yoon and K. J. Kim, *Macromol. Rapid Commun.*, 2011, **32**, 831.

CHAPTER 4

Thermal and Thermoelectric Properties of Nanomaterials

GANG ZHANG

Institute of High Performance Computing, A*Star, Singapore 138632
Email: zhangg@ihpc.a-star.edu.sg

4.1 Introduction

Recently, thermoelectric materials have attracted extensive attention again due to the increasing awareness of the global warming on the planet's environment, a renewed requirement for long-life electrical power sources, and the increasing miniaturization of electronic circuits and sensors. Thermoelectrics is able to make a contribution to meet the requirements of all the above activities. Today, the heat engines used in most thermal power stations typically operate at 30–40% efficiency. Thermoelectric modules can potentially convert part of the wasted heat directly into electricity, reducing the usage of fossil fuels.

In addition to waste heat harvesting, thermoelectric materials can also be applied in temperature management in microelectronics. Power dissipation issues have recently become one of the greatest challenges for integrated electronics, which limits the performance of a wide range of electronics from handheld devices to massive data centers, and has doubled in the past five years. In existing commercial silicon chips, hot spot removal is a key for the future generation of IC chips. As the silicon stacked chips or three-dimensional chips are introduced, this can create even smaller and hotter spots. Thermoelectric cooling is a silent and environment friendly solution.

RSC Nanoscience & Nanotechnology No. 32
Nanofabrication and its Application in Renewable Energy
Edited by Gang Zhang and Navin Manjooran
© The Royal Society of Chemistry 2014
Published by the Royal Society of Chemistry, www.rsc.org

The performance efficiency of thermoelectric material is best measured by the figure of merit: $ZT = \frac{S^2 \times \sigma}{\kappa} T$, here S is the Seebeck coefficient,[1] σ is the electrical conductivity, T is the absolute temperature, and κ is the thermal conductivity; $\kappa = \kappa_e + \kappa_p$, where κ_e and κ_p are the electron and phonon (lattice vibration) contributions to the thermal conductivity, respectively. In order to make a material competitive for thermoelectric purposes, the ZT of the material must be larger than three. There are several ways to do this. The first approach is to increase the Seebeck coefficient S. However, for general materials, simply increasing S will lead to a simultaneous decrease in electrical conductivity. The second approach is to increase the electrical conductivity. This has also proven to be ineffective, because electrons are also carriers of heat and an increase in the electrical conductivity will also lead to an increase in the thermal conductivity. The ideal case is to reduce the thermal conductivity without affecting the electrical conductivity. It is possible to achieve this in nanoscale materials due to their reduced thermal conductivity.

Moreover, the advent of nanotechnology has had a dramatic effect on thermoelectric material development and has resulted in the syntheses of nanostructured materials whose thermoelectric properties surpass the best performance of its bulk counter, such as silicon nanowires (SiNWs).[2,3]

It is obvious that the systematic applications of nanothermoelectric materials depend on how well we understand their material properties. Thermal properties in nanostructures differ significantly from those in macrostructures because the characteristic length scales of phonons are comparable to the characteristic length of nanostructures. A large number of theoretical and experimental works have been done to explore the thermal and thermoelectric properties and applications of nanomaterials. Due to the limit of space, we only address the most fundamental aspects of thermal and thermoelectric properties. Nanoscale thermal conductivity has been reviewed comprehensively.[4-6] In this chapter, it provides an update of recent developments and serves both as an authoritative reference text on thermoelectric property for the professional scientist and engineer, and as a source of general information on thermoelectric application for the well-informed layman. This chapter is a review of the area and is arranged in three sections, entitled 'Introduction'; 'Thermal conduction in nanomaterials' and 'Thermoelectric properties of nanomaterials'.

4.2 Thermal Conductivity of Nanomaterials

4.2.1 High Thermal Conductivity of Carbon Nanotubes

Carbon nanotubes (CNTs) are one of the promising nanoscale materials discovered in 1990s. Depending on their chirality and diameter, the nanotubes can be either metallic or semiconducting. At room temperature, the electronic resistivity is about 10^{-4}–10^{-3} Ω cm for the metallic nanotubes, while the resistivity is about 10 Ω cm for semiconducting tubes.

There has been some theoretical and experimental works on the heat conduction of CNTs. Thermal conductivity of CNTs is about 6600 W/mK at room temperature,[7] and it shows a peaking behavior before falling off at higher temperatures due to the onset of Umklapp scattering.[8] It was demonstrated that even for the metallic nanotubes, thermal conductivity is dominated by phonons at all temperatures.[9,10] Kim *et al.* measured the thermal conductivity of a single CNT and found that thermal conductivity is larger than 3000 W mK^{-1} at room temperature.[11] More interestingly, it is found that thermal conductivity, κ, of CNTs diverges with length as a power law: $\kappa \propto L^{\beta}$, where the exponents β depends on the temperature and CNT diameters and the value of β is between 0.12 to 0.4[12] and between 0.11 to 0.32.[13]

The high thermal conductivity of CNTs is for a pure and defect-free tube. Actually, nanotubes can have natural defects and doping in the process of fabrication. Since the phonon frequency depends on mass, the isotopic doping can lead to increased phonon scattering. Thus, natural CNTs always contain a significant number of scattering centers leading to localization of some phonon modes and reduces thermal conductivity. Figure 4.1 shows the dependence of thermal conductivity on the impurity percentage.[12] The thermal conductivity decreases as the percentage of ^{14}C impurity increases. With 40–50% ^{14}C, the thermal conductivity is reduced to about 40% of that found in a pure ^{12}C CNT. The thermal conductivity decreases more quickly at a low percentage range than at a high percentage range. It is shown that the thermal conductivity decreases about 20% with only 5% ^{14}C isotope impurity. This anomalously large isotopic effect in a low-dimensional system has been verified by Chang *et al.*[14] experimentally, in boron nitride nanotubes as shown in Figure 4.1. Similar to the mechanism of length dependent thermal conductivity, the origin of observable reduction of thermal

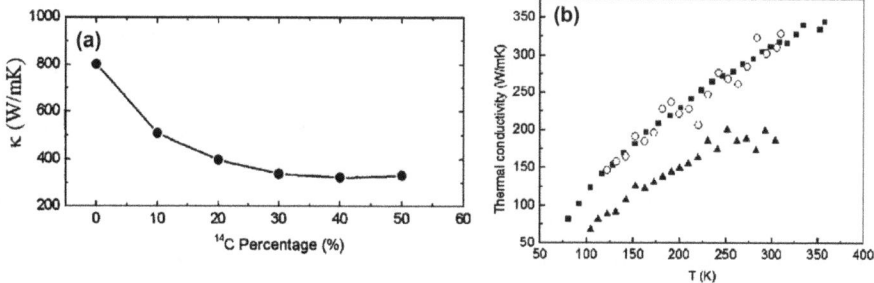

Figure 4.1 (a) Thermal conductivity *vs* ^{14}C impurity percentage for (5,5) SWNT at 300 K.[12] Reprinted with permission from the American Institute of Physics. (b) The thermal conductivity of a boron nitride nanotube (solid triangles), an isotopically pure boron nitride nanotube (solid squares) and a carbon nanotube (open circles). These tubes are with similar outer diameters.[14]
Reprinted with permission from the American Physical Society.

conductivity in isotope-disordered NTs can also be understood from the phonon diffusive process.

In addition to isotope doping, practical nanotubes can have other defects, such as vacancy and conformational defects. A remarkable change in defect scattering, from a quantum to a classical feature is observed. It was found that at room temperature, thermal conductance is critically affected by defect scattering since incident phonons are scattered by localized phonons around the defects. Compared with the vacancy impact on the thermal conductivity of diamond, it is shown that vacancies in nanotubes are not much more influential than in bulk materials.[15] This is probably due to the fact that the strong valence double bond network in carbon nanotubes provides effective additional channels for phonons to bypass the vacancy sites.

Similarly, conformational defects, such as a (5, 7, 7, 5) defect where four hexagons change into two pentagons and two hexagons, can also reduce the thermal conductivity significantly. Compared with vacancies, the (5, 7, 7, 5) defect is a milder form of point defect, since it does not change the basic bonding characteristic and induces much less structural deformation. So the decrease in thermal conductivity is less than in the case of vacancies.

The impact of nitrogen substitutional impurity on the CNT thermal conductivity was studied by using an *ab initio* density functional approach.[16] They found that most of the reduction of transmission takes place at the higher frequency part of the phonon spectrum, thus its effect on thermal conductance is only noticeable at high temperature. The surface chemisorbed molecules can also be treated as one type of doping.[17] It was found that there is a rapid drop in thermal conductivity with chemisorptions, where chemisorptions of as little as 1% of the CNT atoms reduces the thermal conductivity by over a factor of 3.

4.2.2 Low Thermal Conductivity in Silicon Nanowires

In addition to CNTs, silicon nanowires (SiNWs) have attracted great attention in recent years because of their potential applications in many areas. SiNWs are an appealing choice in the novel nanoscale thermoelectric materials because of their small sizes and ideal interface compatibility with conventional Si-based technology. In SiNWs, the electrical conductivity and electron contribution to the Seebeck coefficient are similar to those of bulk silicon, but exhibit an 100-fold reduction in thermal conductivity, showing that the electrical and thermal conductivities are decoupled. Recent experiments have provided direct evidence that an approximately 100-fold improvement of the *ZT* values over bulk Si are achieved in SiNW over a broad temperature range.[18,19] This large increase of *ZT* is contributed by the decrease of thermal conductivity. SiNWs have attracted broad interests in recently years due to their fascinating potential applications.

Due to the size effect and high surface-to-volume ratio, the thermal conductivity of SiNW differs substantially from that of bulk materials. Volz and

Chen[20] have found that the thermal conductivity of individual SiNWs is more than two orders of magnitude lower than the bulk value. Li *et al.*[21] have also reported a significant reduction of thermal conductivity in silicon nanowires compared to the thermal conductivity in bulk silicon experimentally.

Although the thermal conductivity of SiNWs is lower than that of the bulk silicon, it is still larger than the reported ultralow thermal conductivity $(0.05 \text{ W m}^{-1}\text{K}^{-1})$ found in layered materials. So it is indispensable to reduce the thermal conductivity of SiNWs further in order to achieve higher thermoelectric performance. Natural SiNWs always contain a significant number of scattering centers leading to the localization of some phonon modes and a reduction in thermal conductivity. By using molecular dynamics simulations, Yang *et al.*[22] have proposed to dope SiNWs with an isotope impurity randomly for the reduction of thermal conductivity. In silicon isotopes, ^{28}Si is with the highest natural abundance (92%), then ^{29}Si and ^{30}Si with 5% and 3%, respectively. Here the effect of doping ^{29}Si to ^{28}Si NWs are shown.

At a low isotopic percentage, the small ratio of atoms of impurity can induce a large reduction in thermal conductivity. Contrast this to the high sensitivity at the two ends, the thermal conductivity versus isotopic concentration curves are almost flat at the center part as shown in Figure 4.2, where the value of thermal conductivity is only 77% (^{29}Si doping) of that of pure ^{28}Si NW.

Moreover, silicon and germanium can form a continuous series of substitutional solids, offer a continuously variable system with a wide range of crystal lattices and band gaps. The thermal conductivity of $Si_{1-x}Ge_x$ NW depends on the Ge content, the lowest value is 18% of that of pure SiNW. At the two ends of the curves, the thermal conductivity shows a very sensitive dependence on Ge content. It is quite interesting that with only 5% Ge atoms

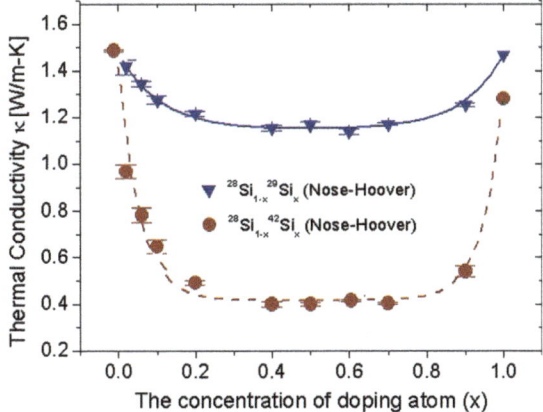

Figure 4.2 Thermal conductivity of SiNWs versus the percentage of randomly doping isotope atoms (^{42}Si and ^{29}Si) at 300 K.[22]
Reprinted with permission from the American Chemical Society.

(Si$_{0.95}$Ge$_{0.05}$ NW), its thermal conductivity can be reduced to 50%. The tunability of thermal conductivity makes this kind of semiconductor system very attractive for industrial application.[23]

The thermal conductivity of NWs decreases with wire diameter. On the one hand, the phonon-phonon interaction increases with size reduction due to the confinement, which causes the decrease of heat conduction. On the other hand, as the surface-to-volume ratio increases, the large surface scattering has great influence on the transport. The phonon thermal conductivity increases remarkably as the diameter increases until the diameter is larger than in the order of hundreds of nanometers. This reflects the fact that scattering at the surface introduces diffuse phonon relaxation in the transverse cross-section. After the diameter reaches large values, the portion of phonons experiencing boundary scattering becomes much smaller, as a result, the thermal conductivity tends to be constant and is close to the value of bulk silicon.[24]

Moreover, thermal conductivity of SiNW can be reduced by introducing more surface scattering: making silicon nanotube (SiNT) structures.[25] Figure 4.3 shows the thermal conductivity of SiNWs and SiNTs versus the cross-sectional area at 300 K. It is interesting to find that only a 1% reduction in cross-sectional area induces the reduction of thermal conductivity of 35%. Moreover, with an increasing size of the hole, a linear dependence of thermal conductivity on cross-sectional area is observed. We can see that with the same cross-sectional area, thermal conductivity of SiNTs is only about 33% of that of SiNWs. This additional reduction is due to the localization of phonon states on the surface.

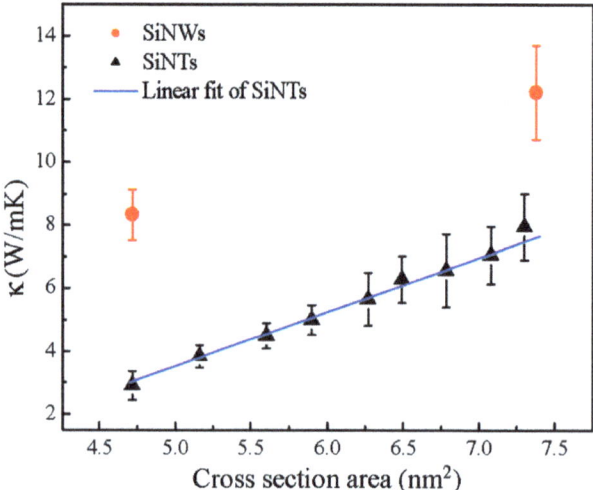

Figure 4.3   Thermal conductivity of SiNWs and SiNTs versus cross-sectional area at 300 K.[25]
Reprinted with permission from the American Chemical Society.

Vibrational eigen-mode analysis is necessary to understand the underlying physical mechanism of thermal conductivity reduction in SiNTs. Mode localization can be quantitatively characterized by the participation ratio P, which measures the fraction of atoms participating in a given mode, and effectively indicates the localized modes with $O(1/N)$ and delocalized modes with $O(1)$. Figure 4.4 compares the participation of each eigen-mode for SiNWs and SiNTs. It shows a reduction of p-ratio in SiNTs for both low frequency phonons and high frequency phonons, compared with SiNWs. Figure 4.4 also shows the normalized energy distribution on the cross-section plane. It is clearly shown that the intensity of localized modes is almost zero in the center of NWs, but with a finite value at the boundary. Due to the inner-surface introduced in SiNTs, energy localization also shows up around the hollow region. These results provide direct numerical evidence that localization takes place on the surface region.

Recently, a new coherent mechanism was proposed to control thermal conductivity of nanowires.[26] The coherence of phonons can be described by the heat current autocorrelation function (HCACF). Figure 4.5(a) shows the typical time dependence of HCACF for Ge/Si core-shell NWs, SiNWs and SiNTs. For both SiNWs and SiNTs, there is a rapid decay of HCACF at the beginning, followed by a long-time tail with a much slower decay. This is the standard two-stage decaying characteristic of HCACF. However, a nontrivial oscillation appears for a long time in the HCACF for core-shell NWs (Figure 4.5a). The long-time region of HCACF reveals that this nontrivial oscillation is not random but shows a periodic manner (Figure 4.5b). This

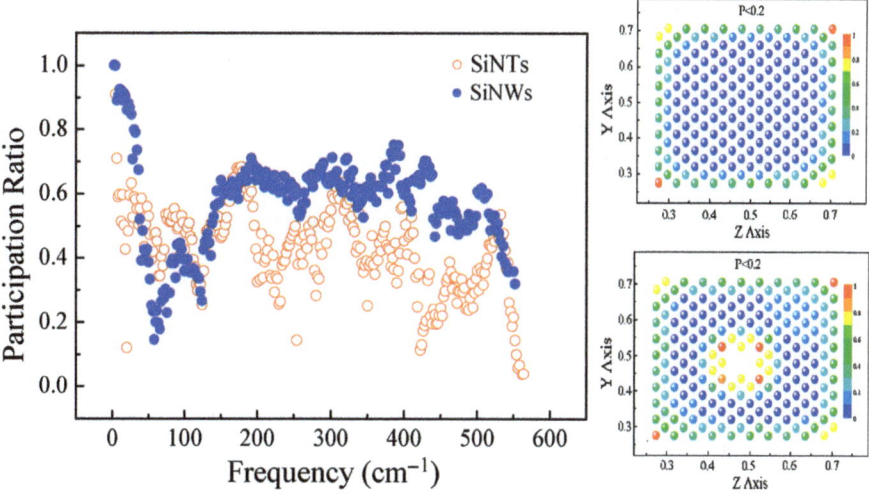

Figure 4.4   Left: Participation ratio of each eigen-mode for SiNTs and SiNWs with the same cross-sectional area. Right: Normalized energy distribution on the cross-section plane for SiNWs and SiNTs. Here are the energy distributions for modes with P < 0.2.[25]
Reprinted with permission from the American Chemical Society.

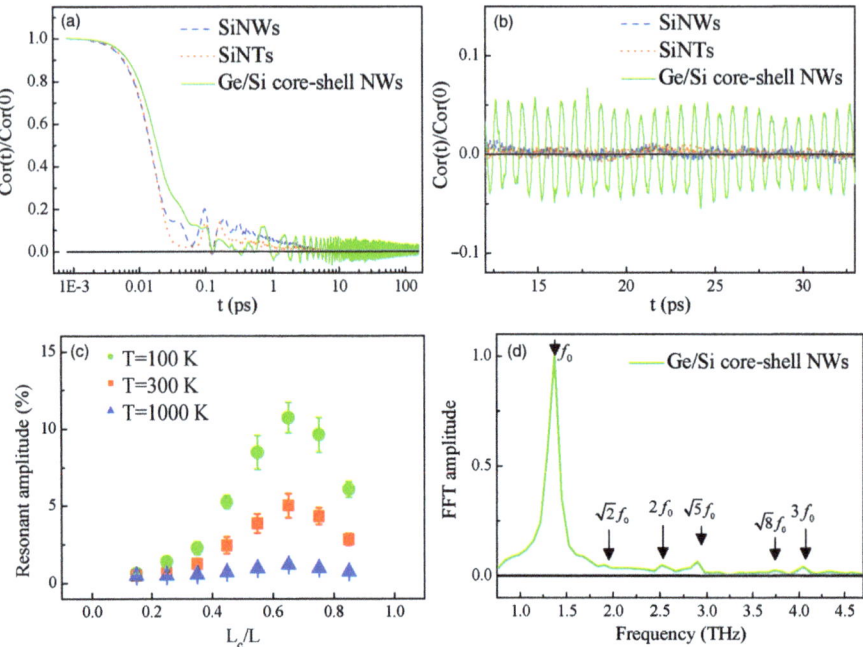

Figure 4.5 (a) Time dependence of normalized heat current autocorrelation function (HCACF). (b) Long-time region of (a). (c) Oscillation amplitude versus core-shell ratio at different temperature. (d) Amplitude of the fast Fourier transform (FFT) of the long-time region of normalized HCACF for Ge/Si core-shell NWs.[26]
Reprinted with permission from the American Institute of Physics.

resonance effect in core-shell NWs is structure- and temperature-dependent. As shown in Figure 4.5(c), when the core size increases, the resonance effect becomes stronger, reaches its maximum amplitude, and then decreases. For a given core-shell structure, the oscillation amplitude becomes larger at a lower temperature.

The fast Fourier transform (FFT) analysis of HCACF suggests that the nontrivial oscillation is caused by the phonon coherent resonance effect in the transverse direction (Figure 4.5(d)). In core-shell NWs, atoms on the same cross-section plane have a different sound velocity in the longitudinal direction. As a result, there is a strong coupling of vibrational modes between the longitudinal and transverse vibrations. When the frequency of the longitudinal phonon mode is close to the eigen-frequency of the transverse mode, coherent resonance occurs. Thus the coherent resonance effect in the transverse direction can indeed manifest itself in HCACF along the longitudinal direction in core-shell NWs.

As the transverse phonons are non-propagating, this resonance effect can significantly slow down the heat transport in the longitudinal direction.[27] As shown in Figure 4.6, in the range of thin coating layers (less than a critical value), the creation of core-shell structures will induce the phonon

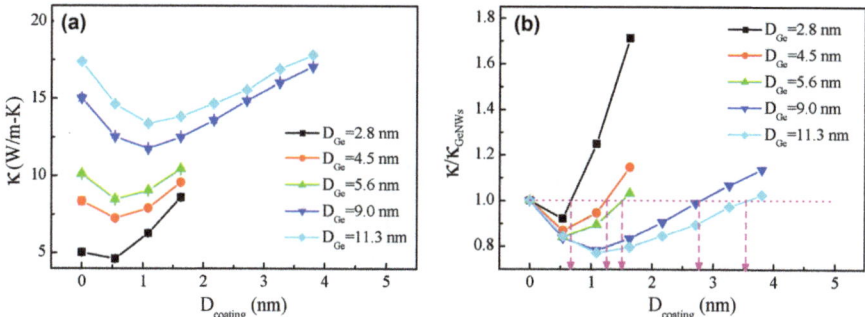

Figure 4.6 (a) Thermal conductivity of GeNWs and Ge/Si core-shell NWs for different D_{Ge} at room temperature. (b) Normalized thermal conductivity versus coating thickness for different D_{Ge}. Thermal conductivity of GeNWs at each D_{Ge} is used as a reference.[27]
Reprinted with permission from the American Chemical Society.

resonance between the transverse and longitudinal modes, thus offering a coherent mechanism to reduce the thermal conductivity. On the other hand, when the coating thickness is greater than a certain critical value, the thermal conductivity of the resultant coated GeNWs becomes larger than that of pristine GeNW without coating. This is because an increase in the total cross-sectional area after coating reduces the strength of the surface scattering. Using the coating method, the low thermal conductivity feature close to a very thin NW can be obtained from a much thicker NW. For example, thermal conductivity of pristine GeNWs with $D_{Ge} = 9.0$ nm is close to that of coated GeNWs with $D_{Ge} = 11.3$ nm and $D_{coating} = 2.7$ nm. The cross-sectional side length of coated NW is almost twice that for pristine GeNW, while they have a similar thermal conductivity. The atomistic coating method offers a novel and flexible route for the design and fabrication of NWs for TE applications. The reduced thermal conductivity in core-shell nanowires has been observed experimentally.[28,29]

Based on the different effects on thermal conductivity of SiNWs, a new concept, phononics engineering, has been proposed.[30] Phononics engineering is the study to control and modify the phonon thermal conductivity of nanomaterials through engineering their internal structures and controlling phonon scattering regimes. As shown in Figure 4.7, the different scattering mechanisms manifest themselves as the dominant ones only in a certain range of frequency regimes. High frequency phonons are sensitive to impurity and vacancy scattering. In comparison to high frequency phonons, low frequency phonons are grossly suppressed by boundary scattering, amorphous disorder scattering and coherent resonance in core-shell structures. In a real nanomaterial, there exists multiple scattering mechanisms at the same time. Hence with further development of theoretical modeling and analysis tools, it is expected that such phononic engineering will provide an efficient and practically useful tool for controlling the thermal conductivity of nanomaterials.

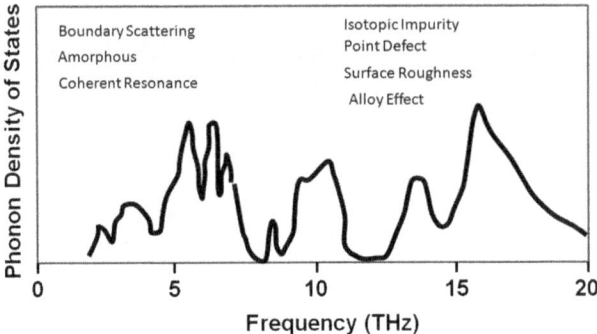

Figure 4.7 Schematic of the dominant range of frequency for various phonon scattering mechanisms.[30]
Reprinted with permission from the Wiley online library.

4.2.3 Ultra-high Thermal Conductivity of Graphenes

Graphene nanoribbons (GNRs) are a new type of quasi–one-dimensional carbon material, which can be obtained either by cutting mechanically exfoliated graphenes,[31] or by patterning epitaxially growth graphenes.[32] As a novel one-atom-thick material, GNRs have numerous unique properties different from those of other nanomaterials, such as high carrier mobility and giant Seebeck coefficient. Interestingly, high thermal conductivity (as high as 5000 W m^{-1} K^{-1})[33,34] has been demonstrated in graphene. This has raised the exciting prospect of using them for thermal management devices.

It has been demonstrated that the thermal conductance of a graphene sheet depends weakly on the direction angle of the thermal flux due to the directional dependence of group velocities of the phonon modes in the graphene.[35] There are rich physical phenomena about the thermal properties of GNRs. The effects of size,[36] defects,[37] doping,[38] shape,[39] stress/strain,[40] substrates,[41] inter-layer interactions,[42,43] chirality,[44] edge effect,[45] hydrogen coverage,[46] and other effects on thermal conductivity of nanoribbons have been widely studied.

As shown in Figure 4.8(a), the room temperature thermal conductance of zigzag GNRs is about thirty percent larger than that of armchair GNRs. This anisotropy phenomenon will disappear when the width is larger than 100 nm. For both armchair and zigzag GNRs, its thermal conductivity depends on the width. As shown in Figure 4.8(b), the thermal conductivity of zigzag GNRs increases firstly and then turns to decrease with the width increasing, while the armchair GNRs thermal conductivity monotonously increases with the width increasing, which comes from the competition between the edge localized phonon effect and the phonon Umklapp scattering effect.

As a combination of stress/strain, inter-layer and substrate effect, the folded GNR is a good candidate for future phonon engineering in graphene derivatives.[47] As shown in Figure 4.8, the percentage of reduction is dependent on the number of folds. The thermal conductivity of a GNR with

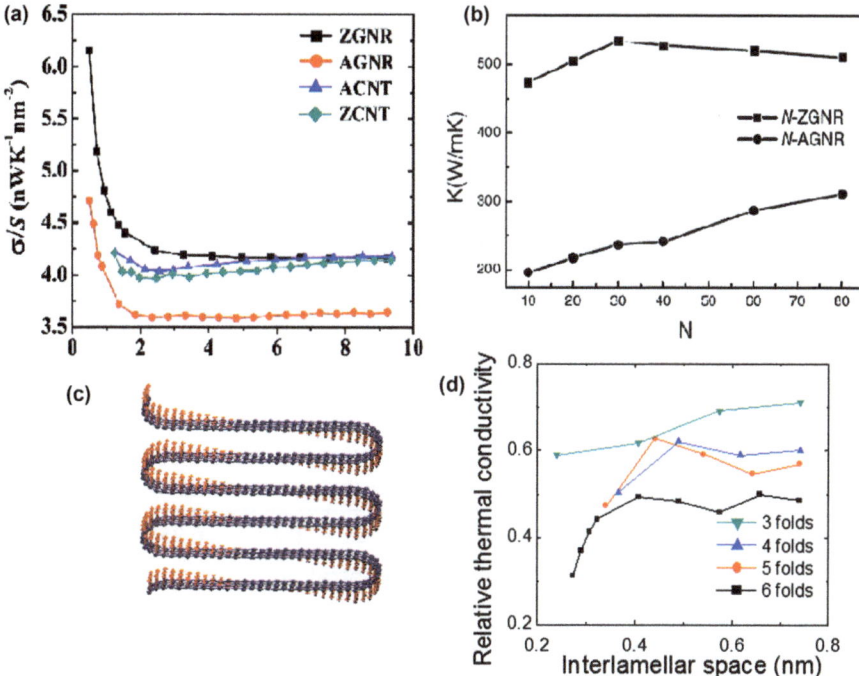

Figure 4.8 (a) The scaled thermal conductance for zigzag graphene nanoribbon (ZGNR), armchair GNR (AGNR), zigzag carbon nanotube (ZCNT) and armchair CNT (ACNT).[44] Reprinted with permission from the American Institute of Physics. (b) Thermal conductivity of N-AGNR and N-ZGNR with variation of N.[45] Reprinted with permission from the American Institute of Physics. (c) The side view of a schematic picture of a folded GNR. (d) Relative thermal conductivity modulation by compressing interlamellar space with different folds in GNRs. The value 1.0 of relative thermal conductivity corresponds to thermal conductivity of the flat zigzag GNR.[47]
Reprinted with permission from the American Institute of Physics.

six folds can be decreased substantially by up to 70% compared to that of the flat GNR. The transmission spectra show the decrease of thermal conductivity comes from strong scattering of low frequency modes at the folds, which is much different from other high frequency phonon scatterings by impurities, dislocation, and boundaries.

It is interesting to compare the isotopic effect on thermal conductivity of graphene and carbon nanotubes.[48] As shown in Figure 4.9, in the low isotopic doping region, the thermal conductivity of CNTs decreases rapidly with increasing isotopic concentration, 5% isotopic doping can yield a 25% reduction in thermal conductivity of CNTs. While in the high isotopic doping region, the thermal conductivity decreases slowly with further increase in doping. A similar phenomenon is also observed in the isotopic dependent thermal conductivity of GNR. More interestingly, the reduction behavior is

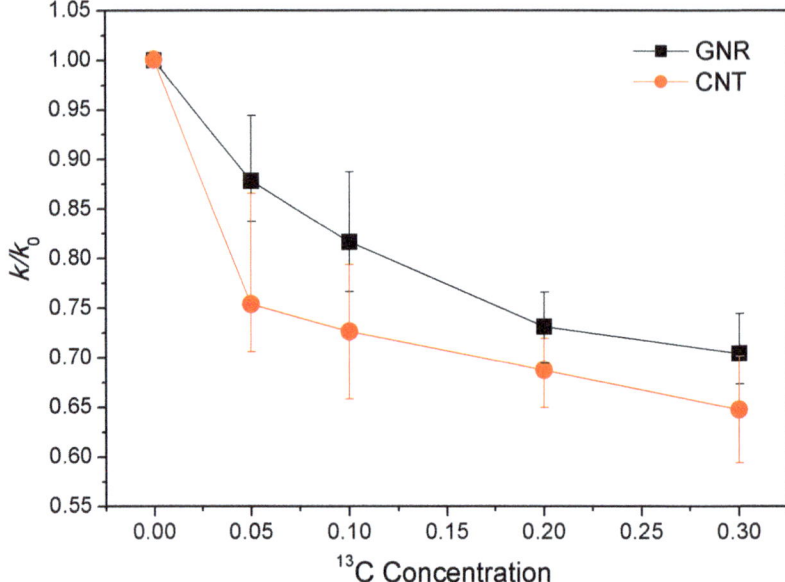

Figure 4.9 Isotopic concentration dependent thermal conductivity of GNR and CNT. The calculated thermal conductivities are normalized by the thermal conductivities of the corresponding pure GNR/CNT.[48] Reprinted with permission from the American Institute of Physics.

much slower in GNRs than that in CNTs. In contrast to the observations on CNTs, thermal conductivity of graphene is reduced by 12% around the isotope concentration of 5%. Such an observation suggests that the effect of isotopic scattering on thermal transport is stronger in CNTs. The random isotopic doping induces localization for phonons with high frequency, while the phonon in this frequency regime is already greatly suppressed by the edge scattering in GNRs. So including isotopic scattering in GNRs will cause much less reduction in thermal conductivity than that in CNTs.

4.2.4   Thermal Conduction of MoS_2

The two-dimensional (2D) suspended monolayer or few-layer molybdenum disulfide (MoS_2) has attracted great attention since it became experimentally accessible, thus opening up new possibilities for nanoelectronics and optoelectronic devices. In comparison with intensive experimental and theoretical research on thermal conductivity of graphene and graphene nanoribbons, the in-plane thermal conductivity of monolayer MoS_2 and MoS_2 nanoribbon has not been well studied. Very recently, the thermal conductivity of a monolayer MoS_2 sheet and nanoribbons (NRs) is investigated by using the molecular dynamics simulations.[49]

One striking feature is the low and isotropic thermal conductivity of a single-layer MoS_2 sheet. At room temperature, the thermal conductivities along both armchair and zigzag directions are essentially the same, about 1.35 W m^{-1}K^{-1}. It was predicted theoretically that 1D and 2D materials will have improved thermoelectric performance compared to the three-dimensional counterparts, due to the quantum confinement in the electron band structure. The high power factor and low thermal conductivity in silicon nanowires have demonstrated that 1D thermoelectric devices can be realized in semiconductor nanowires. Although high power factor is also observed in graphene, the ultra-high thermal conductivity (4800–5600 W m^{-1}K^{-1}) of graphene offsets its advantages in power factor and limits the application of graphene as efficient 2D thermoelectric materials. The low thermal conductivity raises an exciting prospect that monolayer MoS$_2$ can be used as high-performance 2D thermoelectric materials.

The thermal conductivities of MoS$_2$ NRs versus different lengths are shown in Figure 4.10. It is obvious that thermal conductivity increases with NR length. In contrast to the remarkable length dependent thermal conductivity of graphene NRs, thermal conductivity of MoS$_2$ NRs converges to a constant when length is long enough. On the basis of the kinetic theory of phonon transport, it implies that a plot of the inverse of thermal conductivity versus the inverse of the NRs length should be linear and its intercept should be the inverse of thermal conductivity of the infinitely large system. From the intercept and slope values of the plot, the phonon mean free path of the monolayer MoS$_2$ sheet is found to be about 5.2 nm. The intrinsic phonon mean free path according to phonon-phonon Umklapp scattering is about two orders of magnitude lower than that of graphene, which is believed to be responsible to the low thermal conductivity of monolayer MoS$_2$ sheet.

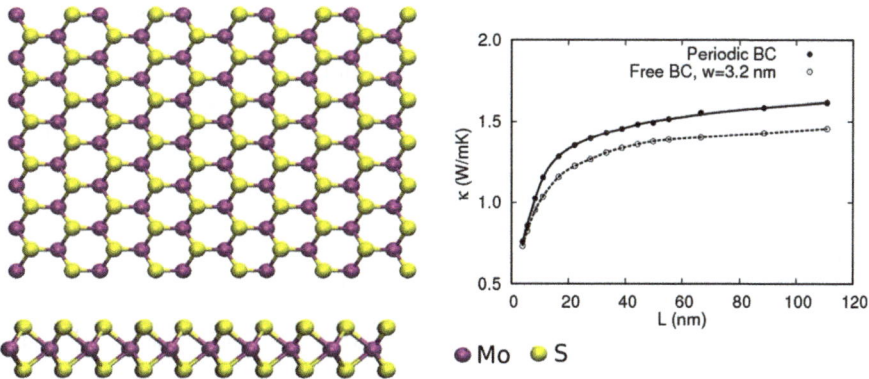

Figure 4.10 Left: Side view and top view of monolayer MoS$_2$. Right: Thermal conductivity of MoS$_2$ NRs as a function of nanoribbon length. Periodic boundary condition (BC) and free BC are used in the width direction of the sheet and NRs, respectively.[49]
Reprinted with permission from the American Institute of Physics.

4.3   Thermoelectric Property of Nanowires

To maximize the performance of a nanoscale thermoelectric cooler, it is indispensable to have a comprehensive understanding of the thermoelectric properties. In this section, we will discuss systematically the impacts of size, isotope concentration and alloy effect on the thermoelectric property of semiconductor nanowires.

4.3.1   Thermoelectric Property of Silicon Nanowires

By using the density functional derived tight-binding method, Shi *et al.*[50] have studied the size effect on the thermoelectric power factor in silicon nanowires. Figure 4.11(a) and (b) show the size effects on electrical conductivity σ and the Seebeck coefficient S with different electron concentration. The size dependence arises from the quantum confinement effect on the electronic band structure of SiNWs. In contrast to the weak size dependence of electronic conductivity, the Seebeck coefficient S decreases remarkably with increasing size. Besides the electronic band gap, the Seebeck

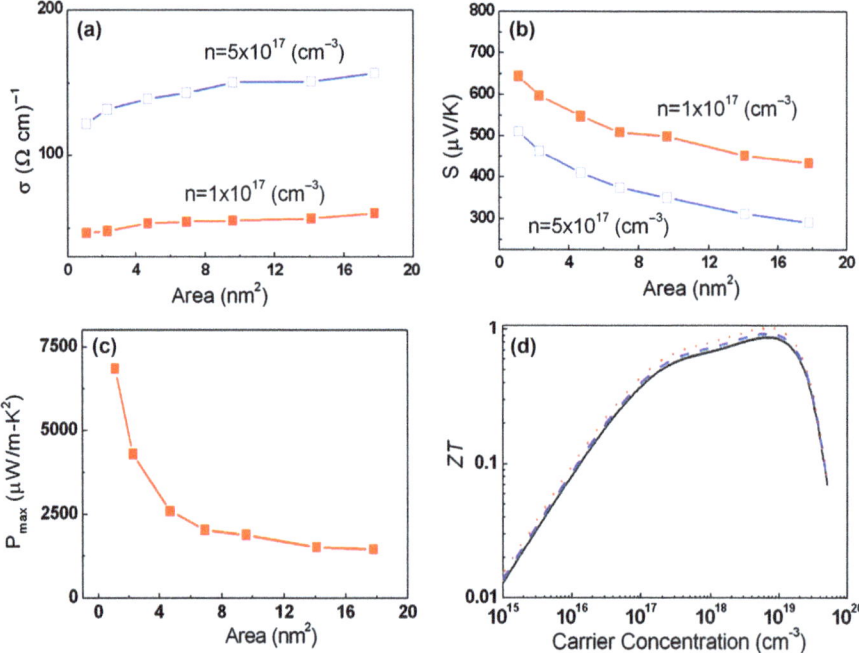

Figure 4.11   (a) and (b) Electrical conductivity and Seebeck coefficient versus cross-sectional area with different carrier concentration. (c) Maximum power factor versus cross-sectional area. (d) *ZT* versus carrier concentration.[50]
Reprinted with permission from the American Institute of Physics.

coefficient S also depends on the detailed band structure, in which the large numbers of electronic stats in the narrow energy ranges can lead to a large S. When the transverse dimension increases, the sharp DOS peaks widen and reduce S.

In thermoelectric application, the power factor P, which is defined as $P = S^2\sigma$, is an important factor influencing the thermoelectric performance directly. Figure 4.11(c) shows the power factor for SiNWs with different cross-sectional areas. When SiNW diameter increases, the maximum attainable power factor, P_{max}, decreases, beacause the slow increase of σ is offset by the obvious decreasing in S. When the area increases from about 1 nm^2 to 18 nm^2, the maximum power factor decreases from about 6800 to 1400 $\mu W\ m^{-1}K^{-2}$. In contrast to the high sensitivity at the small size range, the power factor versus cross-section area curves are almost flat at the large wire range, where the power factor is almost constant when the cross-section area increases from 14.1 to 17.8 nm^2. For SiNWs with a length in the μm scale, the phonon thermal conductivity increases with diameter increases remarkably until the diameter is larger than about a hundred nanometers. Combine the size dependence of the power factor as shown in Figure 4.11(c), we can conclude that ZT will decrease when the NW diameter increases.

The dependence of figure of merit ZT on the carrier concentration is shown in Figure 4.11(d). ZT increases as carrier concentration increases, and there is an optimal carrier concentration yielding the maximum attainable value of ZT. Above this carrier concentration, ZT decreases with increasing carrier concentration. This phenomenon can be understood as below. Increasing carrier concentration has two effects on the power factor and ZT. On the one hand, the increase of the carrier concentration will increase the electrical conductivity. On the other hand, the increase of the carrier concentration will suppress the Seebeck coefficient S. ZT is determined by these two effects that compete with each other. Therefore, there exists an optimal carrier concentration to yield the maximum value of the figure of merit ZT.

4.3.2 Thermoelectric Property of Silicon-Germanium Nanowires

Besides SiNW, it has also been demonstrated that $Si_{1-x}Ge_x$ NW is a promising candidate for high-performance thermoelectric application since its thermal conductivity can be tuned by the Ge contents. The composition dependence of the thermoelectric properties in $Si_{1-x}Ge_x$ NWs has been investigated systematically by Shi *et al.*[51] Figure 4.12 shows the dependence of ZT on the Ge content x. The value of ZT of $Si_{1-x}Ge_x$ NWs increases with Ge content, reaches a maximum and then decreases. It is worth pointing out that the calculated ZT of pure Ge NW is only 1.3 times that of pure SiNW. The remarkable difference in the increased factor of ZT between the respective bulk and nanowires is due to the similarity of the thermal conductivity and

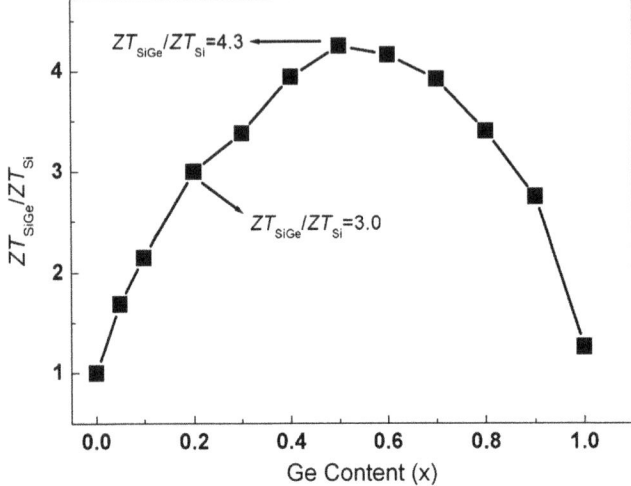

Figure 4.12 The dependence of ZT on Ge content x.[51]
Reprinted with permission from the American Institute of Physics.

electronic property in the nano scale. Moreover, at low Ge content, the small ratio of Ge atoms can induce a large increase in ZT. For instance, in the case of the $Si_{0.8}Ge_{0.2}$ NW, namely, 20% Ge, ZT is about three times of that of pure SiNW. And with 50% Ge atoms ($Si_{0.5}Ge_{0.5}$ NW), the ZT can be as high as 4.3 times that of pure SiNW. Combine with the experimentally measured ZT of n-type SiNW, which is about 0.6–1.0, it is exciting that we may obtain a high ZT value of about 2.5–4.0 in n-type $Si_{1-x}Ge_x$ NWs.

In addition to $Si_{1-x}Ge_x$ NWs, superlattice structured NWs have attracted broad interest due to their fascinating applications in photonics, electronics, and phononics. Figure 4.13 shows the geometry for the $Si_{0.5}Ge_{0.5}$ superlattice NWs with different periodic length. As shown in this figure, the maximum value of ZT (ZT_{max}) first increases with periodic length, reaches the maximum value at the periodic length L of 0.54 nm, and then decreases with further increases of the periodic length. The achievable ZT_{max} for n-type NWs is about twice that of its p-type counterparts. Moreover, the values of ZT_{max} for the superlattice NWs are much larger than those of pristine SiNWs because of their low thermal conductivity. The maximum value of ZT_{max} for n-type superlattice NW is 4.7 at the period length of 0.54 nm, which is a five-fold increase as compared to the equivalent pristine SiNWs ($ZT_{max} = 0.94$). The maximum value of ZT_{max} for p-type NWs is 2.74 with the same periodic length, which is 4.6 times larger than that of the p-type SiNWs ($ZT_{max} \approx 0.6$). The carrier concentration yielding the maximum attainable ZT is 5.3×10^{20} cm^{-3} for n-type and 6.9×10^{20} cm^{-3} for p-type SiGe superlattice NWs, respectively. The optimal carrier concentration yielding the peak ZT is in the same order of magnitude with the maximum achievable free carrier concentration for SiNWs, and demonstrates that superlattice SiGe nanowires are promising thermoelectric materials to achieve high ZT.[52]

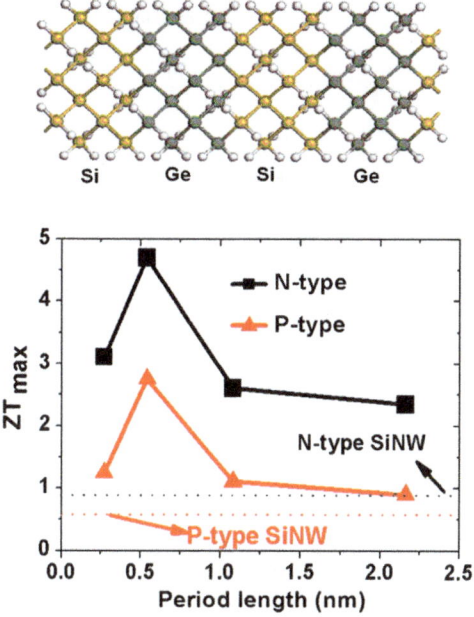

Figure 4.13 Dependence of ZT_{max} on periodic length for p-type and n-type Si$_{0.5}$Ge$_{0.5}$ superlattice NWs.[52]
Reprinted with permission from the American Institute of Physics.

4.3.3 Thermoelectric Properties of ZnO Nanowires

Recently, through the coupling of piezoelectric and semiconducting properties of zinc oxide (ZnO) nanowires, nanoscale mechanical energy can be converted into electrical energy. In addition to piezoelectric and optoelectronic applications, thermoelectric effect is also an important approach to the solution to the energy crisis by converting waste heat into electricity. The study of the thermoelectric property of ZnO ceramics has been motivated by reports of the improved electrical conductivity and reduced thermal conductivity by employing elements (Al, Ga, Mn) as dopants. In the Al and Ga dually doped ZnO ceramics, ZT is about twice that of pure ZnO, which can arrive 0.65 at 1247 K.

Here we discuss the thermoelectric properties of ultrathin Zn$_{1-x}$Ga$_x$O NWs oriented along the (0001) direction.[53] The atomic structure of Zn$_{1-x}$Ga$_x$O NWs is shown in Figure 4.14(a), where the Zn atoms are randomly substituted by Ga atoms. Figure 4.14(b) shows the Ga content dependence of electrical conductance σ and Seebeck coefficient S. σ increases and S decreases as the Ga content increases because there is more charge concentration. At a low carrier concentration range, the Seebeck coefficient is about 400 μV K^{-1}. Figure 4.14(c) shows the power factor $(P = S^2\sigma)$ versus the Ga content. There is an optimal Ga content $(x = 0.04, n = 1.0 \times 10^{17}$ cm$^{-3})$ yielding the maximum attainable value of power factor $(P_{max} = 908 \ \mu$W m^{-1}K$^{-2})$.

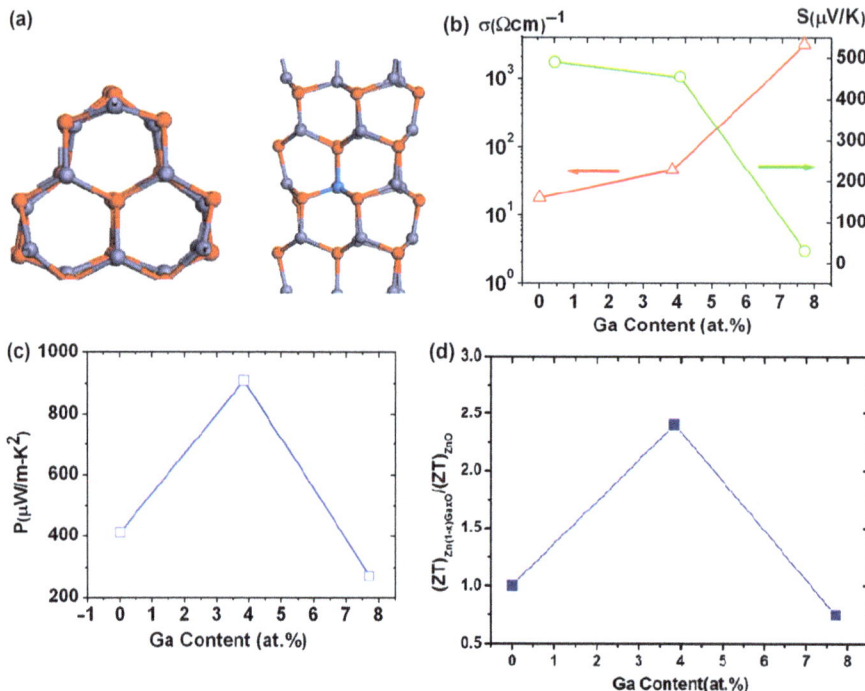

Figure 4.14   (a) The atomic structure for the $Zn_{1-x}Ga_xO$ NW for $x = 0.04$ with the replacement of one Zn atom in the supercell. (b) Electrical conductivity σ and Seebeck coefficient S versus Ga content x. (c) Power factor of $Zn_{1-x}Ga_xO$ NWs versus Ga contents. (d) Figure of merit ZT versus Ga content for $Zn_{1-x}Ga_xO$ NW.[53]
Reprinted with permission from Elsevier.

Combined the power factor and the phonon thermal conductivity, the Ga content dependent ZT is shown in Figure 4.14(d). The value of ZT firstly increases with the Ga content, reaches a maximum value at $x = 0.04$, which is 2.5 times larger than that of pure ZnO wires, and then decreases. This phenomenon can be explained as follows. ZT is contributed by both power factor and thermal conductivity. With Ga content increase, the power factor increases, reaches a maximum value at $x = 0.04$ and then decreases. The curve of the thermal conductivity decreases as the Ga content increases. Thus when $0 \leq x \leq 0.04$, ZT increases with the Ga content. However, when $0.04 \leq x0.08$, the reduction in the thermal conductivity cannot offset the reduction in the power factor, thus resulting in the decrease in ZT with the Ga content. It is worth mentioning that the optimal carrier concentration yielding the maximum attainable value of ZT is 1.0×10^{17} cm^{-3}. This low carrier concentration suggests that the high thermoelectric performance of Ga-doped ZnO NW can be realized by using reliable fabrication technology. This 2.5 times enhancement in ZT provides insights in designing possible ZnO NW array-based piezoelectric, optoelectronic and thermoelectric hybrid energy generator.[53]

References

1. F. J. DiSalvo, *Science*, 1999, **285**, 703.
2. G. Zhang, Q Zhang, D. Kavitha and G.-Q. Lo, *Appl. Phys. Lett.*, 2009, **95**, 243104.
3. G. Zhang, Q. Zhang, C.-T. Bui, G.-Q. Lo and B. Li, *Appl. Phys. Lett.*, 2009, **94**, 213108.
4. A. A. Balandina and D. L. Nika, *Mater. Today*, 2012, **15**, 266.
5. G. Zhang and B. Li, *Nanoscale*, 2010, **2**, 1058.
6. N. Li, J. Ren, L. Wang, G. Zhang, P. Hänggi and B. Li, *Rev. Modern Phys.*, 2012, **84**, 1045.
7. S. Berber, Y.-K. Kwon and D. Tománek, *Phys. Rev. Lett.*, 2000, **84**, 4613.
8. M. A. Osman and D. Srivastava, *Nanotechnology*, 2001, **12**, 21.
9. T. Yamamoto, S. Watanabe and K. Watanabe, *Phys. Rev. Lett.*, 2004, **92**, 075502.
10. J. Hone, M. Whitney, C. Piskoti and A. Zettl, *Phys. Rev. B*, 1999, **59**, R2514.
11. P. Kim, L. Shi, A. Majumdar and P. L. McEuen, *Phys. Rev. Lett.*, 2001, **87**, 215502.
12. G. Zhang and B. Li, *J. Chem. Phys.*, 2005, **123**, 114714.
13. S. Maruyama, *Physica B*, 2002, **323**, 193.
14. C. W. Chang, A. M. Fennimore, A. Afanasiev, D. Okawa, T. Ikuno, H. Garcia, Deyu. Li, A. Majumdar and A. Zettl, *Phys. Rev. Lett.*, 2006, **97**, 085901.
15. J. Che, T. Cagin and W. A. Goddard III, *Nanotechnology*, 2000, **11**, 65.
16. N. Mingo, D. A. Stewart, D. A. Broido and D. Srivastava, *Phys. Rev. B*, 2008, **77**, 033418.
17. C. W. Padgett and D. W. Brenner, *Nano Lett.*, 2004, **4**, 1051.
18. A. I. Hochbaum, R. Chen, R. D. Delgado, W. Liang, E. C. Garnett, M. Najarian, A. Majumdar and P. Yang, *Nature*, 2008, **451**, 163.
19. A. I. Boukai, Y. Bunimovich, J. T. Kheli, J.-K. Yu, W. A. Goddard III and J. R. Heath, *Nature*, 2008, **451**, 168.
20. S. Volz and G. Chen, *Appl. Phys. Lett.*, 1999, **75**, 2056.
21. D. Li, Y. Wu, P. Kim, L. Shi, P. Yang and A. Majumdar, *Appl. Phys. Lett.*, 2003, **83**, 2934.
22. N. Yang, G. Zhang and B. Li, *Nano Lett.*, 2008, **8**, 276.
23. J. Chen, G. Zhang and B. Li, *Appl. Phys. Lett.*, 2008, **95**, 073117.
24. L. H. Liang and B. Li, *Phys. Rev. B*, 2006, **73**, 153303.
25. J. Chen, G. Zhang and B. Li, *Nano Lett.*, 2010, **10**, 3978.
26. J. Chen, G. Zhang and B. Li, *J. Chem. Phys.*, 2011, **135**, 104508.
27. J. Chen, G. Zhang and B. Li, *Nano Lett.*, 2012, **12**, 2826.
28. M. C. Wingert, Z. C. Y. Chen, E. Dechaumphai, J. Moon, J.-H. Kim, J. Xiang and R. Chen, *Nano Lett.*, 2011, **11**, 5507.
29. J. Moon, J.-H. Kim, Z. C. Y. Chen, J. Xiang and R. Chen, *Nano Lett.*, 2013, **13**, 1196.
30. G. Zhang and Y.-W. Zhang, *Phys. Status Solidi RRL*, 2013, **7**, 754.

31. K. S. Novoselov, A. K. Geim, S. V. Morozov, D. Jiang, Y. Zhang, S. V. Dubonos, I. V. Grigorieva and A. A. Firsov, *Science*, 2004, **306**, 666.
32. C. Berger, Z. Song, X. Li, X. Wu, N. Brown, C. Naud, D. Mayou, T. Li, J. Hass, A. N. Marchenkov, E. H. Conrad, P. N. First and W. A. De Heer, *Science*, 2006, **312**, 1191.
33. A. A. Balandin, S. Ghosh, W. Bao, I. Calizo, D. Teweldebrhan, F. Miao and C. N. Lau, *Nano Lett.*, 2008, **8**, 902.
34. S. Ghosh, I. Calizo, D. Teweldebrhan, E. P. Pokatilov, D. L. Nika, A. A. Balandin, W. Bao, F. Miao and C. N. Lau, *Appl. Phys. Lett.*, 2008, **92**, 151911.
35. J.-W. Jiang, J.-S. Wang and B. Li, *Phys. Rev. B*, 2009, **79**, 205418.
36. W. J. Evans, L. Hu and P. Keblinski, *Appl. Phys. Lett.*, 2010, **96**, 203112.
37. Z.-X. Xie, K.-Q. Chen and W. Duan, *J. Phys. Condens. Mat.*, 2011, **23**, 315302.
38. J. W. Jiang, J. H. Lan, J. S. Wang and B. W. Li, *J. Appl. Phys.*, 2010, **107**, 054314.
39. X.-F. Peng, X.-J. Wang, Z.-Q. Gong and K.-Q. Chen, *Appl. Phys. Lett.*, 2011, **99**, 233105.
40. X. Li, K. Maute, M. L. Dunn and R. Yang, *Phys. Rev. B*, 2010, **81**, 245318.
41. Z.-X. Guo, J. W. Ding and X.-G. Gong, *Phys. Rev. B*, 2012, **85**, 235429.
42. G. Zhang and H. Zhang, *Nanoscale*, 2011, **3**, 4604.
43. J. Chen, G. Zhang and B. Li, *Nanoscale*, 2013, **5**, 532.
44. Y. Xu, X. Chen, B.-L. Gu and W. Duan, *Appl. Phys. Lett.*, 2009, **95**, 233116.
45. Z.-X. Guo, D. Zhang and X.-G. Gong, *Appl. Phys. Lett.*, 2009, **95**, 163103.
46. Q.-X. Pei, Z.-D. Sha and Y.-W. Zhang, *Carbon*, 2011, **49**, 4752.
47. N. Yang, X. Ni, J.-W. Jiang and B. Li, *Appl. Phys. Lett.*, 2012, **100**, 093107.
48. X. Li, J. Chen, C. Yu and G. Zhang, *Appl. Phys. Lett.*, 2013, **103**, 013111.
49. X. Liu, G. Zhang, Q.-X. Pei and Y.-W. Zhang, *Appl. Phys. Lett.*, 2013, **103**, 133113.
50. L. Shi, D. Yao, G. Zhang and B. Li, *Appl. Phys. Lett.*, 2009, **95**, 063102.
51. L. Shi, D. Yao, G. Zhang and B. Li, *Appl. Phys. Lett.*, 2010, **96**, 173108.
52. L. Shi, J. Jiang, G. Zhang and B. Li, *Appl. Phys. Lett.*, 2012, **101**, 233114.
53. L. Shi, J. Chen, G. Zhang and B. Li, *Phys. Lett. A*, 2012, **376**, 978.

CHAPTER 5

Nanotubes for Energy Storage

HUI PAN

Faculty of Science and Technology, University of Macau, Macao SAR, China
Email: huipan@umac.mo

5.1 Introduction

Energy has been a key in the development of every sphere of human. Energy demands worldwide, for electricity as well as transportation and heating fuels, are expected to double by 2050, and more than triple the demand by the end of the century, raising critical energy supply issues.[1] Managing the use, transportation, and storage of energy is inevitable in any functional society. The traditional usable energy sources, including fossil fuels and coal, will fall short of this demand over the long term, and their continued use produces harmful side effects such as pollution that threatens human health and greenhouse gases associated with climate change. Sufficient supplies of clean energy are intimately linked with global stability, economic prosperity, and quality of life. Finding energy sources to satisfy the world's growing demand is one of society's foremost challenges for the next half-century. Solar energy and hydrogen are considered to be two of the most important candidates because they are abundant, clean and renewable. A variety of technologies have been developed to take advantage of solar energy, such as solar electricity (photovoltaic cell, PV), solar fuels (photo-electrochemical cell, PEC), and solar thermal systems. The PV cell directly converts the solar energy to electricity, while the PEC splits water into hydrogen and oxygen, so as to convert the solar energy into chemical energy.

RSC Nanoscience & Nanotechnology No. 32
Nanofabrication and its Application in Renewable Energy
Edited by Gang Zhang and Navin Manjooran
© The Royal Society of Chemistry 2014
Published by the Royal Society of Chemistry, www.rsc.org

The thermal energy is obtained from the sun in thermal systems. Without significant investigation and investigation into energy storage technologies, however, the current clean energy technologies will increasingly struggle to provide reliable, affordable electricity, because the excessive energies (electricity and hydrogen) should be discarded and wasted. The use of transient renewable energy sources would be greatly enhanced by improved energy storage technologies.

The energy storage technologies promise to address the growing limitations of energy supply and demand, and enhance the reliability of the electric grid. Current large-scale energy storage systems are both electrochemically based (*e.g.*, advanced lead-carbon batteries, lithium ion batteries, sodium-based batteries, flow batteries, and electrochemical capacitors) and kinetic-energy-based (*e.g.*, compressed-air energy storage and high-speed flywheels).[2] The current energy storage technologies, however, do not adequately meet the wide-ranging needs of the electric power industry due to the high cost, low energy capacity, low efficiency, and current complexity of many of today's storage technologies. Significant advances in materials and devices are needed, therefore, to address many of the challenges associated with energy storage system economics and technical performance, and to realize the potential of large-scale energy storage technologies because the materials composing these technologies determine the majority of their performance specifications.[2] The selection of materials should make the storage technologies affordable, efficient, reliable, and green. It is now widely recognized that technology breakthroughs and truly revolutionary developments are needed.

Nanoscience and nanotechnology arise from the exploitation of new properties, phenomena, processes, and functionalities that matter exhibits at intermediate sizes between isolated atoms or molecules (≈ 1 nm) and bulk materials (over 100 nm). Nanoscience and nanotechnology must play a pivotal role in the development of energy storage technologies because all elementary steps of energy storage (such as charge transfer, molecular rearrangement, chemical reactions, *etc.*) take place on the nanoscale. The development of new nanoscale materials and structures, therefore, creates for an entirely new paradigm for developing new and revolutionary energy storage technologies like advanced batteries, fuel cells, supercapacitors, and biofuels.[3-11]

Nanotubes (NTs) are recognized as a new class of material that has had a profound impact on a wide range of applications since Iijima reported the synthesis of carbon nanotubes (CNTs) in 1991.[12] One of the areas in which nanotubes have demonstrated their advantages is energy storage due to their 'inherently unique' properties,[13-17] such as the unique structure (one-dimensional cylindrical tubule), high surface area, high conductivity, low density, high rigidity, and high tensile strength. In this review, we wish to describe our understanding of the current 'state of the art' energy storage of nanotubes, and discuss recent progresses in their applications to hydrogen storage, supercapacitor, and Li-battery.

5.2 Hydrogen Storage

As an ideal clean energy carrier for the future, hydrogen can be produced from a variety of energy resources, has the highest energy density per unit mass, and produces the least pollution since it can be extracted from natural resources, such as water or biomass, and its use produces water only. A transition from fossil fuels to hydrogen as a major fuel in the next 50 years, therefore, could increase energy security. One of the most difficult challenges for using hydrogen is its very low density which hampers the development of a cost effective, reliable and safe hydrogen storage system.

For transportation, a hydrogen density of >9 wt% (81 g L^{-1}) for the storage system (including materials and tank) is required for H_2-O_2 fuel-cell powered vehicles to travel for 300 miles without refueling. Other requirements include reversibility (1500 cycles), cost ($2 kWh^{-1}), fueling time (30 s (kg-H_2)$^{-1}$) and operating temperature/pressure (−40 to 60 °C; <100 atm).[18–20] The current leading methods of hydrogen storage are compressed gas and liquid. Carbon fiber-reinforced 5000-psi and 10,000-psi compressed hydrogen gas tanks are already in use in prototype hydrogen-powered vehicles. However, the density of hydrogen they can hold is not high enough, less than 4.5 wt% hydrogen (36 g L^{-1}). The volumetric density of hydrogen can be doubled by storing hydrogen in a cryo-compressed tank and 5 wt% (78 g L^{-1}) has been achieved at 21 K. However, liquefying hydrogen would consume about 30% of its energy content and continuous boil-off H_2 loss would further reduce its overall efficiency. Thus the long term plan is to develop reversible solid state hydrogen storage materials, such as high surface area adsorbents.[18–20]

Hydrogen storage represents a difficult but exciting challenge. Success in unlocking the mysteries of hydrogen storage in nanoscale materials could, therefore, open the door to the hydrogen future and meet the 2025 targets for practical applications (Figure 5.1). Ideally, an efficient hydrogen storage material must have: a) a high volumetric/gravimetric capacity, b) a fast sorption rate at relatively low temperatures, and c) a high tolerance to recycling.[21,22] The major strategy in H_2 storage in solid sorbents is (1) to increase the specific surface area; (2) to tail pore structure for increasing the hydrogen molecule binding energy: defects and suitable pore structure can increase the binding energy by 20–30% compared with the perfect or large pore structures; (3) to dope with more electronegative atoms such as N to enhance the bonding energy; (4) to deposit nanosized transition metal catalysts to enhance chemisorption with reversible hydrogen spillover: in this approach, a catalyst dissociates molecular hydrogen to hydrogen atoms, which then diffuse into the sorbent and reversibly bond with carbon atoms or other sorbent atoms.[23]

5.2.1 Carbon Nanotubes

Carbon nanotubes (CNTs) can be considered as hollow tubes rolled up from two-dimensional graphene sheets. CNTs can be single-walled (SWCNT),

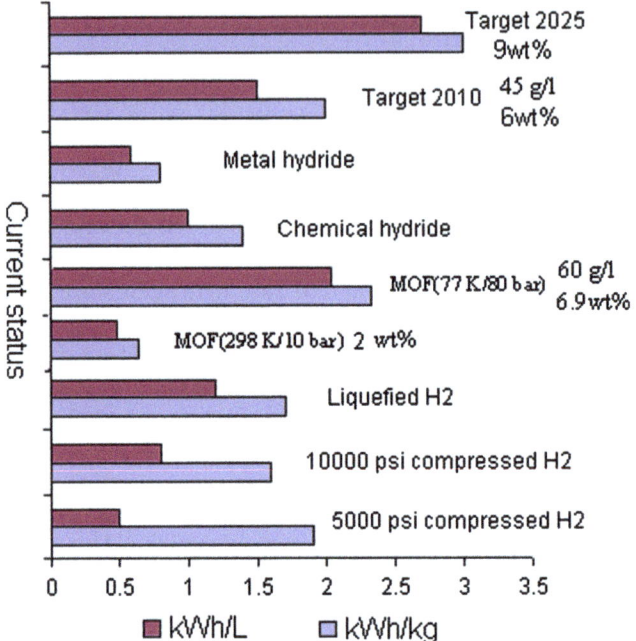

Figure 5.1 The current status of today's hydrogen storage technologies in volumetric and gravimetric terms.

double-walled (DWCNT), and multiwalled (MWCNT), depending on the number of tube walls. SWCNTs are usually 1–2 nm in diameter and up to microns in length. They are often bundled together due to van der Waals attractive forces. MWCNTs normally have a hollow core of tens of nanometers with an interlayer distance of 0.34–0.37 nm. The inner tube diameter of a DWCNT is in the range of 1–3 nm and its interlayer distance is in the range of 0.3–1 nm.

5.2.1.1 Pure Carbon Nanotube

CNTs are of great interest in hydrogen storage because of its high surface area, which may provide adsorption sites for hydrogen atoms/molecules. In 1997, Dillon et al.[24] first reported that the gravimetric storage density in SWCNTs ranges from 5 to 10 wt%H_2. Their exciting and optimistic work stimulated tremendous and intensive research to improve the hydrogen storage ability in carbon-based nanomaterials. Subsequently, their prediction was seemingly realized on SWCNTs,[25,26] and multiwall carbon nanotubes (MWCNTs)[27] under high pressure conditions. However, these results have not been repeated or confirmed independently at other laboratories. Contrarily, Hirscher et al. reported a low hydrogen storage capacity of 0.5 wt% at room temperature and 70 bar for purified SWCNTs and other carbon nanostructures in 2001.[28,29] They clarified the situation and indicated that the desorption of hydrogen from Dillon's samples originates from Ti-alloys introduced during the ultrasonic treatment. If a sonification bar, which was used in pre-treating the CNT samples, was made of stainless steel rather than titanium alloy, the hydrogen storage was below 1 wt%. Lawrence *et al.*[30] also observed that at room temperature adsorption appeared to saturate at 0.9 wt% above a pressure of 200 atm due to monolayer saturation. Bacsa *et al.* further pointed out that the maximum hydrogen storage on their co-containing MWCNTs at room temperature and 10 MPa was 0.5 wt% and the purification cannot improve the hydrogen uptake capacity.[31] Using samples up to 85 g, Ning *et al.* measured only 0.3 wt%H_2 at room temperature and 2.27 wt%H_2 at 77 K and 100 bar on MWCNTs.[32] In their excellent review, Liu *et al.* ascribed the discrepancy in hydrogen storage data of CNTs to the effects of measurement methodology, material species, synthesis techniques, structural perfection, and surface and pore characteristics *etc.*[33] For pure CNTs without any specific treatment, physisorption is the major driving force for hydrogen storage. The ordinary storage capacity by physisorption is limited because of weak hydrogen-hydrogen and hydrogen-carbon interactions. The maximum hydrogen uptake capacity of pure CNTs is in a range of 0.5 wt% to 1.1 wt% at room temperature,[29-40] depending on the structural properties of the CNTs and synthesis methods.

Two different views on how the specific surface area and pore characteristics of CNTs would affect the hydrogen adsorption exist in literature. The hydrogen uptake is a linear function of pressure at 298 K, which can be explained with Henry's law on a sub-monolayer and shows no saturation.[28,29,41,42] In contrast, the isotherm at 77 K shows a saturation at high pressures and can be fitted to a Langmuir type equation, indicating a monolayer formation of hydrogen typical for microporous surfaces. The hydrogen uptake of the different carbon nanostructures including SWCNT and MWCNT (Figure 5.2) shows an almost linear relationship between the storage capacity and specific surface area both at room temperature and at 77 K.[42] It seems that the adsorption is independent of the structures and dependent only on the specific surface area. Nijkamp *et al.*[43] reported that a linear relationship exists between the storage capacity and specific surface area, that is, 1 wt% H_2 is adsorbed for every 500 $m^2\ g^{-1}$ of surface area of sorbent. Zuttle *et al.*[44] also concluded that the reversible hydrogen capacity correlates with the specific surface area of the sample and is 1.5 wt% per 1000 $m^2\ g^{-1}$. Using the inelastic neutron scattering signal associated with rotation of the hydrogen molecule as a sensitive probe for the surroundings of the molecule, Scimmel *et al.*[45,46] concluded that different forms of carbon are essentially the same for hydrogen molecules as long as they possess the characteristic aromatic carbon rings. Thus the amount of hydrogen storage is governed by the number of accessible aromatic C–C bonds in the sample, which is related to the surface area. Zhou *et al.*[47] found that the isotherms of the hydrogen adsorption on MWCNTs at 77 K showed a typical feature of supercritical adsorption. A fundamental feature of supercritical gas is the impossibility of liquefaction at any high pressures. Therefore, monolayer

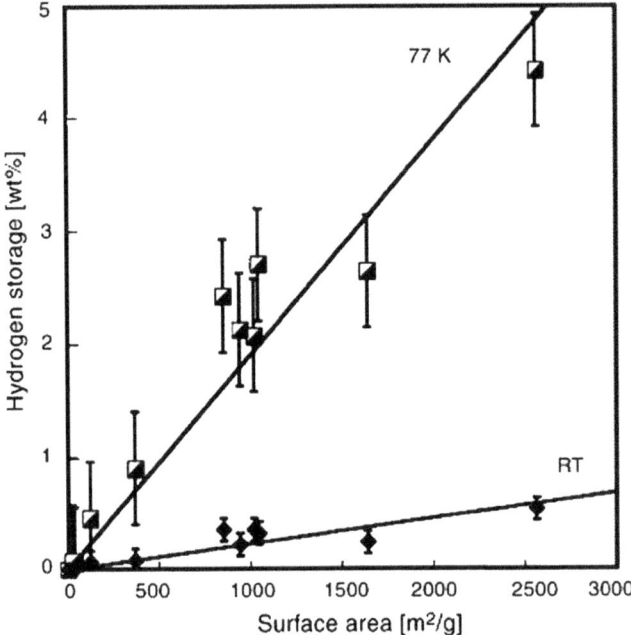

Figure 5.2   Hydrogen storage capacity of various carbon nanostructures versus the specific surface area at RT and at 77 K. the slope of the curves are 0.23×10^{-3} and 1.91×10^{-3} wt% g m^{-2}, respectively.[42] (Copyright: Elsevier).

coverage is the only possible mechanism of adsorption. Since it is a monolayer, the saturated capacity of adsorption depends strongly on the specific surface area of sorbents.

Gao *et al.*,[37] however, reported a different relationship between specific surface area and hydrogen uptake capacity of CNTs and concluded that the higher surface area (characterized by nitrogen cryo-adsorption isotherm) does not always lead to higher hydrogen uptake. In their study, AX-21 (Anderson Development Corp.) has the highest specific surface area $(2790 \text{ m}^2 \text{ g}^{-1})$ with a large mean pore size of 1.5 nm. Its hydrogen uptake is much higher than that of conventional SWCNTs or MWCNTs. However, the hydrogen uptake of AX-21 (0.5 wt% at room temperature and 5 wt% at 77 K at 100 bar) is lower than that of open-tipped multiwalled carbon nanotubes (1.12 wt% at room temperature and 6.34 wt% at 77 K at 100 bar) grown on an anodic aluminium oxide template which has a specific surface area of 840 m^2 g^{-1} and mean pore sizes of 40 nm/0.6 nm (Figure 5.3), suggesting that the molecules of N$_2$, which were used in BET (Brunauer-Emmett-Teller) measurements of specific surface area, are too large to probe some micropores responsible for the H$_2$ adsorption. The high resolution TEM image in Figure 5.4 shows the existence of subnanopores (0.6 nm), such as openended cavities and defective interlayer slits on the wall of the insufficiently

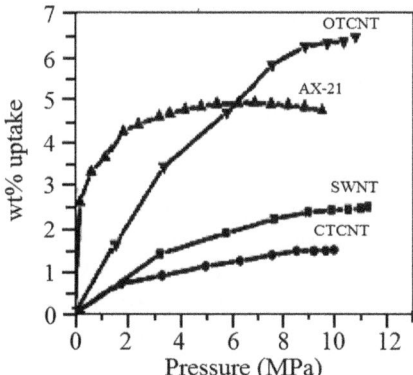

Figure 5.3 Isotherms of wt% H_2 uptake *vs* pressure at 77 K for AX-21, open-tipped CNT (OTCNTs), SWCNTs, and close-tipped CNT (CTCNTs). The measurements were performed on a home-built apparatus based on a volumetric method.[37]
(Copyright: American Institute of Physics).

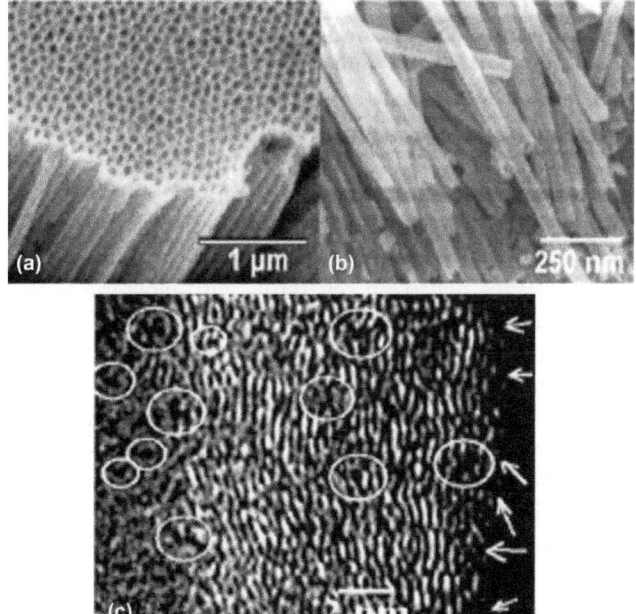

Figure 5.4 (a) SEM image of a highly ordered, open-tipped OTCNT array grown on a both-side-opened AAO template after removal of amorphous carbon. (b) OTCNT completely released from AAO template for hydrogen storage measurement. (c) HRTEM image of the OTCNT wall; the circles indicate subnanopores in the wall body, and the arrows point out open ends on the external wall surface. The internal surface is very rough.[37]
(Copyright: American Institute of Physics).

graphitized open-tipped carbon nanotubes (OTCNTs). A SWCNT sample with a specific surface area of 1160 m^2 g^{-1}, uptake of 3.8 wt% at 77 K and 1 bar while AX-21 with a surface area 2800 m^2 g^{-1}, uptake of only 2.8 wt% under identical conditions. They concluded that defect structures play a dominant role for the hydrogen uptake under high pressures. Bacsa *et al.*[31] also notified that increasing the specific surface area does not necessarily increase the hydrogen storage capacity because the hydrogen storage capacity is related to the pore volume at low pore diameters (<3 nm). Similarly Poirier *et al.* [48] studied hydrogen storage on SWCNTs and activated carbon AX-21 and found no linear relationship between the specific surface area and hydrogen capacity. Consistently, the in-situ Raman experiment showed that H$_2$ on the SWCNT sample generates two extra Raman peaks, one up-shifted and one down-shifted, indicating that there exist at least two distinct sites at which gas molecules can be physically adsorbed.[49] Theoretical calculations[50] suggested four sites in SWCNTs: on the external bundle surface, in a groove formed at the contact between adjacent tubes on the outside of the bundle, within an interior pore of an individual tube, and inside an interstitial channel formed at the contact of three tubes in the bundle interior. The binding energy (E$_B$) of these sites is in the order of E$_{Bchannels}$ (0.119 eV H$_2$$^{-1}$ = 11.47 kJ mol^{-1}) > E$_{Bgrooves}$ (0.089 eV H$_2$$^{-1}$ = 8.58 kJ mol^{-1}) > E$_{Bpores}$ (0.062 eV H$_2$$^{-1}$ = 5.97 kJ mol^{-1}) > E$_{Bsurface}$ (0.049 eV H$_2$$^{-1}$ = 4.72 kJ/mol^{-1}). Access of molecules to the internal tube pores must either be through open SWCNT ends or through defects (holes) in the tube walls. Similarly, Han *et al.*[51] found that oxygen plasma treated MWCNTs may create some 1-nm nanopores in the walls and hence show a new hydrogen desorption peak in the 300–350 K range. Shiraishi *et al.*[52] also suggested that pores with a diameter of less than 1 nm are essential for hydrogen storage. According to the density functional theory calculation[53] pores of 0.5–0.7 nm can adsorb and release hydrogen reversibly with isosteric heat of hydrogen adsorption being around 7–7.5 kJ mol^{-1}.

It can be seen that high surface area is an important factor, but not the essential factor for the high hydrogen uptake capacity of CNTs. The hydrogen uptake capacity of pure CNTs can be enhanced by increasing the surface area. But the capacity should be limited due to monolayer adsorption saturation. One of the optimistic ways to further enhance the hydrogen uptake capacity in CNTs is to create a lot of subnanopores with a diameter of less than 1 nm, which is difficult to access by N$_2$ in the traditional BET method. Several structural modification methods have been proposed to tailor the pore structures of CNTs, including ball milling,[29,54,55] ultrasonication,[28,29,41] chemical-treatment,[56] simple annealing[57] and plasma etching.[58,59] The ball milling can cut short the MWCNTs and open their caps, enhancing hydrogen adsorption due to the increase of defects and surface area.[55] Hischer *et al.*[29] reported that the ball milling of carbon materials in Ar results in little H$_2$ uptake, whereas the ball milling of SWCNTs in H$_2$ atmosphere gives 1 wt% H$_2$ storage capacity at room temperature. Chemical purification treatments are aimed at the removal of amorphous carbons and metal particles present in the samples, and to create defects in the structure of nanotubes in order to provide access to the internal sorption sites. In the HNO$_3$-treated SWCNTs bundle 6 wt%H$_2$ was thus observed at 77 K and 2 bar.[60] Ziegler *et al.* found that a piranha solution (96% H$_2$SO$_4$/30% H$_2$O$_2$ with 4:1 molar ratio) is capable of attacking existing damage sites, generating vacancies in the graphene sidewall, and consuming the oxidized vacancies to yield short, cut nanotubes at high temperatures.[61] Therefore, the room-temperature piranha solutions offer the ability to exploit active damage sites in the sidewalls of the nanotube in a controlled manner to increase H$_2$ storage. Anson *et al.*[53] reported an enhanced adsorption capacity at atmospheric pressure on a KOH-treated SWCNTs, where the specific surface area increased from 262 m^2 g^{-1} for raw SWCNT to 1433 m^2 g^{-1} for the KOH-treated sample. It also increases the volume of micropores in the diameter range 0.5–0.7 nm by 45 times. Shiraishi *et al.*[52] observed that a NaOH-treatment can remove small particles of amorphous carbons, which cover the surface of SWCNTs and is crucial for obtaining clean SWCNTs and for enhancing hydrogen storage. Lee *et al.* reported that the average size of nanopores in MWCNTs decreases from 2.77 to 2.13 nm after the KOH treatment.[62] Shaijumon *et al.*[63–65] reported that the purification process could open the ends of CNTs effectively, and hydrogen molecules could have entered the inner cavities through these ends, leading to an increase in hydrogen sorption capacities. Han *et al.*[58] reported that plasma etching can create nanopores in MWCNTs, providing more sites for the hydrogen adsorption. Mu *et al.*[59] further point out that the defects produced by plasma can supply more hydrogen accesses to the interlayers and hollow interiors of MWCNTs. Etching was found to speed up the adsorption rate and to enhance the hydrogen storage capacity from 0.6 wt% to 1.4 wt% at room temperature and 100 bar. Cao *et al.* reported that the dense alignment of CNTs can improve hydrogen storage by adsorbing molecules on the outer walls as well as the inner walls.[66]

5.2.1.2 *Functionalized/doped Carbon Nanotube*

Functionalization can purposely introduce impurities or functional groups to CNTs to enhance the hydrogen uptake capacity. However, impurities in CNTs is one of the factors reported in the literature[67,68] that induced the discrepancy in hydrogen uptake capacity of CNTs. For example, the K-modified SWCNT can adsorb only <0.69 wt% at room temperature and 70 bar and <1.05 wt% at 77 K and 50 bar.[69] In another case of debates, Hirscher *et al.*[54] and Costa *et al.*[70] had shown the introduction of metal alloy particles in the SWCNTs is solely responsible for the hydrogen uptake of the decorated SWCNTs. These alloys would reversibly hydride/dehydride, accounting therefore for most of the desorption volumes measured in this study. Nevertheless transition metals are often reported to serve as the catalysts and play an important role in increasing the hydrogen storage and improving the kinetics. Yildirim *et al.*[71] demonstrated that a single Ti atom coated on an SWCNT can bind up to four hydrogen molecules based on first-principle calculations, and predicted that SWCNT can adsorb up to 8 wt% hydrogen at high Ti coverage. In Pd-decorated MWCNTs, H_2 is dissociated by metallic Pd into H atoms or protons, which could easily penetrate tube walls and are combined into H_2 in interlayers and hollow interiors.[72] The CNTs decorated with Pd showed a hydrogen storage capability of 3.5 wt% at ambient temperature and pressure of 100 bar, about seven times higher than that (0.6 wt%) of the non-decorated samples. Generally, the hydrogen uptake in metal-decorated CNTs is enhanced.[73–75] The release temperature is, however, also increased due to the high desorption activation energy.[73–75]

To explain and manipulate the hydrogen chemisorption on CNTs with the presence of a metal catalyst, reversible hydrogen spillover has been proposed.[72,76–88] The catalyst in the nanotube wall dissociates molecular hydrogen. Because the nanotube is curved, the attractive forces of the carbon atoms in the lattice overlap. This results in a higher number of carbon atoms interacting with the hydrogen, which lowers the activation energy for hydrogen addition and permits surface diffusion of the hydrogen away from the catalyst. Atomic hydrogen diffuses from the catalyst onto the SWCNT where it is stored. If the carbon-hydrogen binding energy is not too large, the process could be reversible. As shown in Figure 5.5, H_2 molecules are dissociatively adsorbed on the Pt catalyst surface.[82] Atomic hydrogen atoms then spillover to active carbon, which is the support of the Pt catalyst. The physical bridges between the active carbon and SWCNTs promote the secondary spillover of H atoms from the Pt/active-carbon catalyst to SWCNTs. On SWCNTs, the H atoms access different adsorption sites compared to H_2, and the binding energies of the atomic H on SWCNT are sufficiently high as evidenced by the apparent hysteresis. It can be visualized that hydrogen is desorbed first from the interior sites of the SWCNTs, which have lower binding energy than the exterior sites. Since adsorbed H atoms are mobile

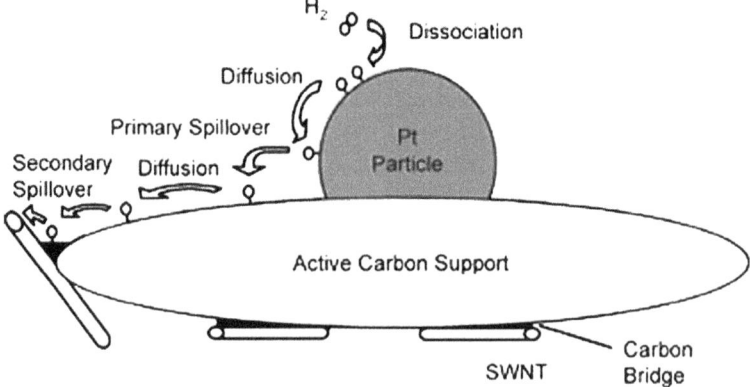

Figure 5.5   Proposed mechanism for primary and secondary spillover on a carbon bridged composite.[82]
(Copyright: American Chemical Society).

and interior-exterior exchange is possible, these interior sites continue to serve as the sites to which hydrogen atoms migrate and from which hydrogen is desorbed as molecular hydrogen. The re-adsorption studies of previously adsorbed samples suggested that nearly 70–85% of the spilled hydrogen occupies the physisorption binding sites. The remaining hydrogen (30–15%) could, however, be chemisorbed on defect sites and could not be removed reversibly.[89]

5.2.2 Inorganic (Non-carbon) Nanotubes

It was proposed that almost any sheet-like material can self-assemble into scrolls by simply exfoliating bulk layered materials into sheets.[90] As long as the two-dimensional (2D) sheet is sufficiently large, the energy gain from non-convalent binding between layers of the scroll is sufficient to compensate for the elastic energy cost of forming the scroll. Many layered materials have demonstrated their ability to form nanotubular or onion-like structures. These include hexagonal BN,[91,92] GaN,[93] transition metal dichalcogenides[94] (*e.g.* TiS_2, WS_2 and MoS_2), metal halides ($NiCl_2$, $MgCl_2$, $CdCl_2$) and metal oxides[95–97] (*e.g.* MnO_2, V_2O_5, and TiO_2). These inorganic nanotubes can store hydrogen between layers of the sheet, on the external surface of the scroll, at the center of the scroll, and in interstitial space between scrolls, and are considered as promising materials for hydrogen storage.

5.2.2.1 Boron Nitride Nanotube

A boron nitride nanotube (BNNT) is isoelectronic and isostructural to a CNT. Unlike CNTs, which can be metallic or semiconducting, BNNT is an insulator with a large bandgap. BNNTs had been successfully synthesized using arc-discharge,[92,98] laser ablation,[99,100] chemical vapor deposition[101–103] and chemical reactions such as $(B_2O_3 + CNT + N_2)$[104] and $(NiB + B$ (or $KBH_4) + N_2)$.[105,106] Ma *et al.*[91] first investigated the hydrogen adsorption properties of two kinds of BNNTs: multiwalled nanotubes and bamboo-like nanotubes. Multiwall nanotubes generally have a narrow diameter distribution from 10 to 30 nm and lengths of only a few micrometers. The tubes are capped at each end. Usually a few nanotubes are bonded together to form bundles. Bamboo-like tubes, in the conformation of joined nanobell segments, occupy a slightly wider diameter range of 10–80 nm with lengths ordinarily exceeding 10 cm (Figure 5.6). Multiwalled and bamboo-like BNNTs can adsorb hydrogen up to 1.8 and 2.6 wt%, respectively, at about 10 MPa and room temperature, much higher than bulk BN powder (0.2 wt%). Subsequently, Oku *et al.*[107,108] also found that a mixture of BNNTs and BN cages can absorb hydrogen up to 3 wt%. The capped feature of BN multiwalled nanotubes prevents the hollow cavity of the tubes from being accessed by hydrogen molecules, which may explain a somewhat lower hydrogen uptake amount. When the nanotube sample was recovered to the ambient pressure, approximately 20% of the physically adsorbed hydrogen was released. A major fraction of the stored hydrogen ($\approx 80\%$) was still

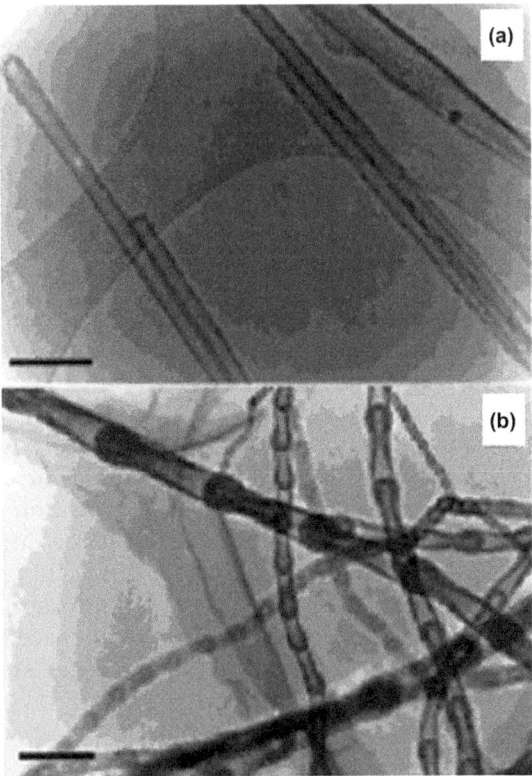

Figure 5.6 The morphologies of BN nanotubes: (a) multiwall nanotubes and (b) bamboo-like nanotubes. Scale bar: 100 nm.[91]
(Copyright: American Chemical Society).

Figure 5.7 TEM image of BN nanotubes after heating at 1500 °C for 6 h in the presence of platimum; the arrow points to platinum nanoparticle.[103]
(Copyright: American Chemical Society).

retained, which is to be released at a slightly higher temperature (≈ 573 K).[109] Tang *et al.*[103] reported the enhancement of hydrogen storage in collapsed BNNTs, which were obtained by heating BN tubes to 1500 °C for 6 h in the presence of a Pt catalyst. All wall layers of the collapsed nanotubes were fractured and protruded outward forming a hairlike structure (Figure 5.7).

The specific surface area of the collapsed BNNTs significantly increases from 254.2 to 789.1 m^2g^{-1}. The absorbed hydrogen in the collapsed BNNTs exceeded that of the raw multiwalled nanotubes by a factor of four. The measured value at 10 MPa is 4.2 wt%. This confirms that the collapsed surfaces of BNNTs effectively increase the hydrogen adsorption capacity due to the high specific surface area and suitable pore structure. Similarly, a substantial amount of chemisorbed H_2 can only be released at a slightly higher temperature.

The higher H_2 storage of BNNTs *vs.* CNTs under identical conditions can be attributed to its nanomorphology as well as the presence of heteropolar B–N bonding.[110] The ionic B–N bonding may induce an extra dipole moment and hence stronger adsorption of hydrogen, since the induced dipole moment of hydrogen molecules is sensitive to local electric fields. The binding energy of H_2 on the BN sheet is calculated to be about 90 meV, significantly higher than 60 meV on graphite.[111] The buckling structure in BNNTs may induce an extra dipole moment so that the calculated binding energy of H_2 on BNNTs can be as high as 110, 85 and 100 meV, respectively, on the top of B_3N_3 hexagon center, boron atom, and nitrogen atom.[110]

5.2.2.2 Silicon Carbide Nanotube

The study on BNNTs for hydrogen adsorption showed that the point charges on the material's surface can improve the storage capacity due to the enhancement of binding energy. Theoretical studies of silicon carbide nanotubes (SiCNTs) reported that between the walls of SiCNT are full of point charges because of the charge transfer of more than half an electron from Si to C.[112,113] These nanotubes with the point charges in their surface make them good candidate materials for hydrogen storage. An *ab initio* density functional theory (DFT) calculation showed an increase of 20% in the binding energy of H in SiC nanotubes compared with pure carbon nanotubes (CNTs). This is explained by the alternative charges that exist in the SiC nanotube walls. A classical Monte Carlo simulation of SiC nanotube bundles showed an even larger increase of the storage capacity in SiC nanotubes, especially in low temperature and high pressure conditions.[114] These results indicated that like other inorganic nanostructures, SiC nanotubes seem to be more suitable materials for hydrogen storage than pure CNTs. Using DFT calculations, Li *et al.*[115] reported that the binding energy can be further increased by Li-doping. The hydrogen molecule adsorption on Li-doped SiCNT is 0.211 eV, higher than that on a pure SiCNT (0.086 eV), due to the charge transfer from Li to the nanotube. They found that up to four molecules can be attached to a Li-adsorbed SiCNT with an average binding energy of 0.165 eV.

Experimentally, many methods have been developed to synthesize SiC nanostrucutres.[116-121] Although the experimental data of the hydrogen storage on SiCNTs is absent, Pol *et al.*[122] reported that 2.5 wt% H_2 can be stored on the high surface area SiC nanorods at room temperature and

60 bar. Similar to other nanostructured compound materials, in the pressure-composition-isotherm the adsorption and desorption curves do not coincide when the pressure is lowered to ambient level. Only ≈ 1.5 wt% H_2 can be reversibly absorbed and released. This suggests some chemisorptions of H_2 on SiC nanorods, which may be responsible for the irreversible nature of the room temperature adsorption.

5.2.2.3   Tungsten Carbide Nanotube

Tungsten carbide has attracted scientific interests for its potential applications due to its unusual properties, such as high melting point, superior hardness, low friction coefficient, high oxidation resistance, and superior electrical conductivity. Tungsten carbide nanotubes (WCNTs) can be synthesized by thermal decomposition of $W(CO)_6$ in the presence of Mg powder.[123] Pan *et al.*[124] reported that WCNTs may be more suitable for hydrogen storage because of the relatively higher but not too high binding energy based on first-principles calculations. The WCNT consisting of alternating C and W atoms is the most stable structure (Figure 5.8). By considering all of the possible adsorption sites and orientation of H_2 molecules, Pan *et al.* found that most sites are unstable for the adsorption of H_2 except the W top site, in the case of H_2 parallel to the tube (Figure 5.9). The bond length of H_2 extended to 0.875 Å, which is larger than that in an isolated H_2 (0.754 Å). The W–H bond length is 1.85 Å. For comparison, the adsorption of a hydrogen molecule on a zigzag single-wall CNT (10, 0) was studied using the same approach. The most stable site for hydrogen molecule is the center of the carbon hexagon. The binding distance and energy are 3.4 Å and 0.01 eV (inset in Figure 5.9), respectively. The binding distance of H_2 on the WCNT is much shorter than that on CNT, BNNT, and SiCNT, where it is about 3.0 Å.[125,126] The calculated binding energy for H_2 on the WCNT is about 0.44 eV, which is much larger than that on CNT (0.06 eV),[126] BNNT (0.09 eV),[125] and

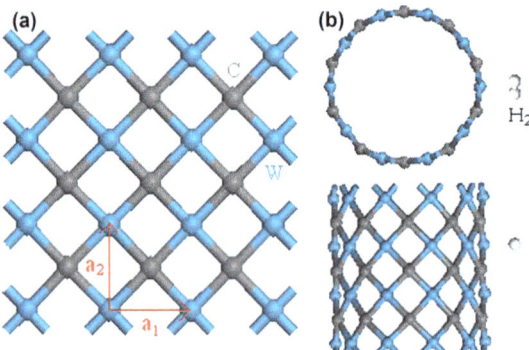

Figure 5.8   Atomic configuration of (a) a tungsten carbide sheet and (b) a tungsten carbide nanotube with a hydrogen molecule.[124] (Copyright: American Institute of Physics).

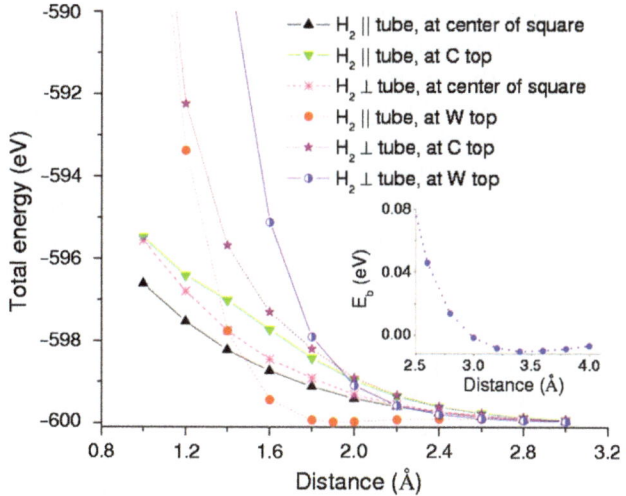

Figure 5.9 Total energy of a hydrogen molecule as a function of its distance from the outer wall of the WCNT (10, 0). The inset shows similar results for a hydrogen molecule approaching the CNT (10, 0).[124] (Copyright: American Institute of Physics).

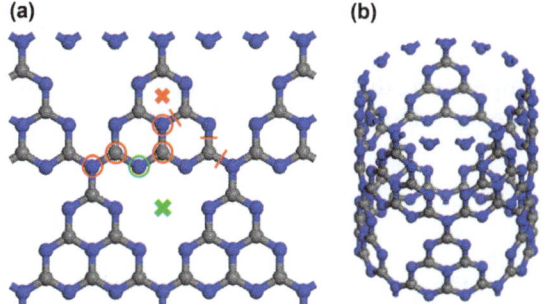

Figure 5.10 Representative structure of a heptazine-based graphitic carbon nitride (a) monolayer; and (b) nanotube. C and N atoms are represented in grey and blue respectively. Positions over which H atoms or H_2 molecules are possibly sited (*i.e.*, out-of-page) are indicated in red and green on (a).[131] (Copyright: Elsevier).

SiCNT (0.07 eV).[126] The shorter binding distance and higher binding energy indicate that the hydrogen storage capability of WCNT is better than that of other nanotubes.

5.2.2.4 Graphitic Carbon Nitride Nanotube

Carbon nitride can exist in various phases, not only depending on the C to N ratio, but also on atomic arrangements.[127] Among these phases, a heptazine-based form of graphitic carbon nitride (g-C_3N_4) (Figure 5.10(a)) is regarded as the most stable structure, and can be realized by thermocondensation of C/N/H-containing precursors.[128] A novel nanotube—g-C_3N_4 nanotube, formed by curling up the g-C_3N_4 monolayer (Figure 5.10(b)), was proposed as a photocalayst for water splitting.[129,130] The unique porous structure of the nanotube, with a pore diameter of ≈0.7 nm (Figure 5.10), should allow greater hydrogen access to the nanotube-interior, thus increasing its specific surface area, allowing hydrogen storage within and improving the storage capacity accordingly. The doubly bonded N atoms at the pore edges provide active sites for H chemisorption and metal functionalization. Koh *et al.*[131] reported a first-principles study on hydrogen storage in g-C_3N_4 nanotubes. They found that the configuration with 1 H atom chemisorbed to each alternate N atom along the pore edges (Figure 5.11(a)) is energetically stable. The calculated binding energy for H-chemisorption in nanotubes increases with the increase of the tube's diameter (Figure 5.11(b)). The binding energy remains at about –0.66 eV as the tube's diameter ranges from 9.5 to 18 Å. Generally, the binding energy for H-chemisorption in nanotubes is significantly lower than that in the monolayer. They further reported that the calculated energy barriers for a molecule penetrating into the tube's interior are about 0.54 eV and 1.1 eV, for the tubes without and with chemisorbed H respectively (Figure 5.12), indicating that the hydrogen molecule can easily pass through the pore and enter the tube interior. The storage and release processes of hydrogen in the graphitic carbon nitride nanotubes should be very fast, and the storage upper-limit should be achievable at a relatively low pressure and temperature, due to the low energy barrier. Correspondingly, it is unlikely for the H_2 molecule to enter the CNT-interior through the tube

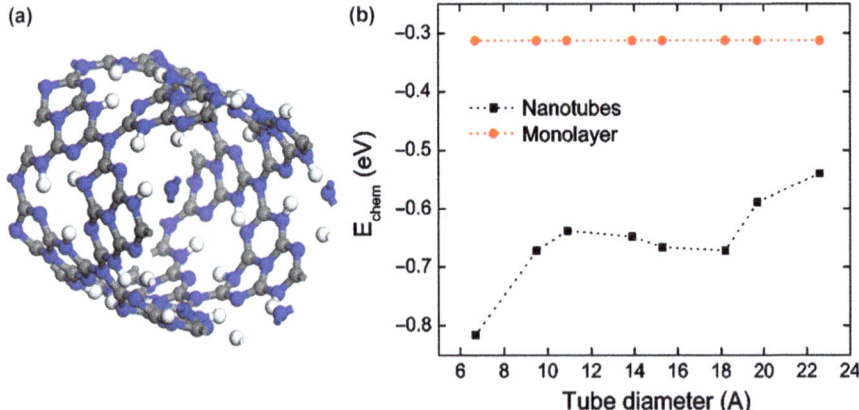

Figure 5.11   (a) Representative structure of a graphitic carbon nitride nanotube (g-C_3N_4 (5, 0)) with three H atoms chemisorbed at each pore. C, N and H atoms are represented in grey, blue and white respectively. (b) Average H–N chemisorption binding energy (E_{chem}) as a function of nanotube diameter. The corresponding binding energy for the g-C_3N_4-sheet system is presented here in red for comparison.[131] (Copyright: Elsevier).

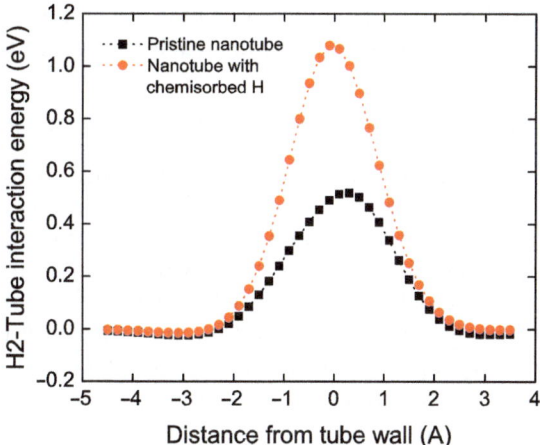

Figure 5.12   Energy profile of a single H_2 molecule as it passes through the pore of a g-C_3N_4 (5, 0) nanotube; the tube wall is at 0 Å, with negative distances representing the tube interior, and vice versa.[131] (Copyright: Elsevier).

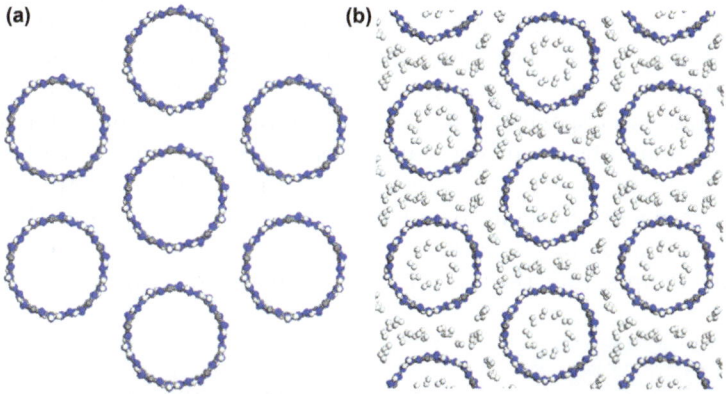

Figure 5.13   (a) Representative structure of a graphitic carbon nitride nanotube (g-C_3N_4 (5, 0)) bundle. (b) Representative structure of a graphitic carbon nitride nanotube (g-C_3N_4 (5, 0)) bundle displaying a hydrogen storage capacity of 5.45 wt. %. C, N and H atoms are represented in grey, blue and white respectively.[131] (Copyright: Elsevier).

wall because of the very high energy barrier (12 eV). The process for CNTs should be slow, and a higher pressure should be required to reach the storage upper-limit, because hydrogen can enter the tube interior only *via* the open ends or defects. An isolated g-C_3N_4 nanotube can store hydrogen up to 4.66 wt. % with an average hydrogen adsorption energy of −0.22 eV per H atom, and a binding distance of 2.4 Å. In the nanotube bundle, the storage capacity can reach 5.45 wt. % (Figure 5.13). At the same time, the binding energy is improved by 10%. This indicates that the storage capacity can be further improved in the bundles and may reach the practical requirement.

5.2.2.5 *Transition Metal Dichalcogenides Nanotubes*

MS_2 (M = Mo, Ti) nanotubes have a nanoscroll structure analogous to carbon nanotubes.[90,132–139] Depending on the preparation methods nanostructured MoS_2 can be single-walled subnanometer-diametered tubes, which are micrometers long and bundled together.[136] They may also occur in the form of nanofibers or fullerene-like nanoparticles. Chen[139] synthesized MoS_2 nanotubes by directly heating $(NH_4)_2MoS_4$ in hydrogen/thiophene. They obtained 0.9 wt% H_2 capacity on the MoS_2 nanotubes, and attributed it to high specific surface area (≈ 60 m^2 g^{-1} *vs* 3.6 m^2 g^{-1} of the bulk) and highly nanoporous structure (0.65 nm layer spacing). The KOH-treated MoS_2 nanotubes showed the highest hydrogen storage capacity of 1.2 wt% at room temperature under 2.5 MPa.[137]

Chen *et al.*[94,137–139] reported three methods (solution chemistry, transport reaction, and gas reaction) that could be used to prepare multiwalled TiS_2 nanotubes. The high-purity TiS_2 nanotubes with openended tips were synthesized by a chemical transport reaction, which involved a mixture of Ti metal sponge and S powder (2:1 atomic ratio) and iodine (>4 mg cm^{-3}) as the transport agent. The TiS_2 nanotubes exhibit an openended tubular structure with an outer diameter of ≈ 30 nm, an inner diameter of ≈ 10 nm and an interlayer spacing of ≈ 0.57 nm (Figure 5.14). Atoms within a S-Ti-S sheet are bound by strong covalent forces, while individual S-Ti-S layers are held together by van der Waals interactions. This enables the introduction of foreign atoms or molecules between the layers by intercalation. The hydrogen concentration in TiS_2 is 2.5 wt% under the hydrogen pressure of 4 MPa at room temperature (Figure 5.15). An increase in temperature causes a

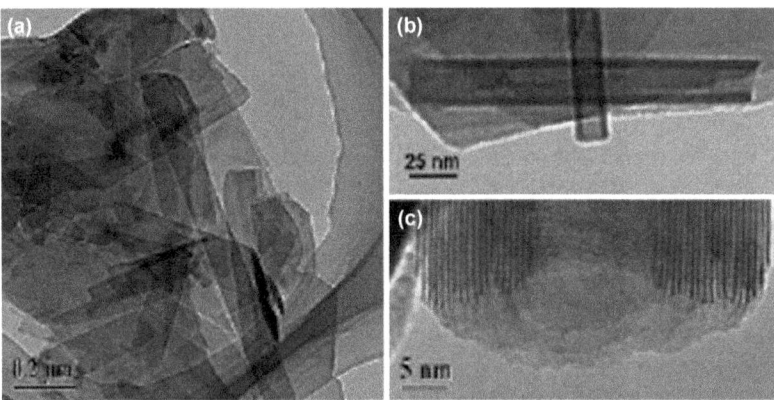

Figure 5.14   TEM (a, b) and HRTEM © images of the as-synthesized TiS_2 nanotubes.[94]
(Copyright: American Chemical Society).

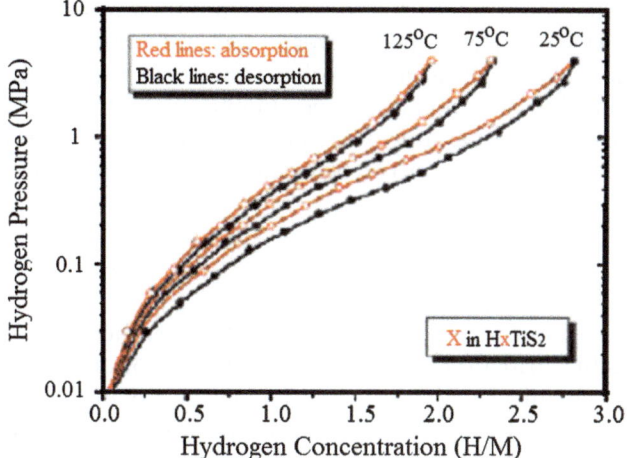

Figure 5.15 PCT curves for hydrogen absorption and desorption of TiS$_2$ nanotubes at 25, 75, and 125 °C.[94]
(Copyright: American Chemical Society).

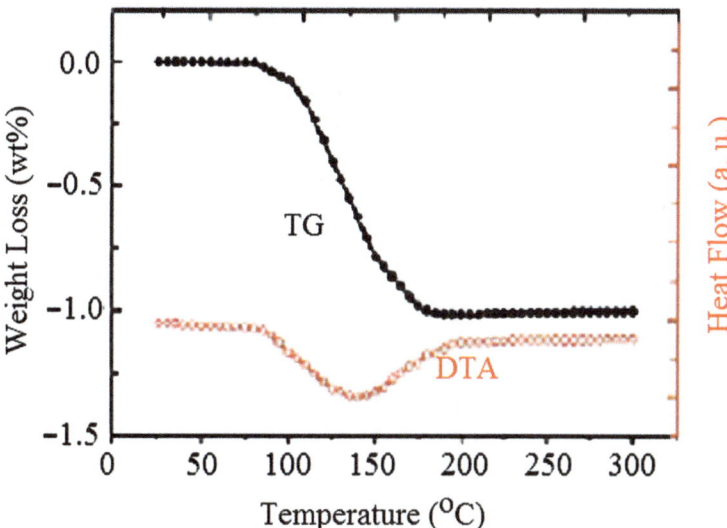

Figure 5.16 TG-DTA characterization of the TiS$_2$ nanotubes in the hydride state (at 25 °C and 4 MPa hydrogen) that was subjected to ambient pressure.[94]
(Copyright: American Chemical Society).

decrease of the hydrogen content, but considerable amounts of hydrogen can still be stored reversibly in the TiS$_2$ nanotubes. The thermal gravimetric analysis (Figure 5.16) shows that the desorption of hydrogen from 75 to 185 °C gives a total loss of about 1.0 wt% H$_2$ (corresponding to H$_{1.2}$TiS$_2$). It means that about 60 wt% of the absorbed hydrogen at 4 MPa was already desorbed when lowering the hydrogen pressure to atmospheric conditions, but that 40 wt% of the absorbed hydrogen was still retained, suggesting that this remaining hydrogen is chemisorbed. And, both the physisorption and the chemisorption play important roles in optimizing the hydrogen storage of TiS_2. The H–S bonding in hydrogenated TiS_2 is weaker than H–O so that the adsorption-desorption on TiS_2 is reversible unlike that in nanostructured oxides.

5.2.2.6   *TiO_2 Nanotube*

Nanostructured metal oxides often possess high surface area and porosity, with accessible oxidation states that may be important in hydrogen adsorption by chemical pathways. Additionally by doping, it is possible to significantly affect the hydrogen uptake capacity of the oxides *via* introducing large amount of defects for gas absorption or by changing the charge distribution or the Fermi level position in an oxide.[140,141]

A TiO_2 nanotube (TiO_2 NT) has been produced by various methods.[142–144] Lim *et al.*[145] had employed a prolonged hydrothermal process in the synthesis of TiO_2 nanotubes. The TiO_2 NTs are entangled together with a hollow core and open tip, very much like a multiwalled carbon nanotube (MWCNT) (Figure 5.17a). The outer diameter of the tubes is uniformly distributed around 10 nm. But the length varies from 200 nm to >1 μm. The N_2 isotherm of TiO_2 NT is typical of type V isotherm. The adsorption-desorption hysteresis loop is observed at $P/P_0 \approx 0.43$, indicating capillary condensation inside the mesoporous tubes. Remarkably smaller adsorption load is observed for the bulk TiO_2 with no hysteresis loop (see Figure 5.17 (b)) The specific

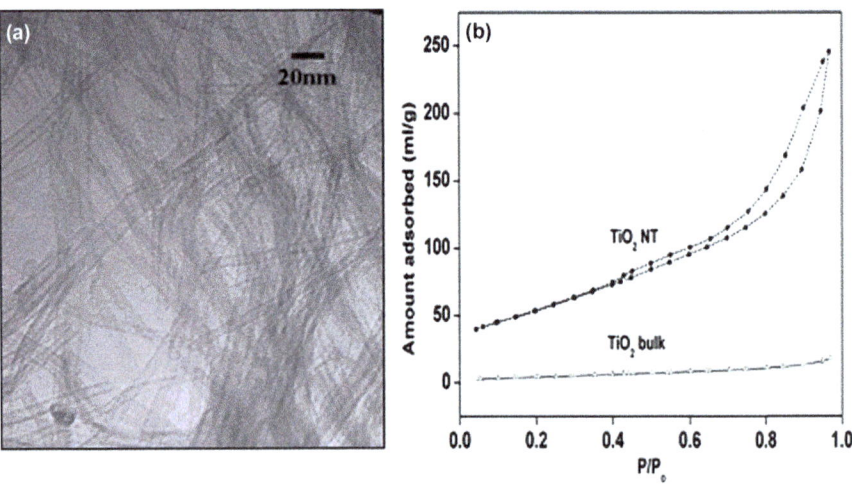

Figure 5.17   (a) TEM images of TiO_2 nanotubes and (b) N_2 isotherms of TiO_2 nanotubes and bulk TiO_2 at 77 K.[145]
(Copyright: American Chemical Society).

surface area (SSA) determined by the BET method in the P/P_0 range between 0.05 and 0.43 is 197 $m^2 g^{-1}$ for TiO$_2$ NT, as compared to 16 $m^2 g^{-1}$ for the bulk TiO$_2$ sample. The mesopore size distributions of the samples are evaluated using Barret-Joyner-Halenda (BJH) methods. The mesopore volumes of bulk TiO$_2$ and TiO$_2$ NT are 0.0301 mL g^{-1} and 0.379 mL g^{-1} respectively, an increment of ≈ 12 times. The average mesopore diameter for TiO$_2$ NT is 7.7 nm, close to the TEM observation. The TiO$_2$ nanotubes have a uniform diameter distribution with ≈ 8 nm diameter. Its specific surface area was 16 times that of anatase powder (196 $m^2 g^{-1}$ *vs* 16 $m^2 g^{-1}$) and its H$_2$ storage capacity was much higher than that of bulk powder at room temperature and 60 bar (2 wt% *vs* 0.9 wt%). Temperature program desorption and various temperature pressure-composition-isotherm studies indicated that 75% of the total H$_2$ is due to physisorption and 25% is due to chemisorptions.[145]

5.2.3 CNT/oxide Composite

As reversible hydrogen storage materials, CNTs can store hydrogen molecules through physical adsorption. The capacity, however, is limited to a small amount (≈ 1 wt%) at ambient temperature and high pressure range. Metal, oxide, and compounds can store hydrogen through chemical reaction and bonding, which may reach high capacity with high release temperature. Nanostructured composite materials formed by CNTs and metal- or metal oxide nanoparticles or oxide should be interesting to hydrogen storage because of the non-classical s-p-d hybridization and spillover phenomena, which may lead to enhanced hydrogen storage capacity.[146]

Two-dimensional (2D) carbon materials, such as graphene, have attracted attention for hydrogen storage applications.[147] The hydrogen storage capacity may be enhanced by architecturally engineering the carbon-based nanostructured materials such as pillared graphene oxide and graphene oxide framework.[148] To achieve the ordering at a small length scale, Aboutalebi *et al.*[149] reported that the self-aligned graphene oxide (GO)-MWCNT hybrid frameworks with a high degree of orientation can be obtained by using liquid crystalline GO, which offers a novel yet simple way of designing GO-based hybrid frameworks with extraordinary hydrogen storage capacities. The GO dispersions show the typical birefringence behavior of liquid crystals (LC) (Figure 5.18(a)). The LC properties of GO sheets induce a spontaneous self-assembly into engineered long range ordered layer-by-layer 3D structures upon simple casting and drying (Figure 5.18(b)).The interlayer spacing of this GO paper, which is in excess of 0.82 nm provides a platform to accommodate and intercalate hydrogen atoms. The SEM image of GO-MWCNTs (Figure 5.18(c)) shows the distribution of MWCNTs on graphene layers. The introduction of MWCNTs to GO clearly does not affect the ordering of GO sheets and MWCNTs act only as spacers between GO layers and also inhibit uncontrolled restacking. This ordering in turn results in a fully intercalated 3D structure (Figure 5.18(d)). The hydrogen uptake of MWCNTs

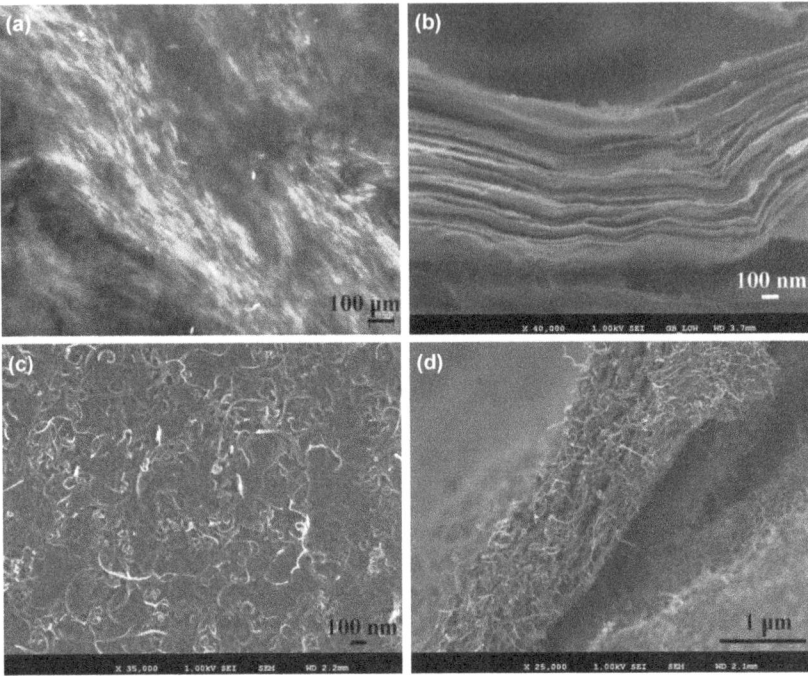

Figure 5.18   a) Polarized optical micrograph of LC graphene oxide dispersion
and SEM micrographs of b) GO paper, c) MWCNTs distributed on
GO layer and d) layer-by-layer assembled GO platelets decorated by
MWCNTs.[149]
(Copyright: Wiley-VCH).

and GO was around 0.9 wt% and 1.4 wt%, respectively, at room temperature
and 5 MPa hydrogen pressure. The hydrogen capacity of GO-MWCNTs
reaches 2.6 wt%, which is even more than the total capacity of the individual
composite constituents, *i.e.* GO and MWCNTs (Figure 5.19). The desorption
process happened instantaneously as it is expected in the case of carbon
materials.

Metal-organic frameworks (MOFs) are a rapidly growing class of porous
crystalline solids that are assembled by the connection of metal ions or
clusters through organic molecular bridges.[150] All porous MOFs to date are
microporous structures with cavities less than 2 nm in size. MOFs have a
high surface area 1000–5900 m^2 g^{-1} and possess a sufficient number of
accessible sites for atomic or molecular adsorption. Compared to con-
ventionally used microporous inorganic materials such as zeolites, MOFs are
more flexible in structural design through control of the architecture and
functionalization of the pores. The ability of H_2 to bind to metal atoms in
MOFs allows the H_2 molecules to pack more closely together and is expected
to provide a major boost in storage capacity over simple H_2 adsorption at
nonmetal sites in previously prepared materials.[151–155] However, H_2 storage

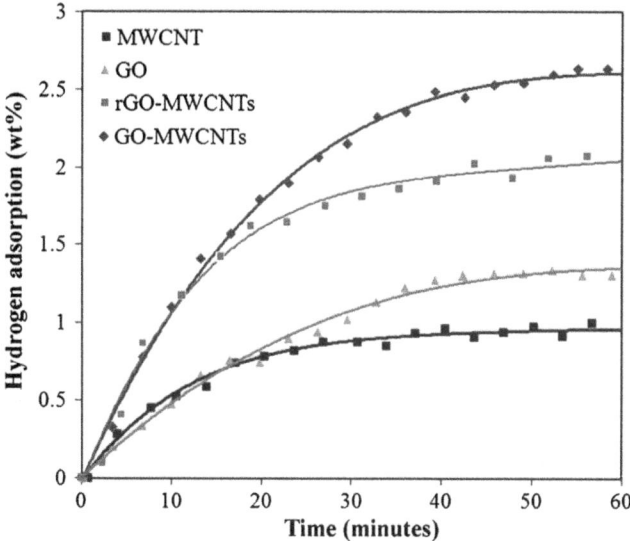

Figure 5.19 Hydrogen adsorption of GO, MWCNTs, GO-MWCNTs and rGO-MWCNTs as a function of time.[149] (Copyright: Wiley-VCH).

by MOFs at room temperature is far below that seen at 77 K and 1 bar (≈ 2.5 wt%). Yang *et al.*[156] studied a new class of H_2 storage materials, formed by a combination of acid-treated MWCNTs and $Zn_4O(bdc)_3$ (bdc = 1,4-benzene-dicarbocylate) (MOF-5). The composite obtained had the same crystal structure and morphology as those of virgin MOF-5, but exhibited a much greater Langmuir specific surface area (increased from 2160 to 3550 $m^2\ g^{-1}$), about a 50% increase in hydrogen storage capacity (from 1.2 to 1.52 wt% at 77 K and 1 bar and from 0.3 to 0.61 wt% at 298 K and 95 bar), and a much improved stability in the presence of ambient moisture.

5.3 Supercapacitor

A supercapacitor (or ultracapacitor) can be viewed as two nonreactive porous plates, or electrodes, immersed in an electrolyte, with a voltage potential applied across the collectors. A porous dielectric separator between the two electrodes prevents the charge from moving between the two electrodes (Figure 5.20). A supercapacitor fills in the gap between batteries and conventional capacitors, covering several orders of magnitude both in energy and power densities. They are an attractive choice for the energy storage applications in portable or remote apparatuses where batteries and conventional capacitors have to be over-dimensioned due to an unfavorable power-to-energy ratio.[13,157,158] In electric, hybrid electric, and fuel cell vehicles, a supercapacitor will serve as a short-time energy storage device with high power capability and will allow storing of the energy from regenerative

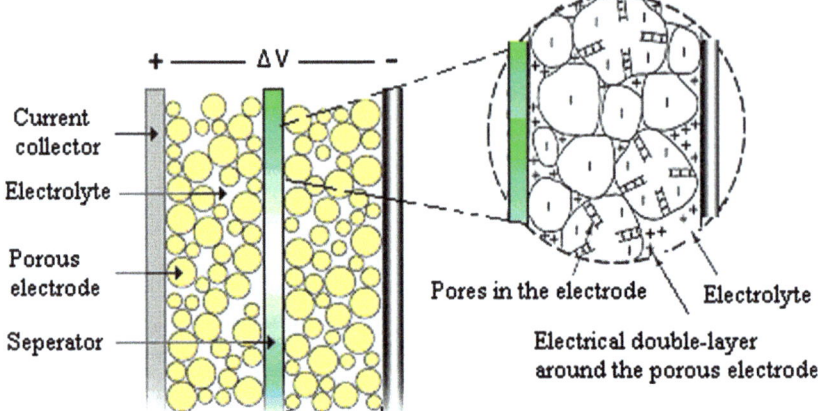

Individual Supercapacitor Cell

Figure 5.20 A presentation of a supercapacitor. http://en.wikipedia.org/wiki/Supercapacitor

braking. Increasing applications also appear in telecommunications, such as cellular phones and personal entertainment instruments.

Supercapacitors are generally classified into two types: pseudocapacitor and electric double-layer capacitor. The electrical charge can be built up in the pseudocapacitor *via* an electron transfer that produces the changes of chemical or oxidization state in the electroactive materials according to Faraday's laws related to electrode potential, *i.e.*, the basis of energy storage in a pseudocapacitor is Faradaic charge transfer. The charge process in the electric double-layer capacitor (EDLC) is non-Faradaic, *i.e.* ideally no electron transfer takes place across the electrode interface and the storage of electric charge and energy is electrostatic.[157] However, it has been recognized now that the EDLC with various high-area carbon electrodes also exhibits a small but significant pseudocapacitance due to electrochemically active redox functionalities.[159]

The electrical energy (E) accumulated in a supercapacitor is related to the capacitance (C) or the stored charges (Q) and voltage (V) by the following formula:

$$E = \frac{CV^2}{2} = \frac{QV}{2} \tag{5.1}$$

The capacitance and stored charge depend essentially on the electrode material used, whereas the operating voltage is determined by the stability window of the electrolyte. The use of high-capacitance materials is a key factor for the improvement of the energy density. Generally, the super-capacitor can provide higher power than most batteries because a large amount of charges (Q) can be stored in the double layers. But, the power density, as indicated in the following formula, is relatively low because of the series resistance.

$$P = \frac{V^2}{4R_s} \tag{5.2}$$

where R_s represents the equivalent series resistance (ERS) of the two electrodes. The improvement of the power and power density requires the development of materials with high capacitance and low resistance, as indicated in eqn (5.1) and (5.2).

A variety of materials, including oxide, polymer, carbon and their composites, can be used as the electrodes of a supercapacitor. Pseudocapacitor utilizes conducting polymers (such as, polyacetylene, polypyrrole, and polyaniline[159–168]), metal oxides (such as RuO_2 and Co_3O_4[169–183]) or polymer-oxide composite[182,184] as electrode materials. The chemical or oxidization state changes of the electrodes induced by the Faradaic charge transfer in the pseudocapacitive behavior may affect the cycling stability and limit their application due to high resistance and poor stability although the specific capacitance can be as high as 900 F g^{-1}.[185] EDLC is normally developed using porous carbon materials (such as activated carbon) as electrodes and the electrical charge is electrostatically accumulated at the electrode/electrolyte interface.[186–198] Generally high surface area in carbon materials is characteristic of a highly developed microporous structure, which is however unfavorable for the electrolyte wetting and rapid ionic motions, especially at high current loads. The combination of high surface area carbon materials with large specific capacity of oxide or polymer would result in high power and power density, and stability, by utilizing both the Faradaic capacitance of the metal oxide or polymer and the double-layer capacitance of the carbon.[199–202] In this section, we focus on the recent development of NT-based supercapacitors, including pure NTs and composites with NTs, oxide and/or polymers.

5.3.1 Carbon Nanotubes

CNTs have a novel structure, a narrow distribution of size in the nanometer range, highly accessible surface area, low resistivity, and high stability,[203–208] suggesting that CNTs are suitable materials for polarizable electrodes. Both SWCNTs and MWCNTs have been studied for electrochemical supercapacitor electrodes due to their unique properties.[209–215] A CNTs-based supercapacitor was first reported by Niu *et al.* in 1997.[209] Their supercapacitor showed a gravimetric capacitance of 102 F g^{-1} and an energy density of 0.5 Wh kg^{-1} obtained at 1 Hz on a single-cell device, using 38 wt% sulfuric acid as the electrolyte. Due to the existence of a catalyst and functional groups in the MWCNTs, both of the Faradaic and non-Faradaic processes were involved in the CNT-based supercapacitors.[209,216] Removing the catalyst from CNTs by thermal oxidization followed by immersion in HCl, however, did not improve the performance because of the formation of amorphous carbon by the thermal oxidization.[217] Experimentally, it is difficult to totally remove the catalyst from the catalyst-assisted CNTs, and, at the same time, keep the graphitization. Here, we focus on the effects of structural properties of pure CNTs (diameter, length, and pore size) on the EDLC.

5.3.1.1 Pure CNTs

Frackowiak *et al.* systematically investigated the effects of structures and diameters of CNTs, and microtexture and elemental composition of the materials on the capacitance.[212] Table 5.1 shows that the capacitance increases with the increase in specific surface area. The smallest value is obtained in CNTs with closed tips and graphitized carbon layers, where the mesopore volume for the diffusion of ions and the active surface for the formation of the electrical double layer are very limited in this material. The nanotubes with numerous edge planes, either due to herringbone morphology (A900Co/Si) or to amorphous carbon coating (A700Co/Si), are the most efficient for the collection of charges. Quite moderate performance is given by straight and rigid nanotubes of large diameter (P800Al) despite a relatively high specific surface area. However taking into account the diameter of the central canal, it is too large in comparison with the size of the solvated ions. On the other hand this particular behavior could be also due to a very hydrophobic character of these tubes as suggested by the very small value of oxygen content (Table 5.1).

The amount of electrical charge accumulated due to electrostatic attraction in EDLC depends on the area of the electrode/electrolyte interface that can be accessed by the charge carriers. The higher surface area of the electrode material could lead to higher capacitance if the area can be fully accessed by the charge carriers. However, the higher surface area does not always result in higher capacitance because the capacitance also depends on the pore size, the size distribution and conductivity. Higher capacitance can be achieved by optimizing all of the related factors. For example, the vertically aligned CNTs with the diameter of about 25 nm and a specific area of 69.5 m^2 g^{-1} had a specific capacitance of 14.1 F g^{-1} and showed excellent rate capability, which was better than those of entangled CNTs due to the

Table 5.1 BET specific surface area, mesopore volume, percentage of oxygen, and capacitance of the analyzed nanotubes.[209] (Copyright: American Institute of Physics).

Type of nanotubes	A700Co/Si	A900Co/Si	A600Co/NaY	P800/Al
V_{meso} (cm^3 STP g^{-1})	435	381	269	643
SBET (m^2 g^{-1})	411	396	128	311
Oxygen (mass %)	10.8	4.6	0.8	<0.3
Capacitance (F g^{-1})	80	62	4	36

larger pore size, more regular pore structure and more conductive paths.[218,219] Reit *et al.*[220] reported that vertically aligned CNTs on metallic substrate can further improve the performance. The power density of the supercapacitor exceeded 3 MW kg^{-1}. Ren *et al.*[221] reported a micro-supercapacitor wire by using aligned multiwalled carbon nanotube (MWCNT) fibers as electrodes. The supercapacitor was fabricated by twisting two aligned MWCNT fibers and showed a mass specific capacitance of 13.31 F g^{-1}, area specific capacitance of 3.01 mF cm^{-2}, or length specific capacitance of 0.015 mF cm^{-1} at 2×10^{-3} mA (1.67 A g^{-1}). Shape engineering of CNTs can also greatly improve the capacitance and power density.[222] When compared with activated carbon cells, the high-densely packed and aligned SWCNTs showed higher capacitance, less capacitance drop at high power operation, and better performance for thick electrodes. The SEM images show that the SWCNTs are high-densely packed after the engineering (Figure 5.21). Cyclic voltammograms of the solid sheet and forest cells were very similar, meaning the two materials have nearly the same capacitance per weight. The capacitance of the SWCNT solid EDLC was larger than that of the forest cell. The energy density was estimated to be 69.4 Wh kg^{-1}. Ion diffusivity plays a key factor to realize compact supercapacitors with high energy density and high power density. Because the electrolyte ions must diffuse through the pores of interstitial regions within the SWCNT packing structure, ion accessibility is limited in the inner region of the solids on the relevant timescale. Superior electrochemical properties of SWCNT solid cells

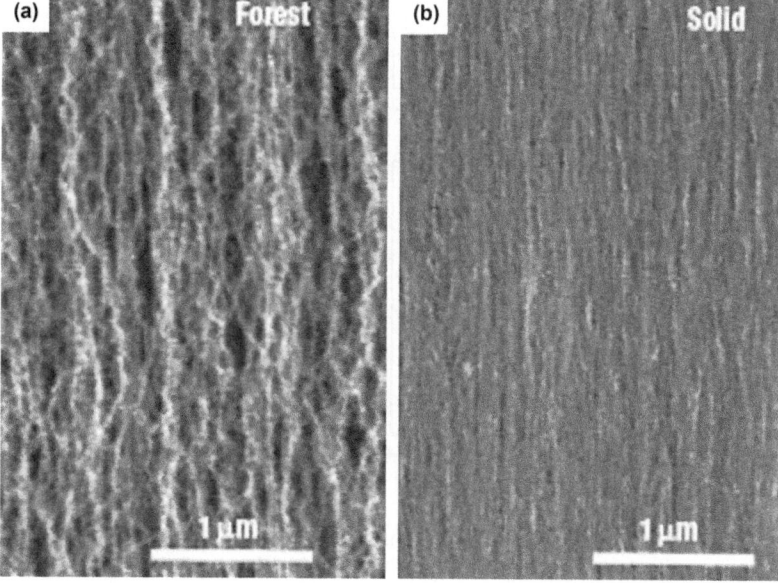

Figure 5.21   SEM images of (a) the as-grown forest and (b) shape engineered SWCNTs.[222]
(Copyright: Nature Publishing Group).

originate from the aligned pore structures compared with activated carbon due to the fast and easy ion diffusivity.[222–224] Izadi-Najafabadi *et al.*[223,224] further reported that the aligned SWCNTs forest can operate at a higher voltage (4 V) while maintaining a durable full charge–discharge cyclability, with an energy density (94 Wh kg^{-1}, 47 Wh L^{-1}) and power density (210 kW kg, 105 kW L^{-1}).

The high power density supercapacitor can also be achieved using electrophoretic deposited (EPD) CNT films and locally aligned CNTs.[211,225,226] The EPD film has a uniform pore structure formed by the open space between entangled nanotubes. Such an open porous structure with a high accessible surface area is unobtainable with other carbon materials, and enables easy access of the solvated ions to the electrode/electrolyte interface, which is crucial for charging the electric double layer. The current response profiles of the CV (cyclic voltammetry) curves at the scan rates of 50 mV s^{-1} and 1000 mV s^{-1} (Figure 5.22) are almost ideally rectangular along the time-potential axis. The excellent CV shape reveals a very rapid current response on voltage reversal at each end potential, and the straight rectangular sides

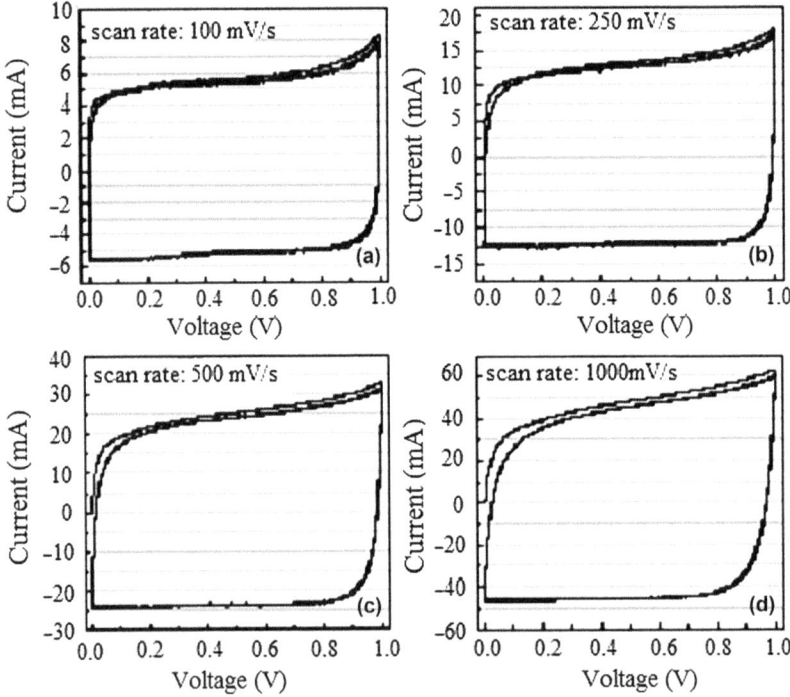

Figure 5.22   (a) CVs of the nanotube thin film supercapacitor cycled from −1 V to +1 V, (b) CVs of the nanotube thin film supercapacitor cycled from 0 V to +1 V for 100 cycles, (c) CVs of a conventional supercapacitor made of carbon particle thin films, and (d) charge/discharge curves of the nanotube thin film supercapacitor.[226]
(Copyright: Institute of Physics).

represent a very small equivalent series resistance (ESR) of the electrodes and also the fast diffusion of electrolyte in the films.[157] The CV plots with different scan rates of the assembled supercapacitor made of high packing and aligned CNTs are close to an ideally rectangular shape even at exceedingly high scan rates of 500 and 1000 mV s^{-1}, indicating an extremely low ESR of the electrodes (Figure 5.23).[226] The E–t responses of the charge process were almost the mirror image of their corresponding discharge counterparts, and no IR drop was observed, again owing to the negligible ESR of the electrodes. The high power density is attributed to the small internal resistance, which results from the coherent structure of the thin films fabricated using a highly concentrated colloidal suspension of carbon nanotubes.

Pan *et al.* systematically investigated the effects of factors, such as diameter, surface area and pore-size distribution, on the capacitance and demonstrated that the supercapacitance can be improved by the shape engineering.[227] The TEM images of AAO-based MWCNTs with a diameter of 50 nm (AM50) and AAO-based tubes-in-tube MWCANTs (ATM50) show clearly that smaller CNTs are confined with a larger one by comparing with the TEM images in Figure 5.24. The CV plots of the samples in the aqueous solution of 0.5 mol L^{-1} H$_2$SO$_4$ at a scan rate of 50 mV s^{-1} show that

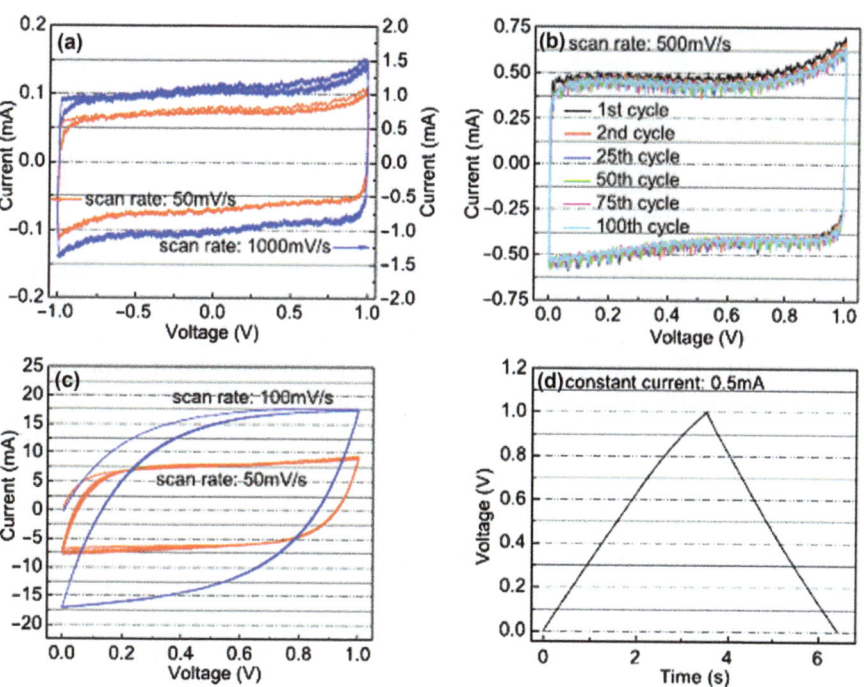

Figure 5.23 Cyclic voltammograms with different scan rates of an assembled supercapacitor using the nanotube thin films as electrodes.[211] (Copyright: Institute of Physics).

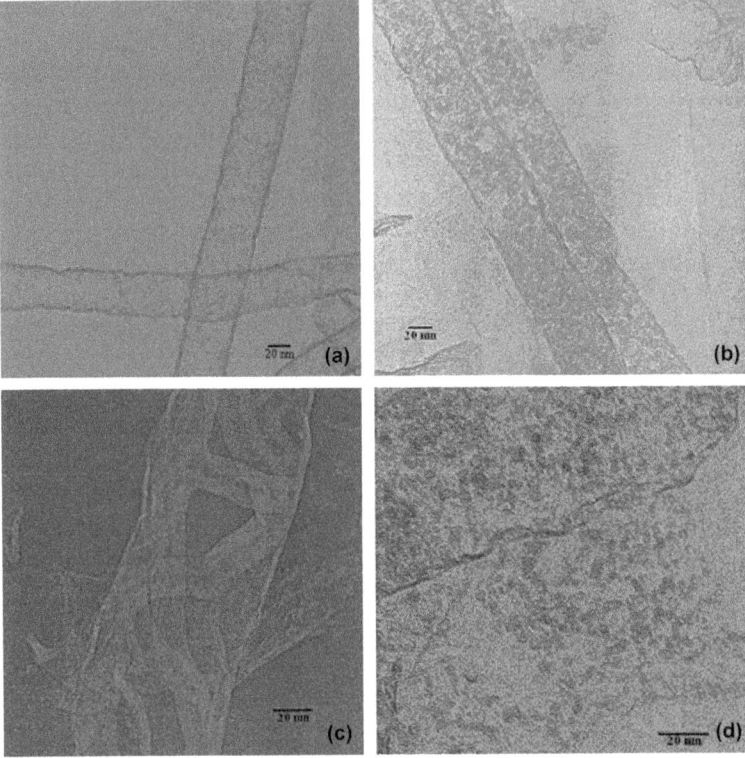

Figure 5.24   TEM images of AAO-based 50 nm MWCNTs after the first- and second-
step pyrolysis of C_2H_4: (a) AM50, (b) ATM50, (c) and (d) fine view
of ATM50.[227]
(Copyright: American Chemical Society).

supercapacitors can be realized due to the existence of the Faradaic processes (Figure 5.25). The redox peaks on the CV plots can be ascribed to
oxygenated groups attached to the surface of the carbon nanostructures,
such as OH^-,[180,181] which leads to the remarkable pseudocapacitance. The
redox reaction (Faradaic process) can be considered as the following:[172,181]

$$> C - OH \Leftrightarrow C = O + H^+ + e^- \tag{5.3}$$

$$> C = O + e^- \Leftrightarrow > C - O^- \tag{5.4}$$

The conductivity of CNTs can also be improved by the OH-functionaliza-
tion because of the bandgap narrowing and carrier (hole) doping.[228] The
better average specific capacitance of ATM50 was attributed to its higher
surface area, better pore-size distribution and conductivity. The amount of
electrical charge accumulated due to electrostatic attraction in EDLC depends on the area of the electrode/electrolyte interface that can be accessed
by the charge carriers. The higher surface area of the electrode material
could lead to higher capacitance if the area can be fully accessed by the

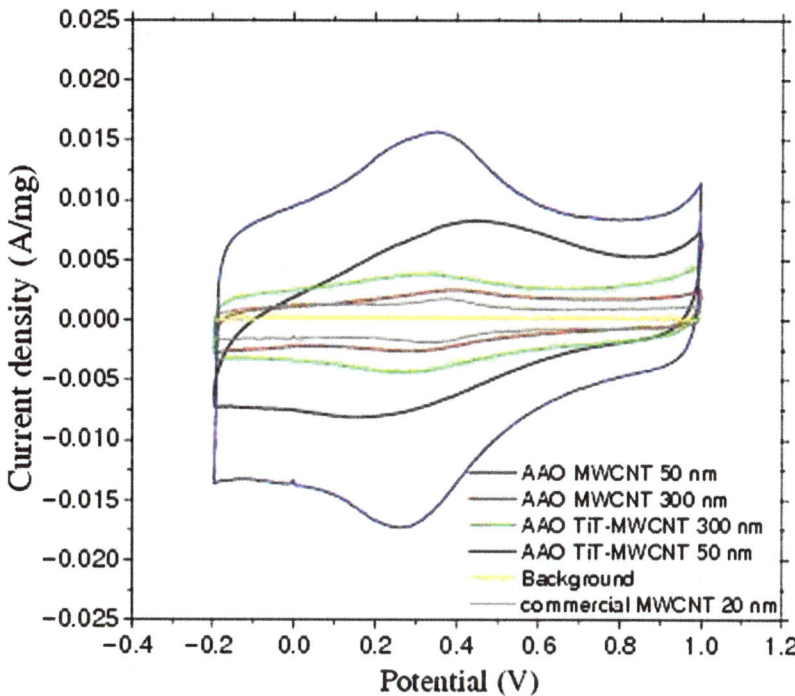

Figure 5.25 The CV plots in 0.5 M H_2SO_4 at a scan rate of 50 mV s^{-1} for the five samples.[227]
(Copyright: American Chemical Society).

Table 5.2 Specific surface area, average pose size, and capacitance of the carbon nanomaterials.[227] (Copyright : American Chemical Society).

	CM20	AM50	AM300	ATM50	ATM300
I_D/I_G	1.03	0.86	0.92	0.74	0.84
Specific area (m^2 g^{-1})	136	649	264	500	390
Average pore diameter (nm)	8.8	3.9	7.4	5.2	9.1
Capacitance (F g^{-1})	17	91	23	203	53

charge carriers. However, higher surface area does not always result in higher capacitance because the capacitance depends on the pore size and its size distribution. The surface area is hardly accessible if it consists of micropores (<2 nm).[194] The average pore diameters of all samples are larger than 2 nm (Table 5.2). The pore-size distributions for AM50 and ATM50 are narrow and show that the dominant pore diameter is about 3.9 nm. However, the pore-size distributions for other samples are broad and extend to a larger size, although the dominant pore diameter is about 2 nm for other samples (Figure 5.26). The average specific capacitance for AM50 and ATM50

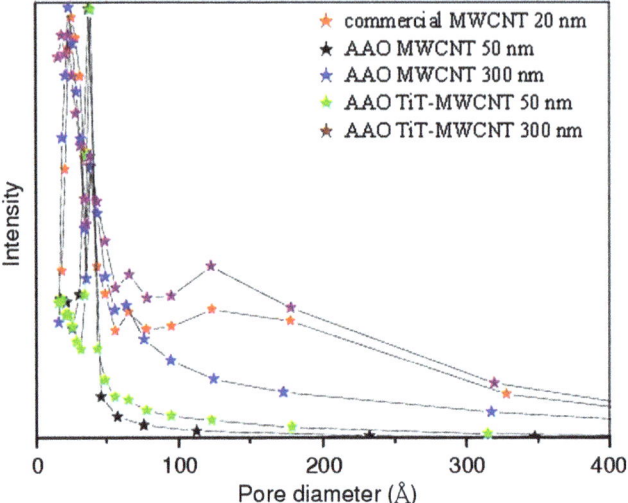

Figure 5.26 The pose size distribution calculated using BJH method.[227] (Copyright: American Chemical Society).

is larger than those of other samples. And, the average specific capacitance increases with the increase of the specific surface area with the exception of ATM50. The electrical conductivity is one of the factors that affect the capacitance. It should be mentioned that the higher the surface area, the poorer the conductivity should be. This should be one of the reasons for the capacitance of ATM50 being larger than that of AM50.[227]

5.3.1.2 Treated CNTs

Annealing is one of the important ways to improve the graphitization of CNTs and remove the amorphous carbon. The effects of heating on the capacitance depend on the heating temperature and the quality of the as-grown CNTs.[229,230] Li *et al.* found that the specific capacitance of MWCNTs was increased by the oxidization up to 650 °C due to the enhanced specific surface area and dispersity.[230] The capacitance, however, decreased with further increasing of the temperature due to reduced surface area. At the same time, the heat treatment reduces the ESR because of the improvement of graphitization, resulting in the enhancement of the power density. With increasing temperature, the specific surface area increases, whereas the average pore diameter decreases and saturates at high temperature (Figure 5.27).[210,231] The raw sample shows a peak at 150 Å and has less distribution in the smaller pore diameter near 20 Å. With increasing heat-treatment temperature, the number of smaller pore diameters increases and reaches the maximum at 1000 °C, whereas the number of pore diameters ranging from 50 ± 250 Å decreases. A maximum specific capacitance of 180 F g^{-1} and a measured power density of 20 kW kg^{-1} for the heat-treated SWCNTs at 1000 °C were obtained. The increased capacitance was well explained by the enhancement of the

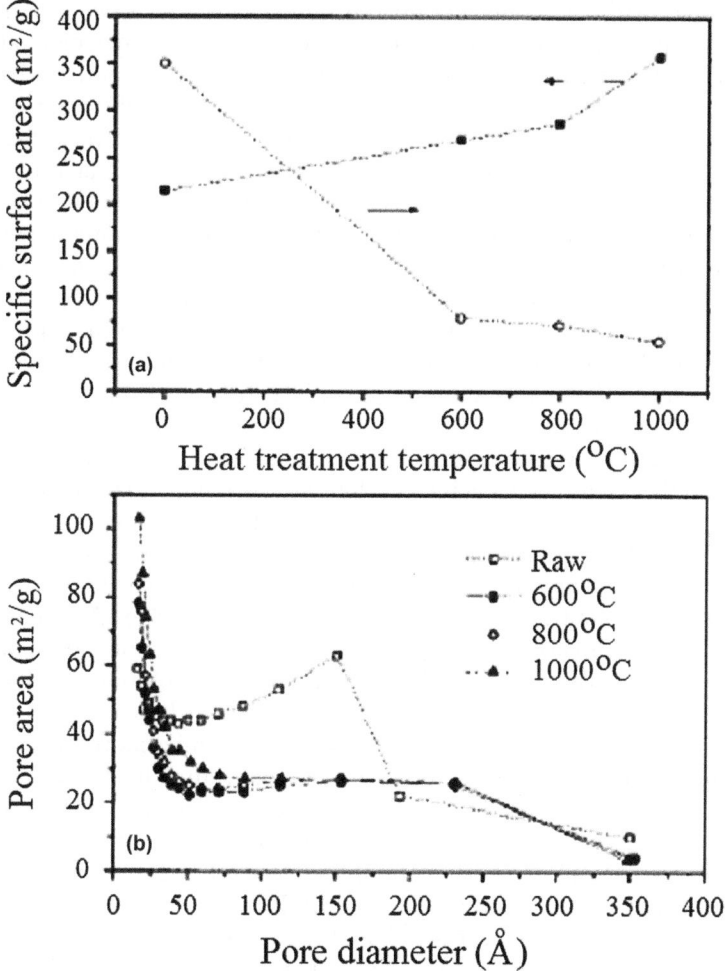

Figure 5.27 (a) The BET (N_2) specific surface areas and the average pore diameters of the CNT electrode as a function of heat-treatment temperature and (b) The pore-size distribution of the CNT electrodes.[210] (Copyright: Wiley-VCH).

specific surface area and the abundant pore distributions at lower pore sizes. It is noted that the effect of the annealing temperature on the capacitance also depends on the raw CNTs used in the experiments.[230,231]

Capacitance of CNT-based supercapacitors can also be enhanced by chemical activation,[212,232,233] functionalization,[234–236] and surface treatment.[237,238] The value of specific capacitance increased significantly after strong oxidation in acid nitride due to the increase of the functional groups on the CNT's surface.[212] Enhanced values of capacitance were observed after activation; in some cases it increased almost seven times because the microporosity of pure MWCNTs can be highly developed using chemical KOH activation.[232] The activated material still possessed a nanotubular morphology with many defects on the outer walls that gave a significant increase of micropore volume, while keeping a noticeable mesoporosity. The electrochemical treatment of CNTs provides an effective and controllable method for changing the pore-size distribution (PSD) of SWCNTs.[233] In particular, a remarkable volume of the small mesopores in the 3.0–5.0 nm diameter range was increased. The SWCNTs treated for 24 h at 1.5 V have a higher specific surface area (109.4 m^2 g^{-1}) and a larger volume of small mesopores (0.048 cm^3 g^{-1} in 3.0–5.0 nm diameter range), compared with the as-grown SWCNTs (46.8 m^2 g^{-1} and 0.026 cm^3 g^{-1}, respectively). The specific capacitance was increased three-fold after electrochemical treatment. The electric double-layer capacitance, depending on the surface functional groups, can be dramatically changed, from a large increase to complete disappearance.[234] The introduction of surface carboxyl groups created a 3.2 times larger capacitance due to the increased hydrophilicity of MWCNTs in an aqueous electrolyte. In contrast, the introduction of alkyl groups resulted in a marked decrease of capacitance. Notably, the complete disappearance of capacitance for samples functionalized with longer alkyl groups, indicating the perfect block of proton access to the carbon nanotubes' surfaces by extreme hydrophobicity. The specific capacitance can also be enhanced by fluorine functionalization with heat treatment.[235] The fluorination of SWCNT walls transformed the nonpolar SWCNTs to the polar ones by forming dipole layers on the walls, resulting in high solubility in deionized water. Fluorinated samples gave lower capacitance than the raw samples before heat treatment due to the increase of the micropore area and the decrease in the average pore diameter. However, after heat treatment, the specific capacitance of the fluorinated samples became higher than those of the raw samples because of the additional redox reaction due to the residual oxygen gases present on the surface of the electrodes. The reduction of ERS was attributed to the improvement of conductivity because of the carrier induced by the functionalization.[236] Pyrrole treated-functionalized SWCNTs have high values of capacitance (350 F g^{-1}), power density (4.8 kW kg^{-1}), and energy density (3.3 kJ kg^{-1}).[237] The high capacitance can also be obtained by the plasma surface treatment with NH_3 due to the enhancement of the total surface area and wettability of the MWCNTs.[238]

5.3.2   CNTs Binary Composites

The composites incorporating a nanotubular backbone coated by an active phase with pseudocapacitive properties, such as CNT/oxide composite, represent an important breakthrough for developing a new generation of supercapacitors based on three basic reasons:[188,239–243] (1) the percolation of the active particles is more efficient with nanotubes than with the traditional carbon materials; (2) the open mesoporous network formed by the entanglement of nanotubes allows the ions to diffuse easily to the active surface of the composite components; and (3) since the nanotubular materials are characterized by a high resiliency, the composite electrodes can easily adapt to the volumetric changes during charge and discharge, which improves drastically the cycling performance. The first two properties are essential to lower the ESR (R_s) and consequently increase the power density.

5.3.2.1 CNTs/graphene

Graphene sheets have been considered as one of the most promising flexible electrode materials for a supercapacitor because of the high specific surface area[244] and extraordinary pliable property.[245] Considerable efforts have been made on the graphene-based supercapacitors,[246–248] and a maximum specific capacitance of 140 F g^{-1} has been achieved.[248] However, a graphene sheet exhibits a lower electrochemical capacity than powder-based graphene electrodes with specific capacitance values ranging from 191 to 279 F g^{-1},[249–252] because graphene sheets with less tacking should exhibit better supercapacitive performance.[253] As a consequence, it is both important and necessary to further improve the electrochemical performance of graphene film. The integration of graphene, including pristine graphene (PG) and graphene oxide (GO), and one-dimensional CNTs into a hybrid material to design hierarchically structured composites is quite a promising way to enhance the dispersion of graphene and CNTs, inherit the virtues of both graphene and CNTs, fabricate an efficient and effective electronic and thermal conductive network, and explore advanced carbon materials because CNTs can bridge the defects for electron transfer and expand the layer distance between graphene sheets.[254–266]

An ideal 3-D structure of the hybrid material should have the graphene sheets separated completely by CNTs in the state of individual layers (Figure 5.28).[261,267] Then, the graphene sheets could be kept completely from restacking if the initial homogeneous distribution of CNTs and graphene sheets is completely kept. In this ideal state, the entire surface of graphene could be exposed, and the electrolyte ions would also use the channels generated by the CNT spacer and thus facilitate the charging/recharging process of the devices. Fan *et al.*[267] reported that the supercapacitor based on graphene/CNT sandwich structures (Figure 5.28(b)) exhibits a specific capacitance of 385 F g^{-1} at 10 mV s^{-1} in 6 M KOH solution. Lu *et al.*[256] obtained a specific capacitance of 265 F g^{-1} at a current of 0.1 A g^{-1} on graphene/CNTs (16% CNTs) electrode in 6 M KOH electrolyte. Li *et al.*[266] reported that the capacitance of the hybrid electrode depends on the graphene to CNT ratio. A specific capacitance of 70–110 F g^{-1} at a low scan rate of 1 mV s^{-1} can be obtained by changing the ratio. Wang *et al.*[261] reported that these ideal 3D hierarchical structures (Figure 5.28(a)) shows a high specific capacitance of 318 F g^{-1} with an energy density of 11.1 Wh kg^{-1}.

5.3.2.2 CNTs/oxides

A hybrid electrode consisting of CNTs and oxide nanoparticles incorporates a nanotubular backbone coated by oxides, which fully utilize the advantages

(a)

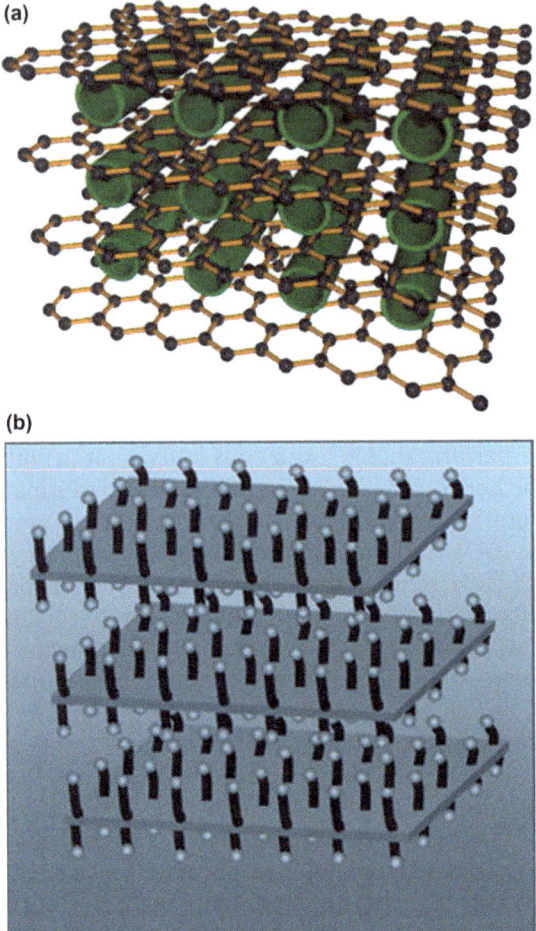

(b)

Figure 5.28 (a) Model for an ideal state of the 3-D hybrid graphene/CNTs.[261] (Copyright: American Chemical Society) (b) Illustration of the formation of hybrid materials with CNTs grown in between graphene nanosheets.[267] (Copyright: Wiley-VCH).

of the pseudocapacitance and EDLC. The open mesoporous network formed by the entanglement of nanotubes may allow the ions to diffuse easily to the active surface of the composite components and to lower the ESR (R_s) and consequently increase the power density. The oxide nanoparticles include RuO_2, Co_3O_4, MnO_2, V_2O_5, ZnO, $Ni(OH)_2$, TiO_2, IrO_2, and oxide composites.

As one of the important materials in oxide supercapacitors, the incorporation of RuO_2 into CNTs had been considered as an efficient way to improve the capacitance. In RuO_2/CNT binary composite electrodes, the surface functionality of CNTs strongly affects the electrostatic charge storage and pseudofaradaic reactions due to the hydrophilicity.[268] Such hydrophilicity induced by the functionalization enables easy access of the solvated ions to the electrode/electrolyte interface, which increases the Faradaic reaction site number of RuO_2 nanoparticles and leads to higher capacitance. The specific capacitance of the RuO_2/pristine-CNTs composite was about 70 F g^{-1} (RuO_2:13 wt% loading). However, the specific capacitance of RuO_2/hydrophilic-CNT (nitric acid treated) composites with the same oxide loading was about 120 F g^{-1}. Kim *et al.* reported that a three-dimensional CNT film substrate with RuO_2 showed both a very high specific capacitance of 1170 F g^{-1} and a high rate capability.[269] To enhance its pseudocapacitance, RuO_2 must be formed with a hydrated amorphous and porous structure, and a smaller size, because this structure provides a large surface area, and forms conduction paths for protons to easily access even the inner part of the RuO_2. The highly dispersed RuO_2 nanoparticles can be obtained on carboxylated carbon nanotubes by preventing agglomeration among RuO_2 nanoparticles through bond formation between the RuO_2 and the surface carboxyl groups of the carbon nanotubes[270] or by treating the CNTs in a concentrated H_2SO_4/HNO_3 (3:1 volume ratio) mixture at 70 °C.[271] The highly dispersed RuO_2 nanoparticles on carbon nanotubes show an increased capacitance because the protons are able to access the inner part of RuO_2 with the decrease in size and its utilization is increased. The high dispersion of RuO_2 is therefore a key factor to increase the capacitance of nanocomposite electrode materials for supercapacitors. A prominently enhanced capacitive performance was also observed in well-dispersed RuO_2 nanoparticles on nitrogen-containing carbon nanotubes.[272,273] The function of nitrogen amalgamation is to create preferential sites on CNTs with lower interfacial energy for attachment of RuO_2 nanoparticles (Figure 5.29). This crucial phenomenon leads to a significant improvement of the overall specific capacitance up to the measured scan rate of 2000 mV s^{-1}, indicating that superior electrochemical performances for supercapacitor applications can be achieved with RuO_2–CNT-based electrodes using the nitrogen incorporation technique. However, the commercialization of RuO_2/CNT composites is very difficult because of the high cost and high toxicity of RuO_2.

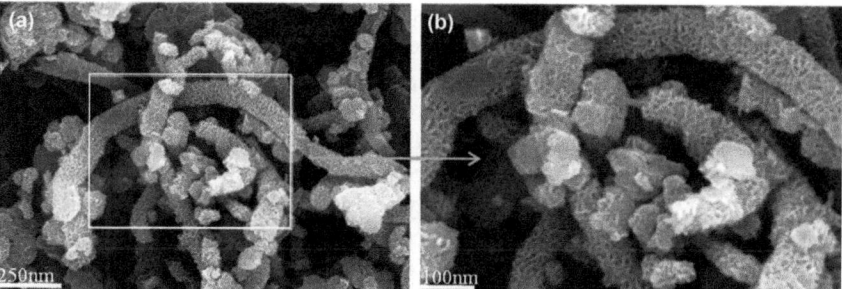

Figure 5.29 SEM images of MWCNTs/MnO_2.[280]
(Copyright: Elsevier).

As one of the most promising pseudocapacitor electrode materials, MnO_2 shows both high specific capacitance and low cost. The performance of real capacitors based on manganese oxide, however, is limited by the two irreversible reactions Mn(IV) to Mn(II) and Mn(IV) to Mn(VII), which potentially depend on the electrolyte pH. In particular, the electrolyte usually leads to the dissolution of the negative electrode. CNTs can help in preserving the electrodes integrity during cycling and keeping the capacitance of oxide.[274–295] It is believed that CNTs not only provide considerable specific surface area for high mass loading of MnO_2 to ensure effective utilization of MnO_2, but also offer an electron pathway to improve electrical conductivity of the electrode materials. Zheng *et al.*[280] reported that the core/shell structure of CNTs/MnO_2 electrode (Figure 5.29) presents excellent electrochemical capacitance properties with the specific capacitance reaching 380 F g^{-1} at the current density of 5 A g^{-1} in 0.5 M Na_2SO_4 electrolyte. In addition, the electrode also exhibits a good performance (the power density: 11.28 kW kg^{-1} at 5 A g^{-1}) and long-term cycling stability (retaining 82.7% of its initial capacitance after 3500 cycles at 5 A g^{-1}). By a facile and scalable asymmetric *in situ* deposition method, Shen *et al.*[281] deposited MnO_2 nanoparticles in vacuum filtrated SWCNT films. The assembled two-electrode SWCNT–MnO_2 electrode showed the optimal specific capacitance, power and energy densities, superior knee frequency at 529.8 F g^{-1}, 73.6 Wh kg^{-1}, 14.6 kW kg^{-1}, and 1318 Hz, respectively, as well as excellent cyclic stability (99.9% capacitance retention after 2000 charge/discharge cycles). The mesopores in CNTs/MnO_2 composite electrodes play a key role in achieving improved electrochemical performance and mechanical flexible.[281,283] Jiang *et al.*[291] found that the carbon conductive agent can improve the performance of the composite. Although the storage performance of the CNTs/MnO_2 composite is improved, its long-term stability still remains unsolved.

Ni(OH)$_2$ is often used in the hybrid supercapacitor with carbon (using KOH solution as an electrolyte). The positive electrode materials (Ni(OH)$_2$) convert to NiOOH with the formation of a proton and electron during the charge process. The rate capability of Ni(OH)$_2$ is associated with the proton diffusion in Ni(OH)$_2$ framework. The Ni(OH)$_2$/CNTs composite provided a shorter diffusion path for proton diffusion and larger reaction surface areas, as well as reducing the electrode resistance due to the high electronic conductivity of CNTs.[296–299] Wang *et al.* reported that the CNTs can reduce the aggregation of Ni(OH)$_2$ nanoparticles, inducing a good distribution of the nanosized Ni(OH)$_2$ particles on the cross-linked, netlike structure CNTs.[296] The rate capability and utilization of Ni(OH)$_2$ were greatly improved, and the composite electrode resistance was reduced. A specific energy density of 32 Wh kg^{-1} at a specific power density of 1500 W kg^{-1} was obtained in the hybrid supercapacitor. The capacitance can be further improved by heating the Ni(OH)$_2$/CNTs composite at 300 °C because of the formation of an extremely thin layer of NiO_x on the CNT film.[297] The specific capacitance decreased with the increase of NiO_x in the composite if the NiO_x percentage was above 8.9 wt%. A specific capacitance of 1701 F g^{-1} was reported for

8.9 wt% NiO$_x$/CNT electrode. Tang *et al.*[298] further reported that the Ni(OH)$_2$/CNTs electrode on Ni foam showed a specific capacitance of 3300 F g^{-1}, an operation window at 1.8 V, and an energy density of 50.6 kW kg^{-1}.

Other oxides, such as ZnO,[300,301] VO$_x$,[302-306] TiO$_2$,[307,308] SnO$_2$,[307,309] IrO$_x$,[310,311] and CrO$_3$,[312] had also been studied for the oxide/CNTs composite electrodes. Zhang *et al.*[301] found that the capacitance of CNTs/ZnO electrode strongly depends on the ZnO loading, and excess oxide loading will deteriorate the capacitive performance by destroying the network structure of the CNT matrix and lowering the conductivity of the electrode. Chen *et al.*[306] reported that asymmetric supercapacitors based on a thick-film CNT/V$_2$O$_5$ nanowire composite anode and commercial active carbons-based cathode in combination with an organic electrolyte exhibits excellent rate capability, high capacity, and cycling stability. A prototype asymmetric supercapacitor showed an energy density of 40 Wh kg^{-1} at a power density of 210 W kg^{-1}, and a maximum power density of 20 kW kg^{-1}. Reddy *et al.*[307] compared the electrochemical properties of RuO$_2$/MWCNT, TiO$_2$/MWCNT, and SnO$_2$/MWCNT nanocrystalline composites for supercapacitor electrodes. The average specific capacitances measured using the three electrochemical techniques of the pure MWCNT, RuO$_2$/MWCNT, TiO$_2$/MWCNT, and SnO$_2$/MWCNT nanocomposite electrodes are 67, 138, 160, and 93 F g^{-1}, respectively. The enhancement of the specific capacitance of metal oxide dispersed MWCNT from pure MWCNT is due to the progressive redox reactions occurring at the surface and bulk of transition metal oxides through Faradaic charge transfer due to the modification of the surface morphology of MWCNT by the nanocrystalline RuO$_2$, TiO$_2$, and SnO$_2$. A CrO$_3$/SWCNT based electrode showed exceptionally quick charge propagation due to the overall physical and textural properties of SWCNT.[312] CrO$_3$ nanoparticles finely dispersed at the nanoscale in the SWCNT make possible the enhanced charging rate of the electrical double layer and allow fast Faradaic reactions. Chromium-containing species present as CrO$_3$ inside SWCNTs as well as in the form of CrO$_2$Cl$_2$ (possibly along with CrO$_3$ too) between SWCNTs within bundles supply redox reactions due to access by the electrolyte in spite of its encapsulated (and intercalated) location because of the numerous side openings created all along the SWCNT defective walls during the filling step.

5.3.2.3 CNTs/polymer

Electronically conducting polymers are promising supercapacitor materials for two main reasons: (1) high specific capacitance, because the charge process involves the entire mass; and (2) high conductivity in the charged state, leading to low resistance and high power density. The main drawback using polymer in supercapacitors is the cycling stability because of typical shrinkage, breaking, and cracks appearing in subsequent cycles. It has already been proved that composites based on CNTs and conducting polymers are very interesting electrode materials because the entangled mesoporous network of nanotubes in the composite can adapt to the volume change.[313-341] That allows the shrinkage to be avoided and hence a more stable capacitance with cycling to be obtained.

The performance of the CNTs/polymer composite electrode strongly depends on the ratio between CNTs and polymer, cell construction, the quality of CNTs, the methods to synthesize the composite, electrolyte type, substrates, and the properties of the polymer.[313-321] The optimal performance of the poly(3,4-ethylenedioxythiophene) (PEDOT) and CNT composite can be achieved with 20–30% of CNTs and 70–80% of PEDOT.[313] The specific capacitance of a polyaniline (PANI)/SWCNT composite electrode increased as the amount of the deposited PANI onto SWCNTs increased up to 73 wt%, where the PANI was wrapped around SWCNT.[317] Beyond 73 wt%, the additional PANI was deposited either in the mesopores between SWCNTs or in the form of film over the surface, which caused a drop in the capacitance (Figure 5.30). In the case of three electrode cells, the capacitance values for the composites (20 wt% of CNTs and 80 wt% of conducting polymers (ECP), such as polyaniline (PANI) and polypyrrole (PPy)) are extremely high (250 to 1100 F g^{-1}). In the two-electrode cell, however, the specific capacitance values are much smaller (190 F g^{-1} for PPy/CNTs and 360 F g^{-1} for PANI/CNTs).[314] It highlights the fact that only two-electrode cells allow a good estimation of materials performance in electrochemical capacitors. An *et al.*[316] demonstrated that the SWCNT/PPy (1/1 in weight) nanocomposite

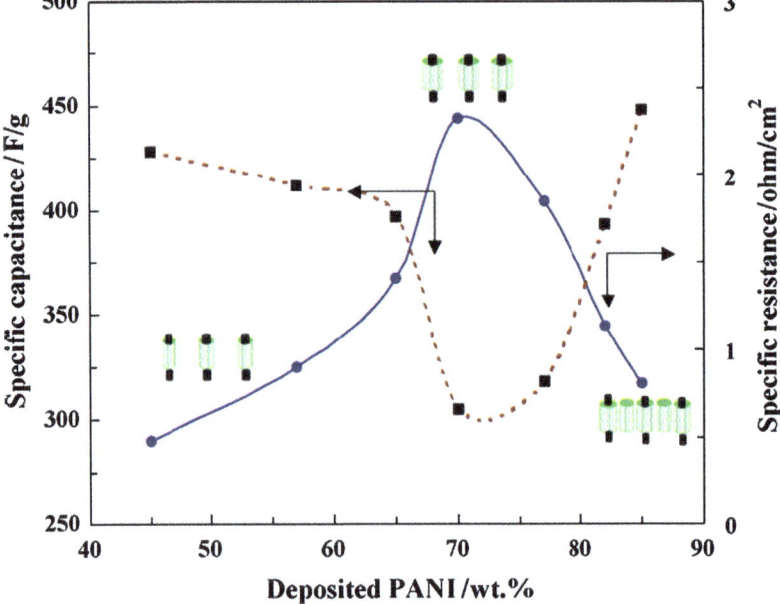

Figure 5.30 Relationship of wt% of deposited PANI on SWCNTs with the (a) specific capacitance (at 10 mA/cm^2 in 1 M H$_2$SO$_4$ electrolyte) and the (b) specific resistance of the PANI/SWCNT composite.[317] (Copyright: Elsevier).

electrode shows much higher specific capacitance than pure PPy and as-grown SWCNT electrodes, due to the uniformly coated PPy on the SWNTs. A maximum specific capacitance of 265 F g^{-1} from the SWNTs/PPy nanocomposite electrode containing 15 wt% of the conducting agent was obtained, where the addition of conducting agent into the SWCNT-PPy nanocomposite electrode gives rise to an increase of the specific capacitance by reducing the internal resistance of the supercapacitor. Ertas *et al.*[334] reported that the substrate for the CNTs/polymer composite strongly affected the capacitance. The capacitance of PProDOT/SWCNTs composite on gold-coated Kapton substrate is much less than that in the pyrene functionalized polyfluorene (Sticky-PF)/SWCNTs substrate. Hu *et al.*[332] reported that the specific capacitance of CNTs/PPy composite electrode can reach up to 587 F g^{-1} at a current of 3 A g^{-1} in 0.1 M NaClO$_4$ solution by introducing a physical defect into super-long CNTs. Similarly, the composite with core-shell structures consisting of PANI (66%) and CNTs prepared *via in situ* polymerization of aniline monomers by using MWCNTs as templates showed a capacitance of 560 F g^{-1}.[338] However, 29% of the capacitance was lost after 700 cycles. Studies showed a steady loss of capacitance to about 57% of the original value after 32 700 charge/discharge cycles.

The stability is one of the important challenges in CNTs/polymer-based supercapacitor. To improve the stability, Rosario-Canales *et al.*[335] fabricated the poly(3,4-propylenedioxythiophene) (PProDOT) and SWCNTs composite that was helically wrapped with ionic, conjugated poly[2,6-{1,5-bis(3-propoxysulfonic acid sodium salt)} naphthylene] ethynylene (PNES). The hybrid PProDOT/PNES/SWCNTs composites retained 90% of their initial charge capacity after 21 000 charge/discharge cycles. Lee *et al.*[336] reported a PPy–CNT composite electrode on a ceramic fabric fabricated by CVD (chemical vapor deposition) and chemical polymerization (Figure 5.31). They found that its capacitance was well retained even after 5000 redox cycles (Figure 5.32). Ertas *et al.*[334] reported that the capacitance of PProDOT/SWCNTs composite on gold-coated Kapton substrate retains 80% of its electroactivity after 32 700 nonstop charge/discharge cycles.

5.3.3 CNTs Ternary Composites

The research on supercapacitors has been aimed at designing and synthesizing hierarchical ternary composite materials composed of nanostructured carbon and pseudo-active materials (polymer and oxide) with high Faradaic capacitance.[342–353] These novel composites with good electrical conductivity and architecturally tailored structure can effectively use both the fast and reversible Faradaic capacitance and indefinitely reversible double-layer capacitance at the electrode–electrolyte interfaces, exhibiting synergistic properties with efficient energy extraction from each component in composite materials. The CNTs ternary composite is composed of CNTs and two other materials from graphene, oxide, and polymer.

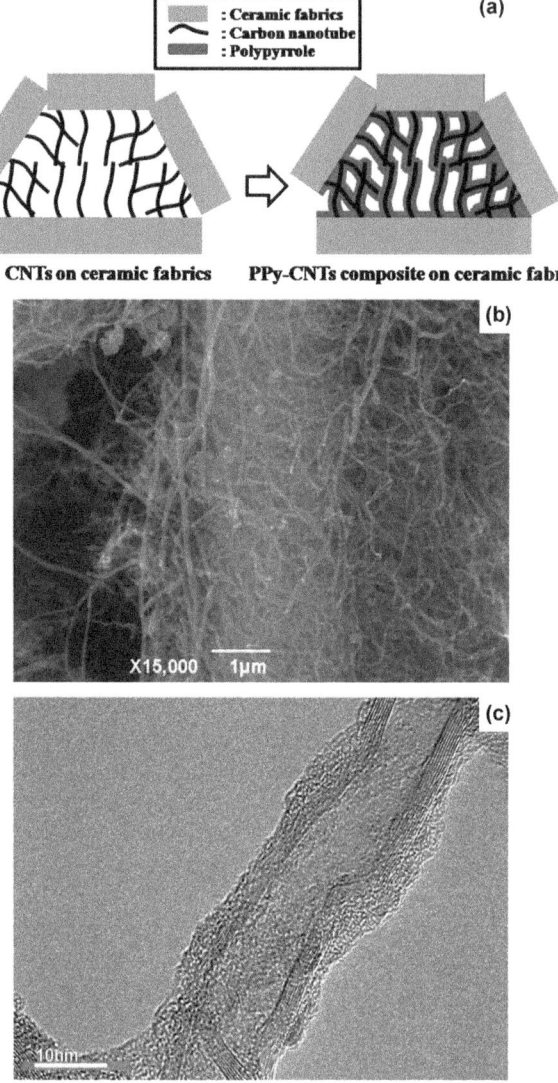

Figure 5.31 Illustration (a), SEM (b) and TEM (c) images of PPy–CNTs composite on ceramic fabric.[336]
(Copyright: Elsevier).

5.3.3.1 *CNTs/graphene/others*

Owing to the graphene and CNTs synergistically accommodating the strain of volume change, the ternary composites of electroactive materials with G and CNT show superior cycle ability that is one of important advantages of supercapacitors relative to that of ordinary batteries. Yan *et al.*[342] reported that a graphene/CNT/PANI composite exhibits the specific capacitance of

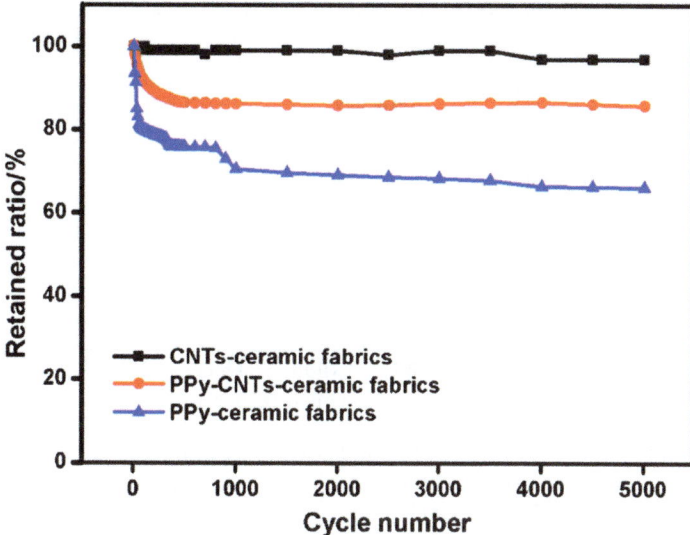

Figure 5.32 Change of capacitance of CNTs, PPy and PPy–CNTs composite electrode on ceramic fabric as a function of the number of charge–discharge cycles at current density of 1 mA cm^{-2}.[336] (Copyright: Elsevier).

1035 F g^{-1} at a scan rate of 1 mV s^{-1} in 6 M of KOH. Though a small amount of CNTs (1 wt%) is added into graphene, the cycle stability of the graphene/CNT/PANI composite is greatly improved due to the maintenance of a highly conductive path, as well as the mechanical strength of the electrode during doping/dedoping processes. After 1000 cycles, the capacitance decreases only 6% of the initial capacitance compared to 52% and 67% for graphene/PANI and CNT/PANI composites. Lu *et al.*[248] reported that graphene/PPy/CNTs ternary composites (8GCPPy) fabricated *via* an *in situ* polymerization method shows three-dimensional hierarchical structure. The prepared graphene/PPy/CNTs composite with graphene : CNT = 8 : 1 exhibits a high specific capacitance (361 F g^{-1}) at a current density of 0.2 A g^{-1}, much higher than that of pure PPy (176 F g^{-1}) and binary composites of CNT/PPy (253 F g^{-1}) and graphene/PPy (265 F g^{-1}). Importantly, the ternary composite shows a stable cycling performance (4% capacity loss after 2000 cycles) (Figure 5.33), due to the graphene and CNTs synergistically releasing the intrinsic differential strain of PPy chains during the charge/discharge processes.

The three-dimensional (3D) carbon nanostructures made by connecting two-dimensional (2D) graphene and one-dimensional (1D) CNTs have attracted great interest, because an ultrahigh surface-to-volume ratio, which is greatly important in energy applications, can be achieved with similar work functions, and can greatly enhanced by reduced agglomeration between nanomaterials.[349–353] By introducing metal, oxide or polymer into the 3D nanoarchitecture, the ternary composite may result in an ultrahigh redox capacitance. Siridar *et al.*[349] reported that the capacitance of 3D graphene/

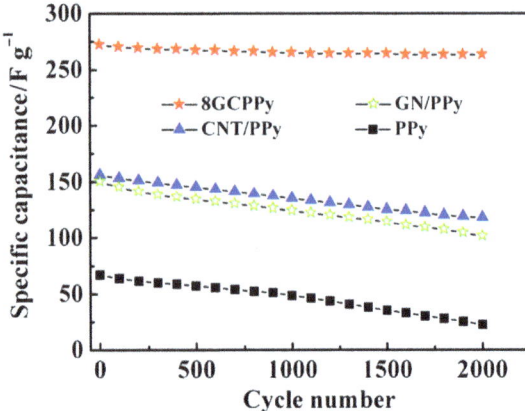

Figure 5.33 SC of CNT/PPy, GN/PPy and 8GCPPy as a function of cycle number at a current density of 6 A g^{-1}.[348] (Copyright: Elsevier).

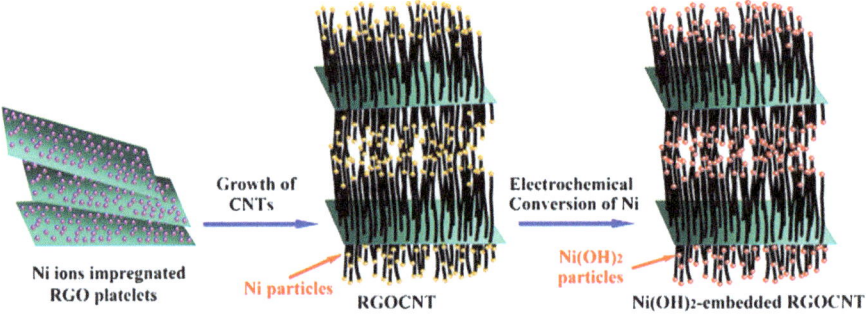

Figure 5.34 Schematic illustration of the experimental steps for the preparation of Ni(OH)$_2$ embedded-carbon-nanotube-graphene composite.[351] (Copyright: Elsevier).

CNT/Pd exhibits an exceptionally high value (1615 F g^{-1}) at a scan rate of 10 mV s even in a weak electrolyte solution of 1 M KOH. After 600 cycles, a capacitance increase of 46% relative to the initial capacitance is observed, indicating excellent electrochemical stability of the electrode. Zhang *et al.*[351] reported that a novel 3D nanostructure consisting of Ni(OH)$_2$ nanoparticles, CNTs, and reduced graphene oxide (Figure 5.34) showed an extremely high electrocapacitance. The specific capacitances are as high as 1235 and 780 F g^{-1} at current densities of 1 and of 20 A g^{-1}, respectively. In addition, the composite retains 80% of its original capacity after 500 cycles at a discharge current density of 10 A g^{-1}.

5.3.3.2 *CNTs/oxide/polymer*

Design and synthesis of new electrode materials for supercapacitors to enhance their electrochemical behaviors has been carried out on the

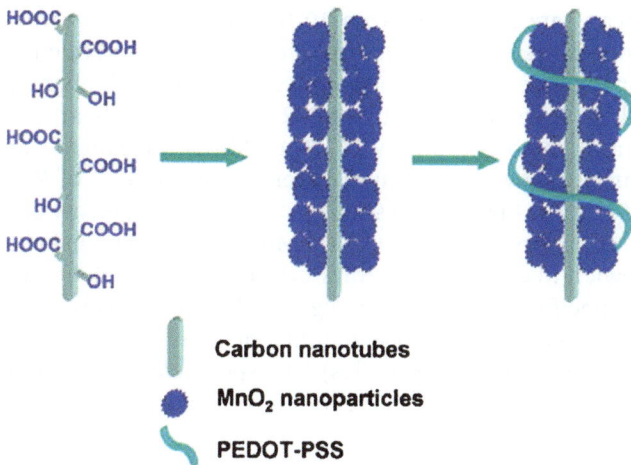

Carbon nanotubes

MnO₂ nanoparticles

PEDOT-PSS

Figure 5.35 Sketch of MnO₂/CNTs/PEDOT-PSS ternary composite.[354] (Copyright: American Chemical Society).

CNT/oxide/polymer ternary composite to effectively utilize the full potential of all the desired functions of each component.[354–357] Hou *et al.*[354] reported a MnO₂/CNT/PEDOT-PSS ternary composite for supercapacitor electrodes. In the composite (Figure 5.35), the MnO₂ nanospheres were directly grown on CNTs, where MnO₂ offers the desired high specific capacitance and the CNT framework provides the improved electrical conductivity and mechanical stability. Furthermore, PEDOT-PSS was added in the composite, which acted not only as a dispersant to stabilize the composite suspension to facilitate the electrode film fabrication, but also offered good interparticle connectivity between the oxide material and the CNTs. The ternary composite showed high specific capacitance (427 F g^{-1} in 1 M Na₂SO₄ solution at a very high scan rate of 500 mV s^{-1}), excellent rate capability (85% of its specific capacitance (from 200 to 168 F g^{-1}) was preserved as the current density increases from 5 to 25 mA cm^{-2}), and long cycling life stability (less than 1% decay in available specific capacity after 1000 cycles). Li *et al.*[355] and Kim *et al.*[356] reported that the MWCNT/PANI/MnO₂ ternary structures showed a specific capacitance of 330–350 F g^{-1} at a low scan rate.

5.3.4 Other Nanotubes

Beside CNTs, other nanotubes, including TiO₂,[358,359] BCN,[360] MnO₂,[361] and polymer,[362] had also been studied for their application as electrodes in supercapacitors. Chen *et al.*[358] reported that the hydrogen-treated bamboo-like TiO₂ NTs showed a specific capacitance of 2 mF cm^{-2} at a scan rate of 1 V s^{-1}. Lu *et al.*[359] reported that the hydrogenated TiO₂ NTs yielded a specific capacitance of 3.24 mF cm^{-1} at a scan rate of 100 mV s^{-1}, and showed a remarkable rate capability with 68% areal capacitance retained when the scan rate increased from 10 to 1000 mV s^{-1}, as well as an out-standing long-term cycling stability with only 3.1% reduction of initial specific capacitance after 10 000 cycles. The vertical aligned BCN NTs showed a high specific capacitance (321.0 F g^{-1}) with an excellent rate capability and high durability.[360] The MnO_2/Mn/MnO_2 sandwich-like nanotube arrays showed maximum specific capacitances of 937 F g^{-1} at a scan rate of 5 mV s^{-1} by CV and 955 F g^{-1} at a current density of 1.5 A g^{-1} by chronopotentiometry in solution of 1.0 M Na_2SO_4.[361] The hybrid MnO_2/Mn/MnO_2 sandwich-like nanotube arrays exhibited an excellent rate capability with a high specific energy of 45 Wh kg^{-1} and a specific power of 23 kW kg^{-1} and excellent long-term cycling stability (less 5% loss of the maximum specific capacitance after 3000 cycles). The specific capacitance of a PEDOT-nanotube-based electrode[362] was reported to be about 132 F g^{-1} at the current density of 5 mA cm^{-2}.

5.4   Lithium Battery

A lithium ion battery (LIB) is a member of a family of rechargeable battery types in which lithium ions move from the anode to the cathode during discharge and back when charging. Li ion batteries use an intercalated lithium compound as the electrode material, compared to the metallic lithium used in a non-rechargeable lithium battery. LIBs are composed of four parts: anode, cathode, separator, and electrolyte (Figure 5.36). The cathode, typically a lithium metal oxide, acts as the positive terminal of the battery (during discharge) and the anode, commercially composed of graphitic carbon, acts as the negative terminal. The cathode and anode react according to the following half reactions:[363-366]

$$\text{Cathode}: LiMO_n = Li_{1-x}MO_n + xLi^+ + xe^- \tag{5.5}$$

$$\text{Anode}: xLi^+ + xe^- + 6C \leftrightarrow Li_xC_6 \tag{5.6}$$

Li$^+$ ions move between the anode and cathode *via* the electrolyte during charging and discharging. The electrolyte is typically a lithium salt, such as $LiPF_6$, dissolved in organic solvent, such as ethylene carbonate, which prevents free electrons from moving between the electrodes. So, the electrons complete the half reaction moving *via* an external wire, leading to a current.

Despite the superior performance of graphite-based LIBs, there are many challenges in the design of LIBs, especially the capacity and cycle life of the anode materials. In many commercial LIBs, the anodes are composed of graphite due to its high electronic (in-plane) conductivity as a consequence of the delocalized π-bonds, low expansion during lithium insertion, and appropriate structure for lithium ion intercalation and diffusion. When lithium intercalates in graphite, it occupies an interstitial site between two planes of graphite, and prevents other lithium ions from binding in directly

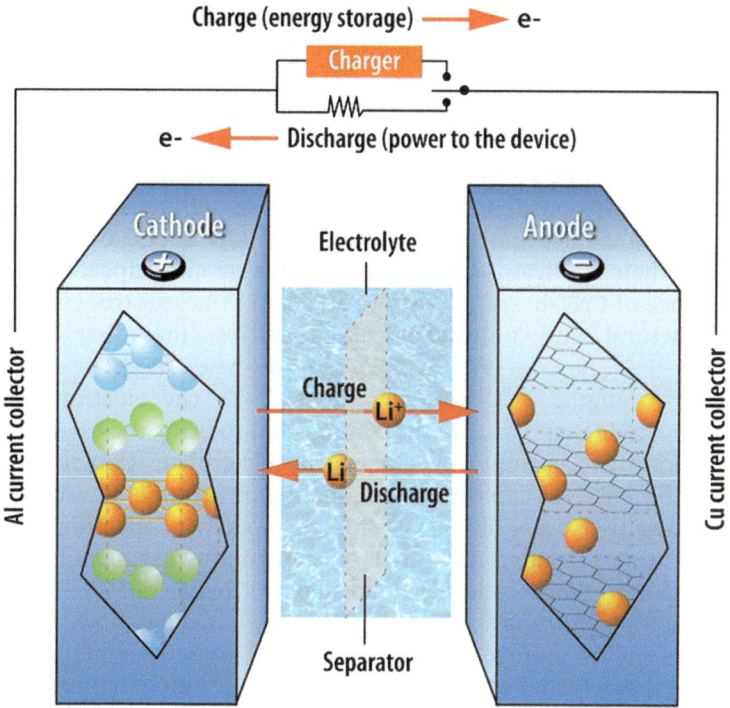

Figure 5.36 Schematic of lithium ion cell. http://www.spacesafetymagazine.com/wp-content/uploads/2013/01/how_a_lithium-ion_battery_works – 4fa2c98-intro.jpg

adjacent interstitial sites, which limits the amount of lithium ions to one for every six carbon atoms, *i.e.* LiC_6, and leads to a limited theoretical specific capacity of 372 mAh g^{-1} and an observed capacity of 280–330 mAh g^{-1} depending on the graphite used.[363–366] As the performance of LIBs strongly depends on the electrode properties, significant improvements in the electrochemical properties of electrode materials are essential to meet the demanding requirements of these applications because the lithium insertion capacity of graphite is relatively low. The hollow nanotubes should be very suitable for the electrodes of LIBs as: (1) their high surface area can improve the electrode/electrolyte contact areas; (2) the reduced size can shorten Li^+ transport distance; (3) the void space in hollow structures prolongs the cycling life by buffering against the local volume change during Li insertion/desertion and alleviating pulverization and aggregation of the electrode materials; and (4) the extra space for the storage of Li^+ can enhance the specific capacity.[365–370] In this part, we provide a comprehensive review on some recent advancements in design, synthesis, and fabrication of a variety of nanotubes as anode–electrode materials in LIBs with enhanced energy and power density, energy efficiencies, reliability, and cycling life. In particular, several effective strategies used to enhance battery performance will be highlighted, including reduced dimensional nanostructuring, surface coating, nanocomposite, and nanoarchitecturing.

5.4.1 Pure CNTs

The unique structures and properties of CNTs, such as a hollow structure, high conductivity, low density, high rigidity, and high tensile strength, have enabled them as a promising candidate for LIB anode electrodes.[365-378] Similar to their applications on hydrogen storage and supercapacitors, the performance of CNT-based anode electrodes in LIBs are strongly affected by their structural and morphological properties, including chirality of SWCNTs,[379-381] type of CNTs (SWCNTs or MWCNTs),[382] defect,[383-387] diameter,[388] length,[389-393] and alignment.[394-396] The electronic conductivity of the anode materials in LIBs is important to bring the carrier as fast as possible from the current collector to the electrolyte. Depending on the chirality, the SWCNTs can be semiconducting or metallic. Kawasaki *et al.*[379] reported that the reversible Li ion storage capacity of metallic SWCNTs is about five times greater than that of semiconducting nanotubes. Chew *et al.*[382] reported that the electrochemical performance of MWCNTs is much better than SWCNTs, and the flexible MWCNT electrodes show stable cycling behavior and allow up to a 10C-rate. Eom *et al.*[384] reported that the reversible capacity of the defective MWCNTs is 681 mAh g^{-1}, much higher than that of the pristine counterpart (351 mAh g^{-1}), because the defects can facilitate the insertion of Li ions into the tube and enhance the reversible lithium capacity. Zhang *et al.*[388] reported that the diameter of MWCNT also affects the electrochemical performance in LIBs. They found that the electrode fabricated from CNTs with diameters of 40–60 nm displayed better performance, as well as long cycle life among all the samples, and excellent coulomb efficiency of up to 101.9%. The length of CNTs strongly affects the electrochemical performance of the lithium storage because short CNTs with a lot of lateral defects facilitate easier intercalation and deintercalation of Li ions.[389-393] The effective diffusion of lithium ion in CNTs is reduced if the tube is long. Wang *et al.*[389,393] reported that the reversible capacity was significantly increased as the length of CNTs decreased (Figure 5.37). The arrangement of CNTs in electrodes of LIBs also affects the activity. Welna *et al.*[394] reported that the aligned MWCNTs provides more sites for the adsorption of lithium, and the intimate contact of the MWCNTs in the aligned structure helps to maintain electrical continuity to the current collector, leading to a high lithium storage capacity of 980 mAh g^{-1} in the first cycle. For the non-aligned MWCNTs, the value is reduced to 158 mAh g^{-1}. The vertically aligned MWCNTs displayed substantially greater rate capability than the non-aligned ones, indicating the important role that structure order plays in electrode performance.

Compared with graphite-based anode materials, the CNT-based anodes show highly irreversible capacity because of the high surface area, hollow structure, and reduced size. However, the electrochemical performance of CNT-based anodes in LIBs strongly depends on their morphology, structure,

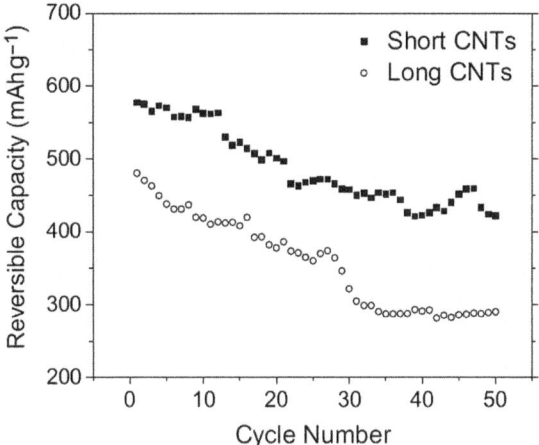

Figure 5.37 Variation of C_{rev} with number of cycles at a current density of 25 mAg^{-1}. Note that the capacity becomes stable after 30 cycles.[389] (Copyright: Wiley-VCH).

and alignment, as well as the synthesis and treatment methods in the fabrication. Their performance as anodes should be improved by optimizing diameter, creating defects, shortening length, aligning the CNTs, and enhancing the conductivity.

5.4.2 CNT-based Hybrid Composites

Although the irreversible capacity of CNT-based anodes is better than that of graphite-based anodes, the improvements in lithium storage capacity and cyclability have not been apparent. The CNT-based hybrid composites, including metals, other carbon nanostructures, transitional metal oxides, and other inorganic materials (such as silicon), show much higher lithium storage capacities of alloying materials and preventing pulverization and crumbling by using CNTs in the anode.[365–370,397–399] The combination of CNTs with other lithium storage materials may show promising at improving LIB anodes with elaborate nanostructure designs.

5.4.2.1 CNTs/silicon

Recently, silicon has attracted interest as an attractive alternative material to replace current graphite anodes for Li ion batteries since it has the highest known theoretical capacity (theoretically 4200 mAh g^{-1} for $Li_{4.4}Si$), low cost, and abundance. However, silicon suffers from severe volume change during charging and discharging that results in a compromised crystallographic structure and ultimately in crumbling and electrode capacity loss. In order to mitigate the effects of pulverization, novel nanoarchitectures have been used. In addition, silicon has a lower conductivity, particularly amorphous silicon, and in the process of lithium alloying and dealloying, the originally crystalline silicon becomes amorphous. The lower conductivity makes it difficult for current to be drawn by the current collector. By incorporating CNTs into silicon, the hybrid composites for anodes in LIBs had shown improvement on electronic conduction and structural buffering.[400–402] The silicon nanostructures in the hybrid composite can be nanoparticles, nanotubes, and nanowires. To form the Si/CNT hybrid composite, there are basically two methods: (1) ball milling of Si powder and CNTs with binder;[403,404] and (2) CVD to deposit Si to CNTs.[400,405–407] The performance of the CNT/Si hybrid anode is dependent on the synthesis method,[400,408] the structures of CNTs and Si,[409–420] and the structure of cell.[421–426]

Forney *et al.*[408] reported that the SWCNTs/Si free-standing anodes fabricated by plasma-enhanced CVD (PECVD) outperform those by low-pressure CVD (LPCVD), with electrode extraction capacities as high as 2500 mAh g^{-1}. When only the Si mass is considered, PECVD–SWCNTs/Si anodes have up to two times higher extraction capacities than LPCVD–SWCNTs/Si anodes at low Si loadings, because PECVD–Si–SWCNT anodes undergo more significant morphological and crystallographic changes during cycling than LPCVD–SWCNTs/Si anodes. Forney's work showed that the CVD method for Si deposition onto SWCNTs current collectors greatly impacts the electrode morphology, and affects the electrochemical performance in turn.

The CNTs/Si hybrid anode is a complicated composite structure by considering the structures of CNTs (SWCNTs, MWCNTs, diameter, and length) and Si (nanoparticle, nanotube, nanowire, crystals, and amorphous state).[409–420] Park *et al.*[409] reported that the MWCNTs embedded with Si nanoparticles (Figure 5.38) exhibited high specific capacity and superior

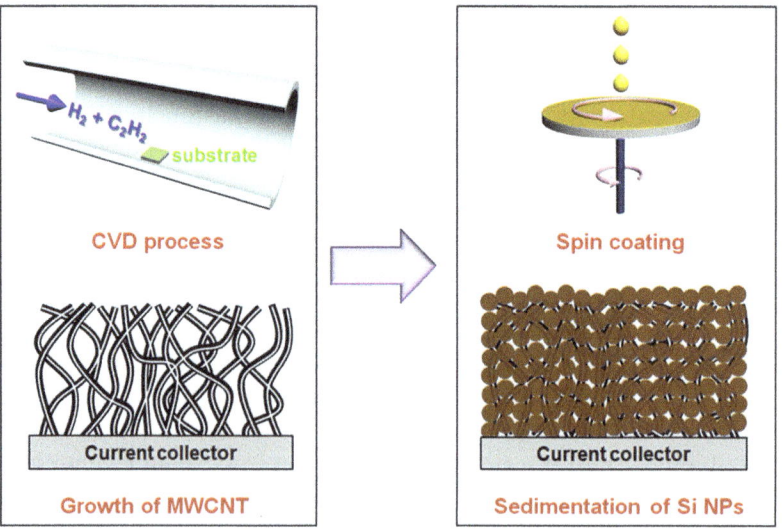

Figure 5.38 Schematic representations of CVD and spin coating processes for preparation of the MWCNTs-embedded Si nanoparticle films.[409] (Copyright: Elsevier).

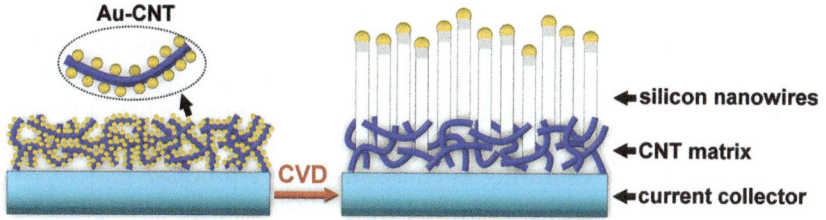

Figure 5.39 Schematic diagram showing the fabrication of the silicon nanowire arrays on carbon nanotube arrays on a current collector (stainless steel substrate).[414]
(Copyright: Wiley-VCH).

capacity retention (2900 and 1510 mAh g^{-1} after 10 and 100 cycles, respectively, at a current density of 840 mA g^{-1}) as well as an outstanding rate capability. Li *et al.*[414] reported that the nanoscale structure composted of CNTs and Si nanowires (Figure 5.39) exhibited a specific capacity of $\approx$3050 mAh g^{-1} for the 24th cycle. Because the electrochemical performance of the anode depends on both the CNTs and the Si, there has not been a clear conclusion on the optimization of the hybrid composite for the improvement of performance. The research on this field should focus on designing the optimized structures *via* the morphological and physical properties of CNT and Si.

Beside the morphologies and structures of CNTs and Si in the hybrid anode, its performance also depends on the Si mass loading per unit area, which, however, is limited by either the fabrication process or the structure design. To improve the mass, novel CNTs/Si structures and cells, such as core/shell structure[411,421–423] and vertically aligned CNTs coated with Si.[424,425] Gohier *et al.*[424] reported that this hierarchical hybrid nanostructure made of vertically aligned CNTs (5 nm in diameter) and Si nanoparticles (10 nm) showed a high reversible Li storage capacity of 3000 mAh g^{-1} at 1.3C, exhibited an impressive rate capability (a capacity of 1900 mAh g^{-1} at 5C and 760 mAh/g at 15C), and can sustain very high C rates without any significant polarization and structural damaging. The key factors for good cycling properties are the perfect adhesion between CNTs directly connected to the current collector and silicon particles thus facilitating the electron and lithium ion transport pathway and limiting the diffusion process occurring in conventional electrodes.

5.4.2.2 CNTs/oxide

Transition metal oxides as anode materials of LIBs have been widely investigated due to their high capacity. Those oxides, however, displayed very poor cycle stability and low cycle efficiency due to stress-induced material failure arising from the volume change in charging and discharging reactions. Similar to the CNT/Si hybrid anode, the incorporation between

CNTs and oxide can improve the electrochemical performance of these anodes because the highly activated surface area of CNTs can not only provide a support for anchoring well-dispersed oxide nanoparticles and work as a highly conductive matrix for enabling good contact between them, but also can effectively prevent the volume expansion/contraction and aggregation of nanoparticles during Li charge/discharge process.[427]

The CNTs/SnO$_2$ hybrid anode of a LIB has been widely studied because of the high theoretical values of discharge and charge capacities of SnO$_2$ (1494 and 782 mAh g^{-1}, respectively).[427–445] Zhang *et al.*[438] reported that the anode made of cross-stacked CNTs with uniformly loaded SnO$_2$ nanoparticles showed over 850 mAh g^{-1} of charge capacity with 100% retention for at least 65 cycles when cycled in the 0.01–3 V potential window versus Li$^+$/Li. Zhu *et al.*[435] reported that the initial discharge capacity and reversible capacity of MWCNTs/SnO$_2$ core-shell structure are up to 1472.7 and 1020.5 mAh g^{-1}, respectively. Moreover, the reversible capacity still remains above 720 mAh g^{-1} over 35 cycles, and the capacity fading is only 0.8% per cycle.

The CNTs/MnO$_2$ composite had been reported as an anode of the LIB because of its high storage capacity, low cost, environmental friendliness, and natural abundance.[446–448] Reddy *et al.*[446] reported that CNT/MnO$_2$ core-shell nanostructures (Figure 5.40) showed an initial discharge capacity of 2170 mAh g^{-1} and a reversible capacity of 500 mAh g^{-1} after 15 cycles.

Other transition metal oxides, such as Co$_3$O$_4$,[449–451] NiO,[452] TiO$_2$,[453,454] MoO$_2$,[455] Fe$_3$O$_4$,[456,457] and V$_2$O$_5$[458] have also been investigated as components as the CNT hybrid anode materials. Du *et al.*[449] reported that the initial capacity of porous Co$_3$O$_4$ on CNT structures is about 1918 mAh g^{-1} and the capacity of the following 19 cycles is almost maintained constant at about 1200 mAh g^{-1}. For CNT/NiO hybrid composites,[452] the as-prepared NiO nanoparticles were uniformly coated onto the surface of MWCNTs. The well crystallized MWCNTs/NiO composites show superior electrochemical performance in LIBs with a lithium storage capacity maintained at

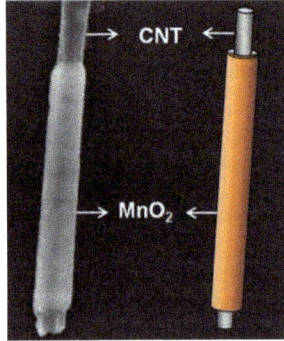

Figure 5.40   SEM image and a schematic representation of a single MnO$_2$/CNT hybrid coaxial nanotube.[446]
(Copyright: American Chemical Society).

≈ 800 mAh g^{-1} after 50 discharge/charge cycles. The CNT/Fe$_3$O$_4$ core-shell composite electrode, with the size of Fe$_3$O$_4$ confined to 5–7 nm and the CNTs aligned, exhibits a high reversible capacity of over 800 mAh g^{-1} based on the total electrode mass, remarkable capacity retention, as well as high rate capability.[456]

5.4.2.3 *CNTs/graphene*

Attempts to combine highly conductive CNTs and chemically reduced graphene have been carried out to obtain hybrid film with superior mechanical, electrical, electrochemical, catalytic or optical properties for applications in LIBs.[459–464] Chen *et al.*[460] reported that multi-layered graphene/CNT composites fabricated by an *in situ* chemical vapor reduction and deposition process delivered a reversible capacity of 518 mAh g^{-1} after 30 cycles. Chen *et al.*[461] reported that the electrochemical performance of CNTs/graphene is dependent on the ratio between the two components. The CNTs/graphene in a weight ratio of 1 : 2 is better than that in 1 : 1. The enhancement of electrochemical performance is attributed to the existence of CNTs as an electrical conducting network, which increases the basal spacing between graphene sheets and facilitates the ionic transportation.

5.4.2.4 *CNTs/metal*

A Sn anode has a high reversible capacity (approximately 994 mAh g^{-1} from Li$_{4.4}$Sn), and is one of the most promising candidates to replace the graphite anode for LIBs. The high energy capacity of a Sn anode results from different amounts of lithium insertion with a range of Li$_x$Sn ($x \leq 4.4$) alloys during the discharge process. However, a large volume expansion (259%) inevitably occurs, when a maximum amount of 4.4 lithium alloy with Sn to form Li$_{4.4}$Sn (a 440% increase in the number of atoms). As a result, Sn anode crumbling causes a loss of electrical contact with the current collectors, thus poor cycle performance, prohibiting Sn anode application in high performance LIBs. CNTs into Sn anodes can significantly mitigate the large volume change problem and relieve mechanical stress in the Sn lattice, thus enhancing anode performance.[465–467] Lee *et al.*[465] reported that SWCNT/Sn nanoparticle composites showed enhanced anode durability (535 mAh g^{-1} in the 50th cycle at 0.2 C). Yu *et al.*[466] reported that the encapsulated Sn/carbon nanoparticles in bamboo-like hollow carbon nanotubes (CNTs) had a high reversible capacity of 737 mAh g^{-1} after 200 cycles at 0.5 C. Kumar *et al.*[467] reported that the reversible capacities of the tin-filled nanotubes were remarkably high, stabilizing in the 720–800 mAh g^{-1} region over the first 20 cycles at 0.1 C.

Intermetallics of Sn alloys are also considered as promising candidates because of relatively smaller volume expansion, high specific capacities (relative to graphite) and good cyclability (relative to Sn).[468–473] Furthermore, the Li intercalation voltages of these alloys are slightly higher than that of

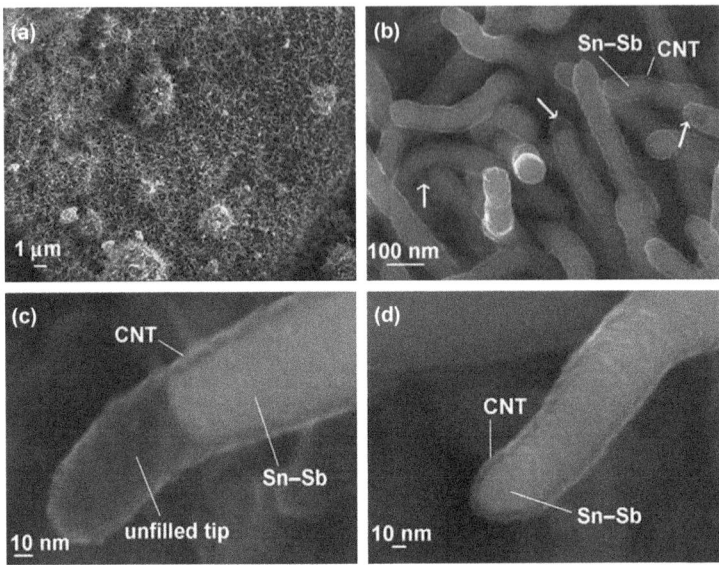

Figure 5.41 FESEM images of the CNT-encapsulated Sn–Sb nanorods showing (a) a large number of rod-like products at low magnification, (b) the thin CNT overlayer of filled nanorods at high magnification, the tips of a few CNTs that have not been completely filled with Sn–Sb are indicated by arrows, (c) the unfilled tip of a filled CNT, and (d) the completely filled tip of a filled CNT.[470] (Copyright: Wiley-VCH).

graphite, which favors operation safety. Chen *et al.*[473] reported that the delithiation capacity of a CNT–56 wt% Sn_2Sb nanocomposite was 372 mAh g^{-1} after 80 deep charge and discharge cycles. Fan *et al.* reported[469] that the nanocomposite with 10 wt% acid functionalized CNTs and SnSb nanoparticles showed a good initial coulombic efficiency of 79% and a reversible capacity of 860 mAh g^{-1} during the 40th cycle at a current density of 160 mA g^{-1}. Wang and Lee reported that the CNT-encapsulated Sn–Sb nanorods (Figure 5.41) are able to sustain high specific capacity about 900 mAh g^{-1} for at least 30 cycles in the 5 mV–2 V range. The CNT-encapsulated Sn–Sb nanorods contain 7.6 wt% of CNTs (*ca.* 90 nm in diameter and 7–10 nm in wall thickness) and 92.4 wt% of Sn–Sb nanorods (as $SnSb_{0.11}$; 70 nm in diameter). Yin *et al.*[472] reported that the milled CNT/$Ag_{36.4}F_{15.6}Sn_{48}$ electrode provided a discharge capacity of 530 mAh g^{-1} in the second cycle, maintaining a rechargeable capacity of, 420 mAh g^{-1} after 300 cycles.

5.4.3 Si Nanotubes

As a most promising and competitive candidate to replace graphite anode material, Si offers a theoretical capacity as high as 4200 mAh g^{-1} with a low working potential. However, the volume change in the course of the battery operation would cause severe mechanical damage of Si electrodes and leads to rapid capacity fading during cycling. One of the strategies is to incorporate CNTs into Si matrix, as discussed in Section 5.4.2.1. Alternatively, reducing the size of Si materials is expected to alleviate the pulverization problem because absolute local volume change is expected to be minimized in nanostructures, especially nanotubes.[474–478] Park *et al.*[478] reported that Si nanotubes showed very high reversible charge capacity of 3247 mAh g^{-1} with coulombic efficiency of 89%, and superior capacity retention even at 5C rate (=15 A g^{-1}). Yoo *et al.*[475] reported that the reversible charge capacity and coulombic efficiency can be enhanced to 3750 mAh g^{-1} and a coulombic efficiency of 97.5%, respectively. Wu *et al.*[476] designed a novel double-walled Si–SiO$_x$ nanotube (DWSiNT) anode, in which the inner wall is active silicon and the outer wall is confining SiO$_x$, which allows lithium ions to pass through. The DWSiNT structure successfully addressed the silicon material and solid–electrolyte interphase stability issues, and showed long cycle life (6000 cycles with 88% capacity retention), high specific charge capacity (2971/1780 mAh g^{-1} at 5C, 940/600 mAh g^{-1} at 12C; capacity based on silicon/total-DWSiNT weight, respectively) and fast charge/discharge rates (up to 20C) (Figure 5.42).

5.4.4 TiO$_2$ Nanotube

TiO$_2$ is also a promising anode material for LIBs because of its high working voltage (more than 1.5 V *vs* Li) and structural stability during lithium insertion/extraction due to small volume expansion (3%), which is free from the decomposition of organic electrolyte and short-circuit caused by the electrode swelling during cycling. However, the inherent poor rate capability of TiO$_2$-based anode limits its practical use. Generally, two approaches – incorporating active/conducting materials into TiO$_2$ (discussed in Section 5.4.2.2), and reducing its size – are employed to overcome the drawback. The porous TiO$_2$ nanostructures, especially nanotubes, demonstrate large lithium flux and short lithium diffusion length, respectively, which leads to an improvement in kinetics associated with lithium, because of the high interface area between active materials and the electrolyte, and the nanometer-sized dimension.[479–491] Bi *et al.*[479] reported that the TiO$_2$-NTs on Ti foam produced by electrochemical anodization techniques were used as anodes in lithium ion batteries; they exhibited high capacities of 103 mAh cm^{-1} at 10 mA cm^{-1} and 83 mAh cm^{-1} at 500 mA cm^{-1}, respectively. They further demostrated that the sample on Ti foam showed better performance than that on Ti foil. Li *et al.*[480] reported that the performance of TiO$_2$ NTs can be improved by using ionic liquids to anodize Ti. In addition to the growth conditions, the post-growth treatment, such as annealing, also strongly affects the performance of the TiO$_2$ NTs-based anode. Yan *et al.*[490] reported that increasing the annealing temperature (up to 500 °C) can improve the coulombic efficiency and lithium-ion transfer rate. Lee *et al.*[481]

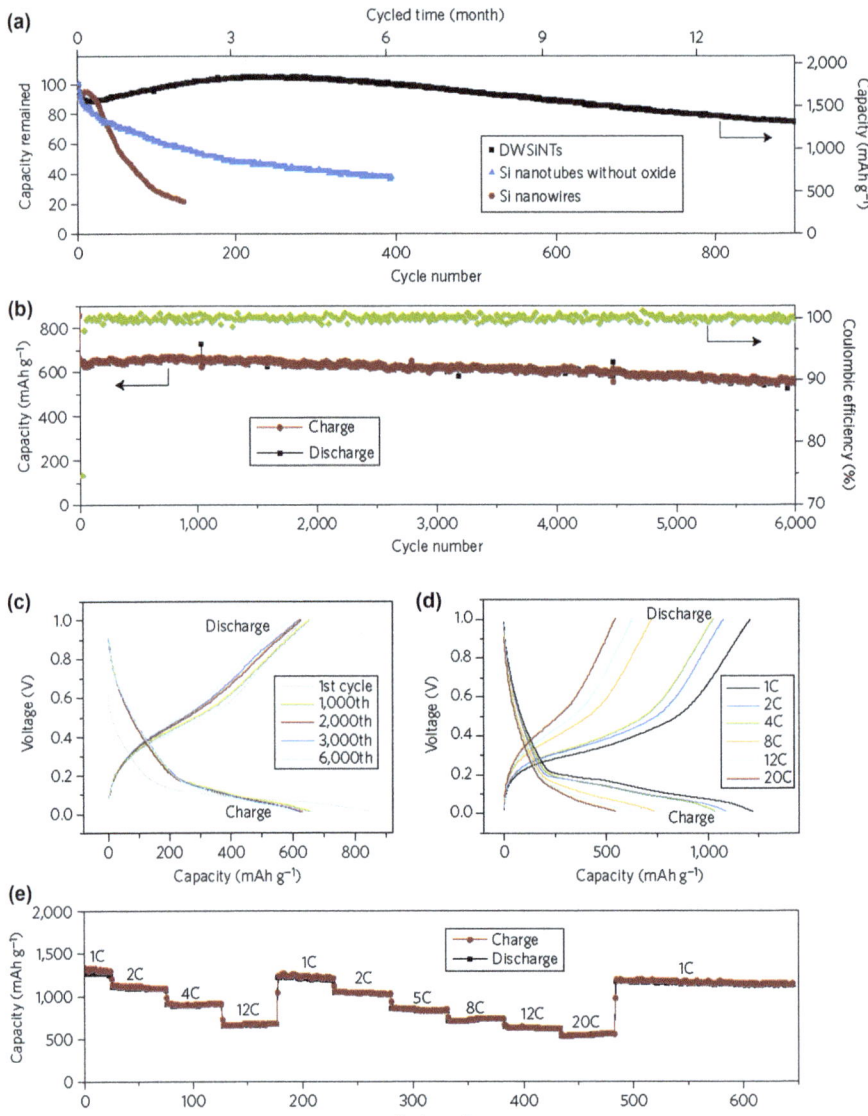

Figure 5.42 Electrochemical characteristics of DWSiNTs tested between 1 V and 0.01 V. (a) Capacity retention of different silicon nanostructures. All samples were cycled at the same charge/discharge rate of C/5. The calendar life and delithiation capacity of DWSiNTs can also be seen in this figure. (b) Lithiation/delithiation capacity and CE of DWSiNTs cycled at 12C for 6000 cycles. There is no significant capacity fading after 6000 cycles. (c) Voltage profiles plotted for the 1st, 1000th, 2000th, 3000th and 6000th cycles. (d),(e), Galvanostatic charge/discharge profiles (d) and capacity (e) of DWSiNTs cycled at various rates from 1C to 20C. All electrochemical measurements (a–e) were carried out at room temperature in two-electrode 2032 coin-type half-cells. All the specific capacities of DWSiNTs are reported based on the total weight of Si–SiO$_x$.[476] (Copyright: Nature publisher group).

reported that the TiO_2 NTs annealed in O_2 atmosphere showed better performance. The noticeably improved electrochemical performances were attributed to the crystalline transformation, and the appropriate morphology, such as large pore, thin wall, and good tube structure.[481,484–491] Han et al.[482] that TiO_2 NTs array grown directly on a current collector showed superior rate capability than conventional randomly oriented electrodes due to the reduction of the contact resistance between the electrode and current collector.

Beside CNTs, other materials, such as oxide, carbon, and graphene, incorporation also can improve the electrochemical performance of TiO_2-NTs-based anodes in LIBs.[492–500] Wang et al.[492] reported that the novel composites presented excellent electrochemical performance with high Li storage capacity (357 mAh g^{-1} at the rate of 10 mA g^{-1}, exceeding the theoretical capacity value 336 mAh g^{-1} of TiO_2) and excellent rate performance due to the synergetic and interactive effects, namely, the 'morphology' and 'electronic' interactions of both components. The TiO_2 nanotube/graphene composite also exhibited excellent rate capacities of 150 mAh g^{-1} (at the rate of 4000 mA g^{-1}) after 50 cycles and 80 mAh g^{-1} (at the rate of 8000 mA g^{-1}) after 2000 cycles; the coulombic efficiency was approximately 99.5%, indicating excellent cycling stability and reversibility. Park et al.[499] reported that carbon-coated TiO_2 nanotubes deliver a remarkable lithium-ion intercalation/deintercalation performance, such as reversible capacities of 286 and 150 mAh g^{-1} at 250 and 7500 mA g^{-1}, respectively, due to improved ionic and electronic kinetics. Similarly, the oxide/TiO_2 nanotube composites exhibit an enhanced performance with a higher capacity and rate capability at various test current densities than bare TiO_2 nanotubes because the oxide layer coating can not only increase the conductivity of TiO_2, but could also contribute to the overall reversible capacity during charge/discharge.[493–496]

5.4.5 SnO_2 Nanotube

SnO_2 has also been considered as a promising anode material for LIBs due to its reported high specific capacity, low lithium ion intercalation potential, non-toxicity, and low production cost. However, the practical application of SnO_2 as an anode material in LIBs has largely been hampered by its poor cycle performance and rate capability, which are caused by the huge volume change (300%) during charge and discharge, giving rise to pulverization of electrode materials and loss of electrical contact with the current collector. Using SnO_2 NTs as anodes in LIBs showed an improvement in the cyclic performance because the hollow interior structures could shorten the pathway lengths of lithium ions and accommodate the large volume change.[501–506] Wang et al. reported[502] that the ordered SnO_2 nanotube arrays showed a high discharge capacity of 790 mAh g^{-1} after 20 cycles with the retention of about 73%, which is higher than the theoretical capacity of SnO_2 anode. Wang et al.[501] reported that the porous SnO_2-NTs grown by a rapid microwave-assisted hydrothermal process showed high lithium ion storage

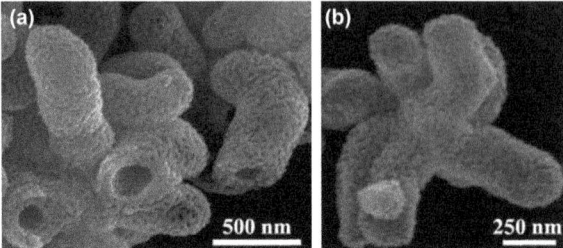

Figure 5.43 (a) and (b) SEM images of as-synthesized porous SnO_2 nanotubes.[502] (Copyright: Elsevier).

capacity (1258, 951, 757, 603, 458, and 288 mAh g^{-1}, at 0.1, 0.2, 0.5, 1, 2, and 4C, respectively) due to their unique geometry and porous structures (Figure 5.43).

5.5 Summary and Perspectives

This article presents a broad overview of the up-to-date development of nanotubes as energy storage materials. The effects of their structures/ morphologies, the functionalization/doping, complex composites, and ex- perimental methods on their performance in hydrogen storage, super- capacitor, and lithium battery, which have been attracting increasing interest from physicists, chemists, materials scientists, and particularly from industry, are systematically discussed based on two general basic pre- requisites on the design of materials for the improvement of performance: (1) high storage capacity and (2) stability or recyclability. For particular ap- plication, the specific criteria are also discussed.

For energy storage, the performance of the pure NT-based storage cells is closely related to the physical properties of the NTs, such as specific surface area, then, strongly depending on the synthesis and post-treatment methods of the NTs. The specific surface area, which can be accessed by H_2 in hydrogen storage, charges in the supercapacitor, and Li^+ in the battery, is one of the critical factors to improve the performance. Other factors, in- cluding pore size, pore-size distribution, alignment of NTs, length, and diameter, also affect their storage capacities. The conductivity of NTs is also very important for their applications in supercapacitors and lithium bat- teries. The performance of the storage cell can be improved by optimizing these factors. Functionalization, doping, and aligning and packing NTs can lead to high capacity through the improvement of pore-size distribution, conductivity, ion diffusivity, and addition of functional groups. Among all of the NTs, the carbon-related NTs (CNTs, BCN, BN, g-C_3N_4, *etc.*) show higher stability because of their super physical and chemical properties. The non- carbon NTs, such as oxide NTs, show high capacity, but much lower stability/ recyclability and lower conductivity (for supercapacitor and lithium battery).

The hybrid/ternary composites by optimally utilizing the advantages of NTs and other active phases represent an important breakthrough to develop new energy storage materials for supercapacitors and lithium batteries. The energy storage cell with high capacity and stability can be obtained by mixing the NTs with oxide, polymer, or both as the electrode. To achieve the purpose, the design of the composite, including the choice of original materials and their ratio, morphology, structure, and experimental methods, is particularly critical to the performance by considering the conductivity, ion diffusion, and charge storage.

It should be pointed out that the NT-based materials for energy storage have not reached our perspectives if considering cost and large-scale industry production, although significant progresses on NT-based energy storage materials have been achieved in the past decade. The design of new materials and composites for energy storage, such as porous NTs and their composites, may provide guidance for the improvement of performance (high capacity and reliability) with reduced cost and easy fabrication.

References

1. Basic research needs for solar energy utilization, Report of Department of Energy, USA. http://science.energy.gov/bes/news-and-resources/reports/abstracts/#SEU.
2. Advanced materials and devices for stationary electrical energy storage applications, Report of Department of Energy, USA. http://energy.gov/oe/downloads/advanced-materials-and-devices-stationary-electrical-energy-storage-applications.
3. L. F. Nazar, G. Goward, F. Leroux, M. Duncan, H. Huang, T. Kerr and J. Gaubicher, *Int. J. Inorg. Mater.*, 2001, **3**, 191.
4. A. S. Arico, P. Bruce, B. Scrosati, J. M. Tarascon and W. Van Schalkwijk, *Nature Mater.*, 2005, **4**, 366.
5. E. Serrano, G. Rus and J. Garcia-Martinez, *Renewable & Sustainable Energy Rev.*, 2009, **13**, 2373.
6. J. N. Tiwari, R. N Tiwari and K. S. Kim, *Prog. Mater. Sci.*, 2012, **57**, 724.
7. Q. F. Zhang, E. Uchaker, S. L. Candelaria and G. Z. Cao, *Chem. Rev. Soc.*, 2013, **42**, 3127.
8. X. B. Chen, C. Li, M. Graetzel, R. Kostecki and S. S. Mao, *Chem. Rev. Soc.*, 2012, **41**, 7909.
9. A. L. M. Reddy, S. R. Gowda, M. M. Shaijumon and P. M. Ajayan, *Adv. Mater.*, 2012, **24**, 5045.
10. F. Y. Cheng and J. Chen, *J. Mater. Res.*, 2006, **21**, 2744.
11. J. Liu, J. G. Zhang, Z. G. Yang, J. P. Lemmon, C. Imhoff, G. L. Graff, L. Y. Li, J. Z. Hu, C. M. Wang, J. Xiao, G. Xia, V. V. Viswanathan, S. Baskaran, V. Sprenkle, X. L. Li, Y. Y. Shao and B. Schwenzer, *Adv. Func. Mater.*, 2013, **23**, 929.
12. S. Iijima, *Nature*, 1991, **354**, 56.
13. H. Pan, J. Lin and Y. P. Feng, *Nanoscale Res. Lett.*, 2010, **5**, 654.

14. Z. L. Xiong, Y. S. Yun and H. J. Jin, *Materials*, 2013, **6**, 1138.

15. M. Paradise and T. Goswami, *Mater. Design*, 2007, **28**, 1477.

16. A. Goncharov, A. Guglya and E. Melnikova, *Int. J. Hydro. Energy*, 2012, **37**, 18061.

17. C. de las Casas and W. Z. Li, *J. Power Sources*, 2012, **208**, 74.

18. https://www1.eere.energy.gov/hydrogenandfuelcells/storage/storage_challenges.html.

19. D. K. Ross, *Vacuum*, 2006, **80**, 1084.

20. A. Züttel, *Naturwissenschaften*, 2004, **91**, 157.

21. A. Züttel, P. Wenger, S. Rentsch, P. Sudan, Ph Mauron and Ch. Emmenegger, *J. Power Sources*, 2003, **118**, 1.

22. S. Louis, in *Topics in Applied Physics*, Springer, Berlin, 1988, **vol. 63**, p. 350.

23. J. Y. Lin, H. Pan, and Y. P. Feng, Hydrogen Storage by Nanostructured Materials, in *Encyclopedia of Nanoscience and Nanotechnology*, ed. H. S. Nalwa, American Scientific Publisher, California, 2011, p. 225.

24. A. C. Dillon, K. M. Jones, T. A. Bekkedahl, C. H. Kiang, D. S. Bethune and M. J. Heben, *Nature*, 1997, **386**, 377.

25. C. Liu, Y. Y. Fan, M. Liu, H. T. Cong, H. M. Cheng and M. S. Dresselhaus, *Science*, 1999, **286**, 1127.

26. Y. Ye, C. C. Ahn, C. Witham, B. Fultz, J. Liu, A. G. Rinzler, D. Colbert, K. A. Smith and R. E. Smalley, *Appl. Phys. Lett.*, 1999, **74**, 2307.

27. P. X. Hou, Q. Yang, S. Bai, S. Xu, M. Liu and H. M. Cheng, *J. Phys. Chem. B*, 2002, **106**, 963.

28. M. Hirscher, M. Becher, M. Haluska, U. Detlaff-Weglikowska, A. Quintel, G. S. Duesberg, Y. M. Choi, P. Downes, M. Hulman, S. Roth, I. Stepanek and P. Bernier, *Appl. Phys. A*, 2001, **72**, 129.

29. M. Hirscher, M. Becher, M. Haluska, A. Quintel, V. Skakalova, Y. M. Choi, U. Dettlaff-Weglikowska, R. Roth, I. Stepanek, P. Bernier, A. Leonhardt and J. Fink, *J. Alloys Compounds*, 2002, **330–332**, 654.

30. J. Lawrence and G. Xu, *Appl. Phys. Lett.*, 2004, **84**, 918.

31. R. Bacsa, C. Laurent, R. Morishima, H. Suzuki and M. L. Lay, *J. Phys. Chem. B*, 2004, **108**, 12718.

32. G. Q. Ning, F. Wei, G. H. Luo, Q. X. Wang, Y. L. Wu and H. Yu, *Appl. Phys. A*, 2004, **78**, 955.

33. C. Liu and H. M. Cheng, *J. Phys. D: Appl. Phys.*, 2005, **38**, R231.

34. K. Murata, K. Kaneko, H. Kanoh, D. Kasuya, K. Takahashi, F. Kokai, M. Yudasaka and S. Iijima, *J. Phys. Chem. B*, 2002, **106**, 11132.

35. M. Ritschel, M. Uhlemann, O. Gutfleisch, A. Leonhardt, A. Graff, Ch. Taschner and J. Fink, *Appl. Phys. Lett.*, 2002, **80**, 2985.

36. G. Gundiah, A. Govindaraj, N. Rajalakshmi, K. S. Dhathathreyan and C. N. R. Rao, *J. Mater. Chem.*, 2003, **13**, 209.

37. H. Gao, X. B. Wu, J. T. Li, G. T. Wu, J. Y. Lin, K. Wu and D. S. Xu, *Appl. Phys. Lett.*, 2003, **83**, 3389.

38. L. Zhou, Y. Zhou and Y Sun, *Int. J. Hydro. Energy*, 2004, **29**, 475.

39. H. Kajiura, S. Tsutsui, K. Kadono, M. Kakuta, M Ata and Y. Murakami, *Appl. Phys. Lett.*, 2003, **82**, 1105.

40. A. Lan and A. Mukasyan, *J. Phys. Chem. B*, 2005, **109**, 16011.
41. M. Hirscher and B. Panella, *J. Alloys Compounds*, 2005, **404–406**, 399.
42. B. Panella, M. Hirscher and S. Roth, *Carbon*, 2005, **43**, 2209.
43. M. G. Nijkamp, J. E. M. J. Raaymakers, A. J. van Dillen and K. P. de Jong, *Appl. Phys. A*, 2001, **72**, 619.
44. A. Züttel, P. Sudan, Ph. Mauron, T. Kiyobayashi, Ch. Emmenegger and L. Schlapbach, *Int. J. Hydro. Energy*, 2002, **27**, 203.
45. H. G. Schimmel, G. J. Kearley, M. G. Nijkamp, C. T. Visser, K. P. de Jong and F. M. Mulder, *Chem. Eur. J.*, 2003, **9**, 4764.
46. H. G. Schimmel, G. Nijkamp, G. J. Kearley, A. Riveraa, K. P. de Jong and F. M. Mulder, *Mater. Sci. Eng. B*, 2004, **108**, 124.
47. Y. Zhou, K. Feng, Y. Sun and L. Zhou, *Chem. Phys. Lett.*, 2003, **380**, 526.
48. E. Poirier, R. Chahine, P. Benard, D. Cosseme, L. Lafile, E. Melanc, O. Bose and S. Desilets, *Appl. Phys. A*, 2004, **78**, 961.
49. B. K. Pradhan, G. U. Sumanasekera, K. W. Adu, H. E. Romero, K. A. Williams and P. C. Eklund, *Physica B*, 2002, **323**, 115.
50. K. A. Williams and P. C. Eklund, *Chem. Phys. Lett.*, 2000, **320**, 352.
51. S. S. Han, H. S. Kim, K. S. Han, J. Y. Lee, H. M. Lee, J. K. Kang, S. I. Woo, A. C. T. van Duin and W. A. Goddard III, *Appl. Phys. Lett.*, 2005, **87**, 213113.
52. M. Shiraishi, T. Takenobu, Y. Yamada, M. Ata and H. Kataura, *Chem. Phys. Lett.*, 2002, **358**, 213.
53. A. Anson, J. Jagiello, J. B. Parra, M. L. Sanjuan, A. M. Benito, W. K. Maser and M. T. Martınez, *J. Phys. Chem. B*, 2004, **108**, 15820.
54. M. Hirscher, M. Becher, M. Haluskaa, F. von Zeppelina, X. Chen, U. Dettlaff-Weglikowska and S. Roth, *J. Alloys Compounds*, 2003, **356/357**, 433.
55. F. Liu, X. Zhang, J. Cheng, J. Tu, F. Kong, W. Huang and C. Chen, *Carbon*, 2003, **41,,** 2527.
56. H. Takagi, H. Hatori, Y. Soneda, N. Yoshizawa and Y. Yamada, *Mater. Sci. Eng. B*, 2004, **108**, 143.
57. W. Z. Huang, X. B. Zhang, J. P. Tu, F. Z. Kong, J. X. Ma, F. Liu, H. M. Lu and C. P. Chen, *Mater. Chem. Phys.*, 2002, **78**, 144.
58. K. S. Han, H. S. Kim, M. S. Song, M. S. Park, S. S. Han, J. Y. Lee, J. K. Kang and Y. K. Kim, *Appl. Phys. Lett.*, 2005, **86**, 263105.
59. S. C. Mu, H. L. Tang, S. H. Qian, M. Pan and R. Z. Yuan, *Carbon*, 2006, **44**, 762.
60. B. K. Pradhan, A. R. Harutyunyan, D. Stojkovic, J. C. Grossman, P. Zhang, M. W. Cole, V. Crespi, H. Goto, J. Fujiwara and P. C. Eklund, *J. Mater. Res.*, 2002, **17**, 2209.
61. K. J. Ziegler, Z. Gu, H. Peng, E. L. Flor, R. H. Hauge and R. E. Smalley, *J. Am. Chem. Soc.*, 2005, **127**, 1541, and references therein.
62. S. M. Lee, S. C. Lee, J. H. Jung and H. J. Kim, *Chem. Phys. Lett.*, 2005, **416**, 251.

63. M. M. Shaijumon, N. Bejoy and S. Ramaprabhu, *Appl. Surf. Sci.*, 2005, **242**, 192.
64. M. M. Shaijumon and S. Ramaprabhu, *Chem. Phys. Lett.*, 2003, **374**, 513.
65. M. M. Shaijumon, N. Rajalakshmi, H. Ryu and S. Ramaprabhu, *Nanotechnol.*, 2005, **16**, 518.
66. A. Cao, H. Zhu, X. Zhang, X. Li, D. Ruan, C. Xu, B. Wei, J. Liang and D. Wu, *Chem. Phys. Lett.*, 2001, **342**, 510.
67. P. Chen, X. Wu, J. Lin and K. L. Tan, *Science*, 1999, **285**, 91.
68. R. T. Yang, *Carbon*, 2000, **38**, 623.
69. C. Ahn, Division of Engineering and Applied Science 2003 Progress Report, http://www1.eere.energy.gov/hydrogenandfuelcells/pdfs/merit03/57_caltech_channing_ahn.pdf.
70. P. M. F. J. Costa, K. S. Coleman and M. L. H. Green, *Nanotechnol.*, 2005, **16**, 512.
71. T. Yildirim and S. Ciraci, *Phys. Rev. Lett.*, 2005, **94**, 175501.
72. S. C. Mu, H. L. Tang, S. H. Qian, M. Pan and R. Z. Yuan, *Carbon*, 2006, **44**, 762.
73. H. S. Kim, H. Lee, K. S. Han, J. H. Kim, M. S. Song, M. S. Park, J. Y. Lee and J. K. Kang, *J. Phys. Chem. B*, 2005, **109**, 8983.
74. Z. Zhong, Z. Xiong, L. Sun, J. Luo, P. Chen, X. Wu, J. Y. Lin and K. L. Tan, *J. Phys. Chem. B*, 2002, **106**, 9507.
75. R. Zacharia, K. Y. Kim, A. K. M. F. Kibria and K. K. Nahm, *Chem Phys. Lett.*, 2005, **412**, 369.
76. A. J. Lachawiec Jr., G. Qi and R. T. Yang, *Langmuir*, 2005, **21**, 11418.
77. F. H. Yang and R. T. Yang, *Carbon*, 2002, **40**, 437.
78. A. Lueking and R. T. Yang, *J. Catal.*, 2002, **206**, 165.
79. A. Lueking and R. T. Yang, *AIChE J.*, 2003, **49**, 1556.
80. A. Lueking and R. T. Yang, *Appl. Catal. A*, 2004, **265**, 259.
81. P. J. Tsai, C. H. Yang, W. C. Hsu, W. T. Tsai and J. K. Chang, *Int. J. Hydro. Energy*, 2012, **37**, 6714.
82. F. H. Yang, A. J. Lachawiec Jr. and R. T. Yang, *J. Phys. Chem. B*, 2006, **110**, 6236.
83. E. Yoo, L. Gao, T. Komatsu, N. Yagai, K. Arai, T. Yamazaki, K. Matsuishi, T. Matsumoto and J. Nakamura, *J. Phys. Chem. B*, 2004, **108**, 18903.
84. H. M. Wu, D. Wexler and H. Liu, *Int. J. Hydro. Energy*, 2011, **36**, 9032.
85. H. Cheng, A. C. Cooper and D. P. Pez, *J. Chem. Phys.*, 2004, **120**, 9427.
86. S. Z. Mortazavi, P. Parvin, A. Reyhani, R. Malekfar and S. Mirershadi, *RSC Adv.*, 2013, **3**, 1397.
87. H. M. Wu, D. Wexler and H. K. Liu, *Int. Hydro. Energy*, 2012, **37**, 5686.
88. A. Reyhani, S. Z. Mortazavi, S. Mirershadi, A. Z. Moshfegh, P. Parvin and A. N. Golikand, *J. Phys. Chem. C*, 2011, **115**, 6994.
89. K. Y. Lin, W. T. Tsai and J. K. Chang, *Int. J. Hydro. Energy*, 2010, **35**, 7555.
90. R. Tenne, L. Margulis, M. Genut and G. Hodes, *Nature*, 1992, **360**, 444.

91. R. Ma, Y. Bando, H. Zhu, T. Sato, C. Xu and D. Wu, *J. Am. Chem. Soc.*, 2002, **124**, 7672.
92. N. G. Chopra, R. J. Luyken, K. Cherrey, V. H. Crespi, M. L. Cohen, S. G. Louie and A. Zettl, *Science*, 1995, **269**, 966.
93. J. Goldberger, R. He, Y. Zhang, S. Lee, H. Yan, H. Choi and P. Yang, *Nature*, 2003, **422**, 599.
94. J. Chen, S. L. Li, Z. L. Tao, Y. T. Shen and C. X. Cui, *J. Am. Chem. Soc.*, 2003, **125**, 5284.
95. X. Wang and Y. Li, *J. Am. Chem. Soc.*, 2002, **124**, 2880.
96. L. Q. Mai, W. Chen, Q. Xu, Q. Y. Zhu, C. H. Han and J. F. Peng, *Solid State Commun.*, 2003, **126**, 541.
97. M. Wei, H. Zhou, H. Sugihara, Y. Konishi and H. Arakawa, *Solid State Commun.*, 2005, **133**, 493.
98. J. Cumings and A. Zettl, *Chem. Phys. Lett.*, 2000, **316**, 211.
99. D. Golberg, Y. Bando, M. Eremets, K. Takemura, K. Kurashima and H. Yusa, *Appl. Phys. Lett.*, 1996, **69**, 2045.
100. D. P. Yu, X. S. Sun, C. S. Lee, I. Bello, S. T. Lee, H. D. Gu, K. M. Leung, G. W. Zhou, Z. F. Dong and Z. Zhang, *Appl. Phys. Lett.*, 1996, **72**, 1966.
101. R. Sen, B. C. Satishkumar, A. Govindaraj, K. R. Harikumar, G. Raina, J. P. Zhang, A. K. Cheetham and C. N. R. Rao, *Chem. Phys. Lett.*, 1998, **287**, 671.
102. J. L. Wang, L. P. Zhang, Y. L. Gu, X. Y. Pan, G. W. Zhao and Z. H. Zhang, *J. Exp. Nanosci.*, 2013, **8**, 669.
103. C. Tang, Y. Bando, X. Ding, S. Qi and D. Golberg, *J. Am. Chem. Soc.*, 2002, **124**, 14550.
104. W. Han, Y. Bando, K. Kurashima and T. Sato, *Appl. Phys. Lett.*, 1998, **73**, 3085.
105. C. Tang, Y. Bando and T. Sato, *Chem. Phys. Lett.*, 2002, **362**, 185.
106. S. Lim, J. Luo, W. Ji and J. Lin, *Catal. Today*, 2007, **120**, 346.
107. T. Oku and M. Kuno, *Diamond Relat. Mater.*, 2003, **12**, 840.
108. T. Oku and I. Narita, *Physica B*, 2002, **323**, 216.
109. R. Ma, Y. Bando, T. Sato, D. Golberg, H. Zhu, C. Xu and D. Wu, *Appl. Phys. Lett.*, 2002, **81**, 5225.
110. S. Jhi and Y. Kwon, *Phys. Rev. B.*, 2004, **69**, 245407.
111. P. Wang, S. Orimo, T. Matsushima, H. Fujii and G. Majer, *Appl. Phys. Lett.*, 2002, **80**, 318.
112. A. Mavrandonakis, G. E. Froudakis, M. Schnell and M. Mühlhaüser, *Nano Lett.*, 2003, **3**, 1481.
113. M. Menon, E. Richter and A. N. Andriotis, *Phys. Rev. B.*, 2004, **69**, 115322.
114. G. Mpourmpakis, G. E. Froudakis and J. Samios, *Nano. Lett.*, 2006, **6**, 1581.
115. X. Wang and K. M. Liew, *J. Phys. Chem. C*, 2011, **115**, 3491.
116. C. Pham-Huu, N. Keller, G. Ehret and M. J. Ledoux, *J. Catal.*, 2001, **200**, 400.

117. X. H. Sun, C. P. Li, W. K Wong, N. B. Wong, C. S. Lee, S. T. Lee and B. K. Teo, *J. Am. Chem. Soc.*, 2002, **124**, 14464.

118. H. Dai, E. W. Wang, Y. Z. Lu, S. S. Fan and C. M. Lieber, *Nature*, 1995, **375**, 769.

119. G. W. Meng, L. D. Zhang, C. M. Mo, S. Y. Zhang, Y. Qin, S. P. Feng and H. J. Li, *J. Mater. Res.*, 1998, **13**, 2533.

120. X. T. Zhou, N. Wang, H. L. Lai, H. Y. Peng, I. Bello, N. B. Wong, C. S. Lee and S. T. Lee, *Appl. Phys. Lett.*, 1999, **74**, 3942.

121. Y. B. Li, S. S. Xie, X. P. Zou, D. S. Tang, Z. Q. Liu, W. Y. Zhou and G. Wang, *J. Cryst. Growth*, 2001, **223**, 125.

122. G. V. Pol, S. V. Pol, A. Gedanken, S. H. Lim, Z. Zhong and J. Lin, *J. Phys. Chem. B*, 2006, **110**, 11237.

123. G. V. Pol, S. V. Pol and A. Gedanken, *Adv. Mater.*, 2006, **18**, 2023.

124. H. Pan, Y. P. Feng and J. Y. Lin, *Appl. Phys. Lett.*, 2007, **90**, 223104.

125. S. H. Jhi and Y. K. Kwon, *Phys. Rev. B*, 2004, **69**, 245407.

126. A. M. Seayad and D. V. Antonelli, *Adv. Mater.*, 2004, **16**, 765.

127. E. Kroke and M. Schwarz, *Coord. Chem. Rev.*, 2004, **248**, 493.

128. M. Liu and J.-G. Wang, *J. Inorganic Mater.*, 2009, **24**, 491.

129. H. Pan, Y. W. Zhang, V. B. Shenoy and H. Gao, *ACS Catal.*, 2011, **1**, 99.

130. H. Pan, Y. W. Zhang, V. B. Shenoy and H. Gao, *Nanoscale Res. Lett.*, 2011, **6**, 97.

131. G. Y. Koh, Y. W. Zhang and H. Pan, *Int. J. Hydro. Energy*, 2012, **37**, 4170.

132. M. Hershfinkel, L. A. Gheber, V. Volterra, J. L. Hutchison, L. Margulis and R. Tenne, *J. Am. Chem. Soc.*, 1994, **116**, 1914.

133. Y. Feldman, E. Wasserman, D. J. Srolovitz and R. Tenne, *Science*, 1995, **267**, 222.

134. M. Remskar, A. Mrzel, Z. Skraba, A. Jesih, M. Ceh, J. Demsar, P. Stadelmann, F. Levy and D. Mihailovic, *Science*, 2001, **292**, 479.

135. M. Nath and C. N. R. Rao, *J. Am. Chem. Soc.*, 2001, **123**, 4841.

136. R. Tenne, *Chem. Eur. J.*, 2002, **8**, 5297.

137. J. Chen, S. L. Li and Z. L. Tao, *J. Alloys Compd.*, 2003, **356–357**, 413.

138. J. Chen, Z. L. Tao and S. L. Li, *Angew. Chem. Int. Ed.*, 2003, **42**, 2147.

139. J. Chen and W. Fu, *Appl. Phys. A*, 2004, **78**, 989.

140. Q. Wan, C. L. Lin, X. B Yu and T. H. Wang, *Appl. Phys. Lett.*, 2004, **84**, 124.

141. H. Pan, J. Z. Luo, H. Sun, Y. P. Feng, C. Poh and J. Y. Lin, *Nanotechnol.*, 2006, **17**, 2963.

142. H. Hoyer, *Langmuir*, 1996, **12**, 1411.

143. M. Adachi, Y. Murata and M. Harada, *Chem. Lett.*, 2000, **8**, 942.

144. T. Kasuga, M. Hiramatsu, A. Hoson, T. Sekino and K. Niihara, *Langmuir*, 1998, **14**, 3160.

145. S. Lim, J. Luo, W. Ji and J. Lin, *Inorg. Chem.*, 2005, **44**, 4124.

146. R. Zacharia, S. U. Rather, S. W. Hwang and K. S. Nahm, *Chem. Phys. Lett.*, 2007, **434**, 286.

147. E. Tylianakis, G. M. Psofogiannakis and G. F. Froudakis, *J. Phys. Chem. Lett.*, 2010, **1**, 2459.

148. J. Hong and S. W. Kang, *Colloids Surf., A*, 2011, **374**, 54.
149. S. H. Aboutalebi, S. Aminorroaya-Yamini, I. Nevirkovets, K. Konstantinov and H. K. Liu, *Adv. Energy Mater.*, 2012, **2**, 1439.
150. N. L. Rosi, J. Eckert, E. Eddaoudi, D. T. Vodak, J. Kim, M. O'Keeffe and O. M. Yaghi, *Science*, 2003, **300**, 1127.
151. O. M. Yaghi, G. Li and H. Li, *Nature*, 1995, **378**, 703.
152. G. B. Gardner, D. Venkataraman, J. S. Moore and S. Lee, *Nature*, 1995, **374**, 792.
153. T. P. Vaid, E. B. Lobkovsky and P. T. Wolczanski, *J. Am. Chem. Soc.*, 1997, **119**, 8742.
154. R. M. Kumar, J. V. Sundar and V. Subramanian, *Int. J. Hydrogen Energy*, 2012, **37**, 16070.
155. L. R. MacGillivray, R. H. Groeneman and J. L. Atwood, *J. Am. Chem. Soc.*, 1998, **120**, 2676.
156. S. J. Yang, J. Y. Choi, H. K. Chae, J. H. Cho, K. S. Nahm and C. R. Park, *Mater. Chem.*, 2009, **21**, 1893.
157. B. E. Conway, *Electrochemical supercapacitors: scientific fundamentals and technological applications*, Kluwer Academic/Plenum, New York, 1999.
158. Website: http://en.wikipedia.org/wiki/Supercapacitor.
159. K. Lota, V. Khomenko and E. J. Frackowiak, *J. Phys. Chem. Solids*, 2004, **65**, 295.
160. S. K. Ryu, X. Wu, Y. G. Lee and S. H. Chang, *J. Appl. Polym. Sci.*, 2003, **89**, 1300.
161. S. Patra and N. Munichandraiah, *J. Appl. Polymer Sci.*, 2007, **106**, 1160.
162. F. Marchioni, J. Yang, W. Walker and F. Wudl, *J. Phys. Chem. B*, 2006, **110**, 22202.
163. K. S. Ryu, Y. G. Lee, Y. S. Hong, Y. J. Park, X. L. Wu, K. M. Kim, M. G. Kang, N. G. Park and S. H. Chang, *Electrochim. Acta*, 2004, **50**, 843.
164. M. D. Ingram, H. Staesche and K. S. Ryder, *Solid State Ionics*, 2004, **169**, 51.
165. K. R. Prasad and N. Munichandraiah, *Electrochem. Solid State Lett.*, 2002, **5**, A271.
166. J. Y. Kim and I. J. Chung, *J. Electrochem. Soc.*, 2002, **140**, A1376.
167. K. S. Ryu, K. M. Kim, N. G. Park, Y. J. Park and S H. Chang, *J. Power Sources*, 2002, **103**, 305.
168. J. C. Lassegues, J. Grondin, T. Becker, L. Servant and M. Hernandez, *Solid State Ionics*, 1995, 77, 311.
169. H. C. Chang and C. C. Hu, *Electrochem. Solid State Lett.*, 2004, 7, A466.
170. D. Rochefort and D. Guay, *J. Alloys Compd.*, 2005, **400**, 257.
171. J. N. Broughton and M. J. Brett, *Electrochim. Acta*, 2005, **49**, 4439.
172. P. Ragupathy, H. N. Vasan and N. Munichandraiah, *J. Power Sources*, 2008, **155**, A34.
173. K. Macounová, I. Jirka, A. Trojánek, M. Makarova, Z. Samec and P. Krtil, *J. Electrochem. Soc.*, 2007, **154**, A1077.
174. M. W. Xu, D. D. Zhao, S. J. Bao and H. L. Li, *J. Solid State Electrochem.*, 2007, **11**, 1101.

175. T. P. Gujar, V. R. Shinde, C. D. Lokhande, W. Y. Kim, K. D. Jung and O. S. Joo, *Electrochem. Commun.*, 2007, **9**, 504.

176. S. L. Kuo, J. F. Lee and N. L. Wu, *J. Electrochem. Soc.*, 2007, **154**, A34.

177. K. Yokoshima, T. Shibutani, M. Hirota, W. Sugimoto, Y. Murakami and Y. Takasu, *J. Power Sources*, 2006, **160**, 1480.

178. A. Jayalakshmi, N. Venugopal, K. P. Raja and M. M. Rao, *J. Power Sources*, 2006, **158**, 1538.

179. S. H. Lee, C. E. Tracy and J. R. Pitts, *Electrochem. Solid State Lett.*, 2004, **7**, A299.

180. Y. B. Mo, M. R. Antonio and D. A. Scherson, *J. Phys. Chem. B*, 2000, **104**, 9777.

181. J. M. Miller, B. Dunn, T. D. Tran and R. W. Pekala, *J. Electrochem. Soc.*, 1997, **144**, L309.

182. B. E. Conway, *J. Electrochem. Soc.*, 1991, **138**, 1539.

183. J. H. Jiang and A. Kucernak, *Electrochim. Acta*, 2002, **47**, 2381.

184. L. M. Huang, H. Z. Lin, T. C. Wen and A. Gopalan, *Electrochim. Acta*, 2006, **52**, 1058.

185. J. W. Long, K. E. Swider, C. I. Merzbacher and D. R. Rolison, *Langmuir*, 1999, **15**, 780.

186. H. I. Becker. US. Pat. 2800616, 1957.

187. E. Frackowiak and F. Beguin, *Carbon*, 2001, **39**, 937.

188. E. Frackowiak, *Phys. Chem. Chem. Phys.*, 2007, **9**, 1774.

189. J. Chmiola, G. Yushin, Y. Gogotsi, C. Portet, P. Simon and P. L. Taberna, *Science*, 2006, **313**, 1760.

190. E. Raymundo-Piñero, F. Leroux and F. Béguin, *Adv. Mater.*, 2006, **18**, 1877.

191. A. B. Fuertes, G. Lota, T. A. Centeno and E. Frackowiak, *Electrochim. Acta*, 2005, **50**, 2799.

192. D. Hulicova, J. Yamashita, Y. Soneda, H. Hatori and M. Kodama, *Chem. Mater.*, 2005, **17**, 1241.

193. K. Leitner, A. Lerf, M. Winter, J. O. Besenhard, S. Villar-Rodil, F. Suarez-Garci, A. Martinez-Alonso and J. M. D. Tascon, *J. Power Sources*, 2000, **153**, 419.

194. P. J. Mahon, G. L. Paul, S. M. Keshishian and A. M. Vassallo, *J. Power Sources*, 2000, **91**, 68.

195. N. L. Wu and S. Y. Wang, *J. Power Sources*, 2002, **110**, 233.

196. C. Portet, P. L. Taberna, P. Simon and C. Laberty-Robert, *Electrochim. Acta*, 2004, **49**, 205.

197. C. Kim and K. S. Yang, *Appl. Phys. Lett.*, 2003, **83**, 1216.

198. C. Vix-Guterl, S. Saadallah, K. Jurewicz, E. Frackowiak, M. Reda, J. Parmentier, J. Patarin and F. Beguin, *Mater. Sci. Eng. B*, 2004, **108**, 148.

199. D. Kalpana, K. S. Omkumar, S. S. Kumar and N. G. Renganathan, *Electrochim. Acta*, 2006, **52**, 2309.

200. M. Min, K. Machida, J. H. Jang and K. Naoi, *J. Electrochem. Soc.*, 2006, **153**, A334.

201. C. Arbizzani, M. Mastragostino, L. Meneghello and R. Paraventi, *Adv. Mater.*, 1996, **8**, 331.
202. J. M. Miller and B. Dunn, *Langmuir*, 1999, **15**, 799.
203. S. Iijima, *Nature*, 1991, **354**, 56.
204. M. S. Dresselhaus, G. Dresselhaus and P. C. Eklund, *Science of fullerenes and carbon nanotubes*, Academic, New York/San Diego, 1996.
205. R. Saito, G. Dresselhaus and M. S. Dresselhaus, *J. Appl. Phys.*, 1993, **73**, 494.
206. M. S. Dresselhaus, D. Dresselhaus and R. Saito, *Solid State Commun.*, 1992, **84**, 201.
207. J. P. Issi, L. Langer, J. Heremans and C. H. Olk, *Carbon*, 1995, **33**, 941.
208. T. W. Ebbesen, H. J. Lezec, H. Hiura, J. W. Bennett, H. F. Ghaemi and T. Thio, *Nature*, 1996, **382**, 54.
209. C. Niu, E. K. Sichel, R. Hoch, D. Moy and H. Tennent, *Appl. Phys. Lett.*, 1997, **70**, 1480.
210. K. H. An, W. S. Kim, Y. S. Park, Y. C. Choi, S. M. Lee, D. C. Chung, D. J. Bae, S. C. Lim and Y. H. Lee, *Adv. Mater.*, 2001, **13**, 497.
211. C. S. Du and N. Pan, *Nanotechnol.*, 2006, **17**, 5314.
212. E. Frackowiak, K. Metenier, V. Bertagna and F. Beguin, *Appl. Phys. Lett.*, 2000, 77, 2421.
213. A. L. M. Reddy, F. E. Amitha, I. Jafri and S. Ramaprabhu, *Nanoscale Res. Letts.*, 2008, **3**, 145.
214. C. G. Liu, M. Liu, F. Li and H. M. Cheng, *Appl. Phys. Lett.*, 2008, **92**, 143108.
215. J. N. Barisci, G. G. Wallace, D. Chattopadhyay, F. Papadimitrakopoulos and R. H. Baughman, *J. Electrochem. Soc.*, 2003, **150**, E409.
216. J. N. Barisci, G. G. Wallace and R. H. Baughman, *J. Electrochem. Soc.*, 2000, **147**, 4580.
217. S. Shiraishi, H. Kurihara, K. Okabe, D. Hulicova and A. Oya, *Electrochem. Commun.*, 2002, **4**, 593.
218. H. Zhang, G. P. Cao and Y. S. Yang, *Nanotechnol.*, 2007, **18**, 195607.
219. H. Zhang, G. P. Cao, Y. S. Yang and Z. N. Gu, *J. Electrochem. Soc.*, 2008, **155**, K19.
220. R. Reita, J. Nguyen and W. J. Ready, *Electrochim. Acta*, 2013, **91**, 96.
221. J. Ren, L. Li, C. Chen, X. L. Chen, Z. B. Cai, L. B. Qiu, Y. G. Wang, X. R. Zhu and H. S. Peng, *Adv. Fun. Mater.*, 2013, **25**, 1155.
222. D. N. Futaba, K. Hata, T. Yamada, T. Hiraoka, Y. Hayamizu, Y. Kakudate, O. Tanaike, H. Hatori, M. Yumura and S. Iijima, *Nature Mater.*, 2006, **5**, 987.
223. A. Izadi-Najafabadi, S. Yasuda, K. Kobashi, T. Yamada, D. N. Futaba, H. Hatori, M. Yumura, S. Iijima and K. Hata, *Adv. Mater.*, 2010, **22**, E235.
224. A. Izadi-Najafabadi, D. N. Futaba, S. Iijima and K. Hata, *J. Am. Chem. Soc.*, 2010, **132**, 18017.
225. C. S. Du and N. Pan, *J. Power Sources*, 2006, **160**, 1487.
226. C. S. Du, J. Yeh and N. Pan, *Nanotechnol.*, 2005, **16**, 350.

227. H. Pan, C. K. Poh, Y. P. Feng and J. Lin, *Chem. Mater.*, 2007, **19**, 6120.

228. H. Pan, Y. P. Feng and J. Lin, *Phys. Rev. B*, 2004, **70**, 245425.

229. E. Frackowiak, K. Jurewicz, S. Delpeux and F. Beguin, *J. Power Sources*, 2001, **97–98**, 822.

230. C. Li, D. Wang, T. Liang, X. Wang, J. Wu, X. Hu and J. Liang, *Powder Technol.*, 2004, **142**, 175.

231. K. H. An, W. S. Kim, Y. S. Park, J. M. Moon, D. J. Bae, S. C. Lim, Y. S. Lee and Y. H. Lee, *Adv. Func. Mater.*, 2001, **11**, 387.

232. E. Frackowiak, S. Delpeux, K. Jurewicz, K. Szostak, D. Cazorla-Amoros and F. Beguin, *Chem. Phys. Lett.*, 2002, **261**, 35.

233. C. G. Liu, H. T. Fang, F. Li, M. Liu and H. M. Cheng, *J. Power Sources*, 2006, **160**, 758.

234. Y. T. Kim, Y. Ito, K. Tadai, T. Mitani, U. S. Kim, H. S. Kim and B. W. Cho, *Appl. Phys. Lett.*, 2005, **87**, 234106.

235. J. Y. Lee, K. H. An, J. K. Heo and Y. H. Lee, *J. Phys. Chem. B*, 2003, **107**, 8812.

236. H. Pan, Y. P. Feng and J. Lin, *J. Phys.: Condens. Matter.*, 2006, **18**, 5175.

237. C. Zhou, S. Kumar, C. D. Doyle and J. M. Tour, *Chem. Mater.*, 2005, **17**, 1997.

238. B. J. Yoon, S. H. Jeong, K. H. Lee, H. S. Kim, C. G. Park and J. H. Han, *Chem. Phys. Lett.*, 2004, **380**, 170.

239. L. H. Su, X. G. Zhang, C. Z. Yuan and B. Gao, *J. Electrochem. Soc.*, 2008, **155**, A110.

240. C. Y. Lee, H. M. Tsai, H. J. Chuang, S. Y. Li, P. Lin and T. Y. Tseng, *J. Electrochem. Soc.*, 2005, **152**, A716.

241. G. Arabale, D. Wagh, M. Kulkarni, I. S. Mulla, S. P. Vernekar, K. Vijayamohanan and A. M. Rao, *Chem. Phys. Lett.*, 2003, **376**, 207.

242. K. Liang, K. An and Y. Lee, *J. Mater. Sci. Eng.*, 2005, **21**, 292.

243. M. Hughes, G. Z. Chen, M. S. P. Shaffer, D. J. Fray and A. H. Windle, *Chem. Mater.*, 2002, **14**, 1610.

244. H. K. Chae, D. Y. Siberio-Perez, J. Kim, Y. Go, M. Eddaoudi, A. J. Matzger, M. O'Keeffe and O. M. Yaghi, *Nature*, 2004, **427**, 523.

245. C. G. Lee, X. D. Wei, J. W. Kysar and J. Hone, *Science*, 2008, **321**, 385.

246. D. W. Wang, F. Li, J. P. Zhao, W. C. Ren, Z. G. Chen, J. Tan, Z. S. Wu, I. Gentle, G. Q. Lu and H. M. Cheng, *ACS Nano.*, 2009, **3**, 1745.

247. Q. Wu, Y. X. Xu, Z. Y. Yao, A. R. Liu and G. Q. Shi, *ACS Nano.*, 2010, **4**, 1963.

248. A. P. Yu, I. Roes, A. Davies and Z. W. Chen, *Appl. Phys. Lett.*, 2010, **96**, 253105.

249. Y. Wang, Z. Q. Shi, Y. Huang, Y. F. Ma, C. Y. Wang, M. M. Chen and Y. S. Chen, *J. Phys. Chem. C*, 2009, **113**, 13103.

250. W. Lv, D. M. Tang, Y. B. He, C. H. You, Z. Q. Shi, X. C. Chen, C. M. Chen, P. X. Hou, C. Liu and Q. H. Yang, *ACS Nano.*, 2009, **3**, 3730.

251. Y. W. Zhu, S. Murali, M. D. Stoller, A. Velamakanni, R. D. Piner and R. S. Ruoff, *Carbon*, 2010, **7**, 2118.

252. Q. L. Du, M. B. Zheng, L. F. Zhang, Y. W. Wang, J. H. Chen, L. P. Xue, W. J. Dai, G. B. Ji and J. M. Cao, *Electrochim. Acta*, 2010, **30**, 3897.

253. M. D. Stoller, S. J. Park, Y. W. Zhu, J. H. An and R. S. Ruoff, *Nano Lett.*, 2008, **8**, 3498.

254. Z. J. Fan, J. Yan, L. J. Zhi, Q. Zhang, T. Wei, J. Feng, M. L. Zhang, W. Z. Qian and F. Wei, *Adv. Mater.*, 2010, **22**, 3723.

255. D. S. Yu and L. M. Dai, *J. Phys. Chem. Lett.*, 2010, **1**, 467.

256. X. J. Lu, H. Dou, B. Gao, C. Z. Yuan, S. D. Yang, L. Hao, L. F. Shen and X. G. Zhang, *Electrochim. Acta*, 2011, **56**, 5115.

257. M. Q. Zhao, Q. Zhang, J. Q. Huang, G. L. Tian, T. C. Chen, W. Z. Qian and F. Wei, *Carbon*, 2013, **54**, 403.

258. L. Xu, N. Wei, Y. Zheng, Z. Fan, H. Q. Wang and J. C. Zheng, *J Mater. Chem.*, 2011, **22**, 1435.

259. F. Du, D. Yu, L. Dai, S. Ganguli, V. Varshney and A. K. Roy, *Chem. Mater.*, 2011, **23**, 4810.

260. Y. P. Wu, T. F. Zhang, F. Zhang, Y. Wang, Y. F. Ma, Y. Huang, Y. Y. Liu and Y. S. Chen, *Nano Energy*, 2012, **1**, 820.

261. Y. Wang, Y. Wu, Y. Huang, F. Zhang, Y. Yang, Y. F. Ma and Y. S. Chen, *J. Phys. Chem. C*, 2011, **115**, 23192.

262. L. L. Zhang, Z. G. Xiong and X. S. Zhao, *J. Power Sources*, 2013, **222**, 326.

263. H. C. Gao, F. Xiao, C. B. Ching and H. W. Duan, *ACS Appl. Mater. Interfaces*, 2012, **4**, 7020.

264. H. R. Byon, S. W. Lee, S. Chen, P. T. Hammond and Y. Shao-Horn, *Carbon*, 2011, **49**, 457.

265. Z. W. Xu, Z. Li, C. M. B. Holt, X. H. Tan, H. L. Wang, B. S. Amirkhiz, T. Stephenson and D. Mitlin, *J. Phys. Chem. Lett.*, 2012, **3**, 2928.

266. J. J. Li, Y. W. Ma, X. Jiang, X. M. Feng, Q. L. Fan and W. Huang, *IEEE Tran. Nanotechnol.*, 2012, **11**, 3.

267. Z. J. Fan, J. Yan, L. Zhi, Q. Zhang, T. Wei, J. Feng, M. L. Zhang, W. Z. Qian and F. Wei, *Adv. Mater.*, 2010, **22**, 3723.

268. J. H. Park, J. M. Ko and O. O. Parka, *J. Electrochem. Soc.*, 2003, **150**, A864.

269. H. Kim, J. H. Kim and K. B. Kim, *Electrochem. Solid State Lett.*, 2005, **8**, A369.

270. Y. T. Kim, K. Tadai and T. Mitani, *J. Mater. Chem.*, 2005, **15**, 4914.

271. Y. T. Kim and T. Mitani, *Appl. Phys. Lett.*, 2006, **89**, 033107.

272. W. C. Fang, M. S. Leu, K. H. Chen and L. C. Chen, *J. Electrochem. Soc.*, 2008, **155**, K15.

273. W. C. Fang, K. H. Chen and L. C. Chen, *Nanotechnol.*, 2007, **18**, 485716.

274. E. Raymundo-Pinero, V. Khomenko, E. Frackowiak and F. Beguin, *J. Electrochem. Soc.*, 2005, **152**, A229.

275. V. Subramanian, H. Zhu and B. Wei, *Electrochem. Commun.*, 2006, **8**, 827.

276. S. B. Ma, K. W. Nam, W. S. Yoon, X. Q. Yang, K. Y. Ahn, K. H. Oh and K. B. Kim, *J. Power Sources*, 2008, **178**, 483.

277. X. Xie and L. Gao, *Carbon*, 2007, **45**, 2365.

278. Y. J. Kang, H. Chung and W. Kim, *Synthetic Metals*, 2013, **166**, 40.

279. L. Li, C. Chen, J. Xie, Z. H. Shao and F. X. Yang, *J. Nanomater.*, 2013, 821071.
280. H. J. Zheng, J. X. Wang, Y. Jia and C. A. Ma, *J. Power Sources*, 2012, **216**, 508.
281. J. M. Shen, A. D. Liu, Y. Tu, H. Wang, R. R. Jiang, J. Ouyang and Y. Chen, *Electrochim. Acta*, 2012, **78**, 122.
282. Q. Li, J. M. Anderson, Y. Q. Chen and L. Zhai, *Electrochim. Acta*, 2012, **59**, 548.
283. L. B. Hu, W. Chen, X. Xie, N. A. Liu, Y. Yang, H. Wu, Y. Yao, M. Pasta, N. H. Alshareef and Y. Cui, *ACS Nano.*, 2011, **5**, 8904.
284. R. Amade, E. Jover, B. Caglar, T. Mutlu and E. Bertran, *J. Power Sources*, 2011, **196**, 5779.
285. K. Okamura, R. Inoue, T. Sebille, K. Tomono and M. Nakayama, *J. Electrochem. Soc.*, 2011, **158**, A711.
286. D. D. Zhao, Z. Yang, L. Y. Zhang, X. L. Feng and Y. F. Zhang, *Electrochem. Solid State Lett.*, 2011, **14**, A93.
287. S. W. Zhang, C. Peng, K. C. Ng and G. Z. Chen, *Electrochim. Acta*, 2010, **55**, 7447.
288. M. V. Kiamahalleh, S. A. Sata, S. Buniran and S. H. S. Zein, *Curr. Nanosci.*, 2012, **6**, 553.
289. L. Li, Z. Y. Qin, L. F. Wang, H. J. Liu and M. F. Zhu, *J. Nanopart. Res.*, 2010, **12**, 2349.
290. A. L. M. Reddy, M. M. Shaijumon, S. R. Gowda and P. M. Ajayan, *J. Phys. Chem. C*, 2010, **114**, 658.
291. R. R. Jiang, T. Huang, Y. Tang, J. Liu, L. G. Xue, J. H. Zhuang and A. H. Yu, *Electrochim. Acta*, 2009, **54**, 7173.
292. J. M. Ko and K. M. Kim, *Mater. Chem. Phys.*, 2009, **114**, 837.
293. K. W. Nam, C. W. Lee, X. Q. Yang, B. W. Cho, W. S. Yoon and K. B. Kim, *J. Power Sources*, 2009, **188**, 323.
294. J. Li, Q. M. Yang and I. Zhitornirsky, *J. Power Sources*, 2008, **185**, 1569.
295. S. B. Ma, K. W. Nam, W. S. Yoon, X. Q. Yang, K. Y. Ahn, K. H. Oh and K. B. Kim, *J. Power Sources*, 2008, **178**, 483.
296. Y. G. Wang, L. Yu and Y. Y. Xia, *J. Electrochem. Soc.*, 2006, **153**, A743.
297. K. Nam, K. Kim, E. Lee, W. Yoon, X. Yang and K. Kim, *J. Power Sources*, 2008, **182**, 642.
298. Z. Tang, C. H. Tang and H. Gong, *Adv. Funct. Mater.*, 2012, **22**, 1272.
299. L. L. Zhang, Z. G. Xiong and X. S. Zhao, *J. Power Sources*, 2013, **222**, 326.
300. L. S. Aravinda, K. K. Nagaraja, H. S. Nagaraja, K. U. Bhat and B. R. Bhat, *Electrochim. Acta*, 2013, **95**, 119.
301. Y. P. Zhang, X. W. Sun, L. K. Pan, H. B. Li, Z. Sun, C. P. Sun and B. K. Tay, *Solid State Ionics*, 2009, **180**, 32.
302. G. Wang, M. Qu, Z. Yu and R. Yuan, *Mater. Chem. Phys.*, 2007, **105**, 169.
303. L. Y. Liang, H. M. Liu and W. S. Yang, *J. Alloys Compd.*, 2013, **559**, 167.
304. M. Sathiya, A. S. Prakash, K. Ramesha, J. M. Tarascon and A. K. Shukla, *J. Am. Chem. Soc.*, 2011, **133**, 16291.

305. I. H. Kim, J. H. Kim, B. W. Cho, Y. H. Lee and K. B. Kim, *J. Electrochem. Soc.*, 2006, **153**, A989.
306. Z. Chen, V. Augustyn, J. Wen, Y. W. Zhang, M. Q. Shen, B. Dunn and Y. F. Lu, *Adv. Mater.*, 2011, **23**, 79.
307. A. L. M. Reddy and S. Ramaprabhu, *J. Phys. Chem. C*, 2007, **111**, 7727.
308. Q. Wang, Z. H. Wen and J. H. Li, *J. Nanosci. Nanotechnol.*, 2007, **7**, 3328.
309. Z. J. Li, T. X. Chang, G. Q. Yun and Y. Jia, *Powder Technol.*, 2012, **224**, 306.
310. Y. M. Chen, J. H. Cai, Y. S. Huang, K. Y. Lee, D. S. Tsai and K. K. Tiong, *Thin Solid Film*, 2012, **520**, 2409.
311. Y. M. Chen, J. H. Cai, Y. S. Huang, K. Y. Lee, D. S. Tsai and K. K. Tiong, *Nanotechonol.*, 2011, **22**, 355708.
312. L. Su, S. Zhang, C. Yuan and B. Gao, *J. Electrochem. Soc.*, 2008, **155**, A110.
313. K. Lota, V. Khomenko and E. Frackowiak, *J. Phys. Chem. Solids*, 2004, **65**, 295.
314. V. Khomenko, E. Frackowiak and F. Beguin, *Electrochim. Acta*, 2005, **50**, 2499.
315. M. Hughes, G. Z. Chen, M. S. P. Shaffer, D. J. Fray and A. H. Windle, *Chem. Mater.*, 2002, **14**, 1610.
316. K. H. An, K. K. Jeon, J. K. Heo, S. C. Lim, D. J. Bae and Y. H. Lee, *J. Electrochem. Soc.*, 2002, **149**, A1058.
317. V. Gupta and N. Miura, *Electrochim. Acta*, 2006, **52**, 1721.
318. Y. K. Zhou, B. L. He, W. J. Zhou and H. L. Li, *J. Electrochem. Soc.*, 2004, **151**, A1052.
319. B. Dong, B. L. He, C. L. Xu and H. L. Li, *Mater. Sci. Eng. B*, 2007, **143**, 7.
320. L. B. Kong, J. Zhang, J. J. An, Y. C. Luo and L. Kang, *J. Mater. Sci.*, 2008, **43**, 3664.
321. H. M. Zhang, S. An, X. Ye and S. Yang, *Electrochem. Commun.*, 2007, **9**, 2859.
322. K. Ryu, H. Xue and J. Park, *J Chem. Technol. Biotechnol.*, 2013, **88**, 788.
323. F. Beguin, K. Szostak, G. Lota and E. Frackowiak, *Adv. Mater.*, 2005, **7**, 2380.
324. H. J. Lin, L. Li, J. Ren, Z. B. Cai, L. B. Qiu, Z. B. Yang and H. S. Peng, *Sci. Rep.*, 2013, **3**, 1353.
325. M. Moniruzzaman, S. Sahoo, D. Ghosh, C. K. Das and R. Singh, *J. Appl. Polymer Sci.*, 2013, **128**, 698.
326. P. Sivaraman, A. R. Bhattacharrya, S. P. Mishra, A. P. Thakur, K. Shashidhara and A. B. Samui, *Electrochim. Acta*, 2013, **94**, 182.
327. S. Dhibar, S. Sahoo and C. K. Das, *Polym. Compos.*, 2013, **34**, 517.
328. X. X. Bai, X. J. Hu, S. Y. Zhou, J. Yan, C. H. Sun, P. Chen and L. F. Li, *Electrochim. Acta*, 2013, **87**, 394.
329. R. S. Hastak, P. Sivaraman, D. D. Potphode, K. Shashidhara and A. B. Samui, *J. Solid State Electrochem.*, 2012, **16**, 3215.
330. S. Paul, K. S. Choi, D. J. Lee, P. Sudhagar and Y. S. Kang, *Electrochim. Acta*, 2012, **78**, 649.

331. Y. J. Kang, S. J. Chun, S. S. Lee, B. Y. Kim, J. H. Kim, H. Chung, S. Y. Lee and W. Kim, *ACS Nano.*, 2012, 7, 6400.
332. Y. Hu, Y Zhao, Y Li, H. Li, H. B. Shao and L. T. Qu, *Electrochim. Acta*, 2012, **66**, 279.
333. S. Hu, R. Rajamani and X. Yu, *Appl. Phys. Lett.*, 2012, **100**, 104103.
334. M. Ertas, R. M. Walczak, R. K. Das, A. G. Rinzler and J. R. Reynolds, *Chem. Mater.*, 2012, **24**, 433.
335. M. R. Rosario-Canales, P. Deria, M. J. Therien and J. J. Santiago-Aviles, *ACS Appl. Mater. Interfaces*, 2012, **4**, 102.
336. H. Lee, H Kim, M. S. Cho, J. Choi and Y. Lee, *Electrochim. Acta*, 2011, **56**, 7460.
337. Q. A. Liu, M. H. Nayfeh and S. T. Yau, *J. Power Sources*, 2010, **195**, 7480.
338. Y. Zhou, Z. Y. Qin, L. Li, Y. Zhang, Y. L. Wei, L. F. Wang and M. F. Zhu, *Electrochim. Acta*, 2010, **55**, 3904.
339. B. Gao, Q. B. Fu, L. H. Su, C. Z. Yuan and X. G. Zhang, *Electrochim. Acta*, 2010, **55**, 2311.
340. Y. P. Fang, J. W. Liu, D. J. Yu, J. P. Wicksted, K. Kalkan, C. O. Topal, B. N. Flanders, J. D. Wu, J. Li, i, 2012, **195**, 674.
341. J. Y. Kim, K. H. Kim and K. B. Kim, *J. Power Sources*, 2008, **176**, 396.
342. J. Yan, T. Wei, Z. J. Fan, W. Z. Qian, M. L. Zhang, X. D. Shen and F. Wei, *J. Power Sources*, 2010, **195**, 3041.
343. Q. Li, J. H. Liu, J. H. Zou, A. Chunder, Y. Q. Chen and L. Zhai, *J. Power Sources*, 2011, **196**, 565.
344. S. R. Sivakkumar, J. M. Ko, D. Y. Kim, B. C. Kim and G. G. Wallace, *Electrochim. Acta*, 2007, **52**, 7377.
345. P. K. Sharma, A. Karakoti, S. Seal and L. Zhai, *J. Power Sources*, 2010, **195**, 1256.
346. Y. Hou, Y. W. Cheng, T. Hobson and J. Liu, *Nano Lett.*, 2010, **10**, 2727.
347. K. S. Kim and S. J. Park, *Electrochim. Acta*, 2011, **56**, 1629.
348. X. J. Lu, F. Zhang, H. Dou, C. Z. Yuan, S. D. Yang, L. Hao, L. F. Shen, L. J. Zhang and X. G. Zhan, *Electrochim. Acta*, 2012, **69**, 160.
349. V. Sridhar, H. J. Kim, J. H. Jung, C. G. Lee, S. L. Park and I. K. Oh, *ACS Nano.*, 2012, **6**, 10562.
350. B. You, N. Li, H. Y. Zhu, X. L. Zhu and J. Yang, *ChemSusChem*, 2013, **6**, 474.
351. L. L. Zhang, Z. G. Xiong and X. S. Zhao, *J. Power Sources*, 2013, **222**, 326.
352. Y. W. Cheng, S. T. Lu, H. B. Zhang, C. V. Varanasi and J. Liu, *Nano Lett.*, 2012, **12**, 4206.
353. Z. B. Lei, F. H. Shi and L. Lu, *ACS Appl. Mater. Interfaces*, 2012, **4**, 1058.
354. Y. Hou, Y. W. Cheng, T. Hobson and J. Liu, *Nano Lett.*, 2010, **10**, 2727.
355. Q. Li, J. H. Liu, J. H. Zou, A. Chunder, Y. Q. Chen and L. Zhai, *J. Power Sources*, 2011, **196**, 565.
356. K. S. Kim and S. J. Park, *J. Solid State Electrochem.*, 2012, **16**, 275.
357. J. M. Ko, K. S. Ryu, S. Kim and K. M. Kim, *J. App. Electrochem.*, 2009, **39**, 1331.
358. B. Chen, J. B. Hou and K. Lu, *Langmuir*, 2013, **29**, 5911.

359. X. H. Lu, G. M. Wang, T. Zhai, M. B. Yu, J. Y. Gan, Y. X. Tong and Y. Li, *Nano Lett.*, 2012, **12**, 1690.
360. E. Iyyamperumal, S. Y. Wang and L. M. Dai, *ACS Nano.*, 2012, **6**, 5259.
361. Q. Li, Z. L. Wang, G. R. Li, R. Guo, L. X. Ding and Y. X. Tong, *Nano Lett.*, 2012, **12**, 3803.
362. S. I. Cho and S. B. Lee, *Acc. Chem. Res.*, 2008, **41**, 699.
363. C. de las Casas and W. Z. Li, *J. Power Sources*, 2012, **208**, 74.
364. C. S. Wang, G. T. Wu and W. Z. Li, *J. Power Sources*, 1998, **76**, 1.
365. N. A. Kaskhedikar and J. Maier, *Adv. Mater.*, 2009, **21**, 2664.
366. Z. L. Xiong, Y. S. Yun and H. J. Jin, *Materials*, 2013, **6**, 1138.
367. P. G. Bruce, B. Scrosati and J. M. Tarascon, *Angew. Chem. Int. Ed.*, 2008, **47**, 2930.
368. C. H. Jiang, E. Hosono and H. S. Zhou, *Nano Today*, 2006, **1**, 28.
369. M. K. Song, S. J. Park, F. M. Alamgir, J. Cho and M. L. Liu, *Mater. Sci. Eng., R*, 2011, **72**, 203.
370. J. Liu and D. F. Xue, *Nanoscale Res. Lett.*, 2010, **5**, 1525.
371. L. Chang and H. M. Cheng, *Mater. Today*, 2013, **16**, 19.
372. B. Gao, A. Kleinhammes, X. P. Tang, C. Bower, L. Fleming, Y. Wu and O. Zhou, *Chem. Phys. Lett.*, 1999, **307**, 153.
373. B. Gao, C. Bower, J. D. Lorentzen, A. Kleinhammes, X. P. Tang, L. E. McNeil, Y. Wu and O. Zhou, *Chem. Phys. Lett.*, 2000, **327**, 69.
374. H. Shimoda, B. Gao, X. P. Tang, A. Kleinhammes, L. Fleming, Y. Wu and O. Zhou, *Phys. Rev. Lett.*, 2002, **88**, 015502.
375. Z. H. Yang and H. Q. Wu, *Mater. Lett.*, 2001, **50**, 108.
376. G. Maurin, Ch. Bousquet, F. Henn, P. Bernier, R. Almairac and B. Simon, *Solid State Ionics*, 2000, **136–137**, 1295.
377. F. Leroux, K. Metenier, S. Gautier, E. Frackowiak, S. Bonnamy and F. Beguin, *J. Power Sources*, 1999, **81–82**, 317.
378. A. S. Claye, J. E. Fischer, C. B. Huffman, A. G. Rinzler and R. E. Smalley, *J. Electrochem. Soc.*, 2000, **147**, 2845.
379. S. Kawasaki, T. Hara, Y. Iwai and Y. Suzuki, *Mater. Lett.*, 2008, **62**, 2917.
380. C. M. Schauerman, M. J. Ganter, G. Gaustad, C. W. Babbitt, R. P. Raffaelle and B. J. Landi, *J. Mater. Chem.*, 2012, **22**, 12008.
381. S. H. Ng, J. Wang, Z. P. Guo, G. X. Wang and H. K. Liu, *Electrochim. Acta.*, 2005, **51**, 23.
382. S. Y. Chew, S. H. Ng, J. Z. Wang, P. Novak, F. Krumeich, S. L. Chou, J. Chen and H. K. Liu, *Carbon*, 2009, **47**, 2976.
383. K. N. Ishidate and M. Hasegawa, *Phys. Rev. B*, 2005, **71**, 245418.
384. J. Y. Eom, H. S. Kwon, J. Liu and O. Zhou, *Carbon*, 2004, **42**, 2589.
385. Z. H. Yang and H. Q. Wu, *Solid State Ionics*, 2001, **143**, 173.
386. R. S. Morris, B. G. Dixon, T. Gennett, R. Raffaelle and M. J. Heben, *J. Power Sources.*, 2004, **138**, 277.
387. S. Klink, E. Ventosa, W. Xia, F. La Mantia, M. Muhler and W. Schuhmann, *Electrochem. Commun.*, 2012, **15**, 10.
388. Y. Zhang, T. Chen, J. Wang, G. Min, L. Pan, Z. Song, Z. Sun, W. Zhou and J. Zhang, *Appl. Surf. Sci.*, 2012, **258**, 4729.

389. X. X. Wang, J. N. Wang, H. Chang and Y. F. Zhang, *Adv. Funct. Mater.*, 2007, **17**, 3613.
390. Y. T Yu, C. J. Cui, W. Z. Qian, Q. Xie, C. Zheng, C. Y. Kong and F. Wei, *Asia-Pacific J. Chem. Eng.*, 2013, **8**, 234.
391. S. Yang, J. Huo, H. Song and X. Chen, *Electrochim. Acta*, 2008, **53**, 2238.
392. S. Yang, H. Song and X. Chen, *Electrochem. Commun.*, 2006, **8**, 137.
393. X. Wang, J. Wang and L. F. Su, *J. Power Sources*, 2009, **186**, 194.
394. D. T. Welna, L. Qu, B. E. Taylor, L. Dai and M. F. Durstock, *J. Power Sources*, 2011, **196**, 1455.
395. K. Wang, S. Luo, Y. Wu, X. F. He, F. Zhao, J. P. Wang, K. L. Jiang and S. S. Fan, *Adv. Funct. Mater.*, 2013, **23**, 846.
396. W. Lu, A. Goering, L. T. Qu and L. M. Dai, *Phys. Chem. Chem. Phys.*, 2012, **14**, 12099.
397. X. M. Liu, Z. Huang, S. W. Oh, B. Zhang, P. M. Ma, M. M. F. Yuen and J. K. Kim, *Compos. Sci. Technol.*, 2012, **72**, 121.
398. J. M. Schnorr and T. M. Swager, *Chem. Mater.*, 2011, **23**, 646.
399. B. J. Landi, M. J. Ganter, C. D. Cress, R. A. DiLeo and R. P. Raffaelle, *Energy Environ. Sci.*, 2009, **2**, 638.
400. L. F. Cui, L. Hu, J. W. Choi and Y. Cui, *ACS Nano.*, 2010, **4**, 3671.
401. W. Wang and P. N. Kumta, *ACS Nano.*, 2010, **4**, 2233.
402. L. F. Cui, Y. Yang, C. M. Hsu and Y. Cui, *Nano Lett.*, 2009, **9**, 3370.
403. J. Y. Eom and H. S. Kwon, *ACS Appl. Mater. Interfaces*, 2011, **3**, 1015.
404. J. Lee, J. Bae, J. Heo, I. T. Han, S. N. Cha, D. K. Kim, M. Yang, H. S. Han, W. S. Jeon and J. Chung, *J. Electrochem. Soc.*, 2009, **156**, A905.
405. C. Martin, O. Crosnier, R. Retoux, D. Bélanger, D. M. Schleich and T. Brousse, *Adv. Funct. Mater.*, 2011, **21**, 3524.
406. W. Wang, R. Epur and P. N. Kumta, *Electrochem. Commun.*, 2011, **13**, 429.
407. Y. Fan, Q. Zhang, Q. Z. Xiao, X. H. Wang and K. Huang, *Carbon*, 2013, **59**, 264.
408. M. W. Forney, R. A. DiLeo, A. Raisanen, M. J. Ganter, J. W. Staub, R. E. Rogers, R. D. Ridgley and B. J. Landi, *J. Power Sources*, 2013, **228**, 270.
409. K. S. Park, K. M. Min, S. D. Seo, G. H. Lee, H. W. Shim and D. W. Kim, *Mater. Res. Bulletin*, 2013, **48**, 1732.
410. L. G. Xue, G. J. Xu, Y. Li, S. L. Li, K. Fu, Q. Shi and X. W. Zhang, *ACS Mater. Interfaces*, 2013, **5**, 21.
411. Y. Fan, Q. Zhang, C. X. Lu, Q. Z. Xiao, X. H. Wang and B. K. Tay, *Nanoscale*, 2013, **5**, 1503.
412. K. Evanoff, J. Benson, M. Schauer, I. Kovalenko, D. Lashmore, W. J. Ready and G. Yushin, *ACS Nano.*, 2012, **6**, 9837.
413. J. Zang and Y. P. Zhao, *Composite Part B: Eng.*, 2012, **43**, 76.
414. X. L. Li, J. H. Cho, N. Li, Y. Y. Zhang, D. Williams, S. A. Dayeh and S. T. Picraux, *Adv. Energy Mater.*, 2012, **2**, 87.
415. H. W. Liao, K. Karki, Y. Zhang, J. Cumings and Y. H. Wang, *Adv. Mater.*, 2011, **23**, 4318.

416. C. Martin, O. Crosnier, R. Retoux, D. Belanger, D. M. Schleich and T. Brousse, *Adv. Funct. Mater.*, 2011, **21**, 3524.

417. Y. Fan, Q. Zhang, Q. Z. Xiao, X. H. Wang and K. Huang, *Carbon*, 2013, **59**, 264.

418. S. L. Chou, Y. Zhao, J. Z. Wang, Z. X. Chen, H. K. Liu and S. X. Dou, *J. Phys. Chem. C*, 2010, **114**, 15862.

419. J. Shu, H. Li, R. Z. Yang, Y. Shi and X. J. Huang, *Electrochem. Commun.*, 2006, **8**, 51.

420. J. Y. Eom, J. W. Park, H. S. Kwon and S. Rajendran, *J. Electrochem. Soc.*, 2006, **153**, A1678.

421. L. B. Hu, H. Wu, Y. F. Gao, A. Y. Cao, H. B. Li, J. McDough, X. Xie, M. Zhou and Y. Cui, *Adv. Energy Mater.*, 2011, **1**, 523.

422. C. L. Zhao, Q. Li, W. Wan, J. M. Li, J. J. Li, H. H. Zhou and D. S. Xu, *J. Mater. Chem.*, 2012, **22**, 12193.

423. N. S. Choi, Y. Yao, Y. Cui and J. Cho, *J. Mater. Chem.*, 2011, **21**, 9825.

424. A. Gohier, B. Laik, K. H. Kim, J. L. Maurice, J. P. Pereira-Ramos, C. S. Cojocaru and P. T. Van, *Adv. Mater.*, 2012, **24**, 2592.

425. M. W. Forney, R. A. DiLeo, A. Raisanen, M. J. Ganter, J. W. Staub, R. E. Rogers, R. D. Ridgley and B. J. Landi, *J. Power Sources*, 2013, **228**, 270.

426. C. Kang, I. Lahiri, R. Baskaran, W. G. Kim, Y. K. Sun and W. Choi, *J. Power Sources*, 2012, **219**, 364.

427. D. Ahn, X. C. Xiao, Y. W. Li, A. K. Sachdev, H. W. Park, A. P. Yu and Z. W. Chen, *J. Power Sources*, 2012, **21**, 66.

428. L. Noerochim, J. Z. Wang, S. L. Chou, D. Wexler and H. K. Liu, *Carbon*, 2012, **50**, 1289.

429. H. K. Zhang, H. H. Song, X. H. Chen, J. S. Zhou and H. J. Zhang, *Electrochim. Acta*, 2012, **59**, 160.

430. S. J. Ding, J. S. Chen and X. W. Lou, *Adv. Funct. Mater.*, 2011, **21**, 4120.

431. J. G. Ren, J. B. Yang, A. Abouimrane, D. P. Wang and K. Amine, *J. Power Sources*, 2011, **196**, 8701.

432. C. X. Zhai, N. Du, H. Zhang, J. X. Yu and D. R. Yang, *ACS Appl. Mater. Interfaces*, 2011, **3**, 4067.

433. L. Noerochim, J. Z. Wang, S. L. Chou, H. J. Li and H. K. Liu, *Electrochim. Acta*, 2010, **56**, 314.

434. C. Q. Feng, L. Li, Z. P. Guo and H. Li, *J. Alloys Compd.*, 2010, **504**, 457.

435. C. L. Zhu, M. L. Zhang, Y. J. Qiao, P. Gao and Y. J. Chen, *Mater. Res. Bulletin*, 2010, **45**, 437.

436. G. D. Du, C. Zhong, P. Zhang, Z. P. Guo, Z. X. Chen and H. K. Liu, *Electrochim. Acta*, 2010, **55**, 2582.

437. C. H. Xu, J. Sun and L. A. Gao, *J. Power Sources*, 2011, **196**, 5138.

438. H. X. Zhang, C. Feng, Y. C. Zhai, K. L. Jiang, Q. Q. Li and S. S. Fan, *Adv. Mater.*, 2009, **21**, 2299.

439. N. Du, H. Zhang, B. D. Chen, X. Y. Ma, X. H. Huang, J. P. Tu and D. R. Yang, *Mater. Res. Bulletin*, 2009, **44**, 211.

440. G. Chen, Z. Y. Wang and D. G. Xia, *Chem. Mater.*, 2008, **20**, 6951.

441. Z. Y. Wang, G. Chen and D. G. Xia, *J. Power Sources*, 2008, **184**, 432.

442. Y. J. Chen, C. L. Zhu, X. Y. Xue, X. L. Shi and M. S. Cao, *Appl. Phys. Lett.*, 2008, **92**, 223101.
443. Z. H. Wen, Q. Wang, Q. Zhang and J. H. Li, *Adv. Funct. Mater.*, 2007, **17**, 2772.
444. Y. Wang, H. C. Zeng and J. Y. Lee, *Adv. Mater.*, 2006, **18**, 645.
445. J. Xie and V. K. Varadan, *Mater. Chem Phys.*, 2005, **91**, 274.
446. A. L. M. Reddy, M. M. Shaijumon, S. R. Gowda and P. M. Ajayan, *Nano Lett.*, 2009, **9**, 1002.
447. H. Xia, M. O. Lai and L. Lu, *J. Mater. Chem.*, 2010, **20**, 6896.
448. H. Xia, Y. Wang, M. O. Lai, J. Y. Lin and L. Lu, *Crystal Growth Design*, 2011, **11**, 3306.
449. N. Du, H. Zhang, B. Chen, J. B. Wu, X. Y. Ma, Z. H. Liu, Y. Q. Zhang, D. Yang, X. H. Huang and J. P. Tu, *Adv. Mater.*, 2007, **19**, 4505.
450. L. H. Zhuo, Y. Q. Wu, J. Ming, L. Y. Wang, Y. C. Yu, X. B. Zhang and F. Y. Zhao, *J. Mater. Chem. A*, 2013, **1**, 1141.
451. F. L. Lou, H. T. Zhou, F. Huang, F. Vullum-Bruer, T. D. Tran and D. Chen, *J. Mater. Chem. A*, 2013, **1**, 3757.
452. C. H. Xu, J. Sun and L. Gao, *J. Phys. Chem. C*, 2009, **113**, 20509.
453. H. B. Wu, X. W. Lou and H. H. Hng, *Chem. Eur. J.*, 2012, **12**, 3132.
454. Y. X. Wang, J. Xie, G. S. Cao, T. J. Zhu and X. B. Zhao, *J. Mater. Res.*, 2012, **27**, 417.
455. A. Bhaskar, M. Deepa and T. N. Rao, *ACS Appl. Mater. Interfaces*, 2013, **5**, 2555.
456. Y. Wu, Y. Wei, J. P. Wang, K. L. Jiang and S. S. Fan, *Nano Lett.*, 2013, **13**, 818.
457. Y. He, L. Huang, J. S. Cai, X. M. Zheng and S. G. Sun, *Electrochim. Acta*, 2010, **55**, 1140.
458. X. W. Zhou, G. M. Wu, G. H. Gao, C. J. Cui, H. Y. Yang, J. Shen, B. Zhou and Z. H. Zhang, *Electrochim Acta*, 2012, **74**, 32.
459. E. J. Yoo, J. Kim, E. Hosono, H. S. Zhou, T. Kudo and I. Honma, *Nano. Lett.*, 2008, **8**, 2277.
460. S. Q. Chen, P. Chen and Y. Wang, *Nanoscale*, 2011, **3**, 4323.
461. S. Q. Chen, W. K. Yeoh, Q. Liu and G. X. Wang, *Carbon*, 2012, **50**, 4557.
462. T. Q. Chen, L. K. Pan, K. Yu and Z. Sun, *Solid State Ionics*, 2012, **229**, 9.
463. H. R. Byon, B. M. Gallant, S. W. Lee and Y. Shao-Horn, *Adv. Funct. Mater.*, 2013, **23**, 1037.
464. B. P. Vinayan, R. Nagar, V. Raman, N. Rajalakshmi, K. S. Dhathathreyan and S. Ramaprabhu, *J. Mater. Chem.*, 2012, **22**, 9949.
465. J. H. Lee, B. S. Kong, S. B. Yang and H. T. Jung, *J. Power Sources*, 2009, **194**, 520.
466. Y. Yu, L. Gu, C. L. Wang, A. Dhanabalan, P. A. van Aken and J. Maierm, *Angew. Chem.Int. Ed.*, 2009, **48**, 6485.
467. T. P. Kumar, R. Ramesh, Y. Y. Lin and G. T. K. Fey, *Electrochem. Commun.*, 2004, **6**, 520.
468. X. F. Li, Y. Zhong, M. Cai, M. P. Balogh, D. N. Wang, Y. Zhang, R. Y. Li and X. L. Sun, *Electrochim. Acta*, 2013, **89**, 387.

469. S. F. Fan, T. Sun, X. H. Rui, Q. Y. Yan and H. H. Hng, *J. Power Sources*, 2012, **201**, 288.
470. Y. Wang and J. Y. Lee, *Angew. Chem. Int. Ed.*, 2006, **45**, 7039.
471. L. Huang, J. S. Cai, Y. He, F. S. Ke and S. G. Sun, *Electrochem. Commun.*, 2009, **11**, 950.
472. J. T Yin, M. Wada, Y. Kitano, S. Tanase, O. Kajita and T. Sakai, *J. Electrochem. Soc.*, 2005, **152**, A1341.
473. W. X. Chen, J. Y. Lee and Z. L. Liu, *Electrochem. Commun.*, 2002, **4**, 260.
474. Z. H. Wen, G. H. Lu, S. Mao, H. Kim, S. M. Cui, K. H. Yu, X. K. Huang, P. T. Hurley, O. Mao and J. H. Chen, *Electrochem. Commun.*, 2013, **29**, 67.
475. J. K Yoo, J. Kim, Y. S. Jung and K. Kang, *Adv. Mater.*, 2012, **24**, 5452.
476. H. Wu, G. Chan, J. W. Choi, I. Ryu, M. T. Yao, S. W. McDowell, A. Lee, Y. Jackson, L. B. Yang, Hu and Y. Cui, *Nature Nanotechnol.*, 2012, **7**, 309.
477. T. Song, J. Xia, J. H. Lee, D. H. Lee, M. S. Kwon, J. M. Choi, J. Wu, S. K. Doo, H. Chang, W. I. Park, D. S. Zang, H. Kim, Y. Huang, H. C. Hwang, J. A. Rogers and U. Paik, *Nano Lett.*, 2010, **10**, 1710.
478. M. H. Park, M. G. Kim, J. Joo, K. Kim, J. Kim, S. Ahn, Y. Cui and J. Cho, *Nano Lett.*, 2009, **9**, 3844.
479. Z. H. Bi, M. P. Paranthaman, P. A. Menchhofer, R. R. Dehoff, C. A. Bridges, M. F. Chi, B. K. Guo, X. G. Sun and S. Dai, *J. Power Sources*, 2013, **222**, 461.
480. H. Q. Li, S. K. Martha, R. R. Unocic, H. M. Luo, S. Dai and J. Qu, *J. Power Sources*, 2012, **218**, 88.
481. B. G. Lee, S. C. Nam and J. Choi, *Curr. Appl. Phys.*, 2012, **12**, 1580.
482. H. Han, T. Song, E. K. Lee, A. Devadoss, Y. Jeon, J. Ha, Y. C. Chung, Y. M. Choi, Y. G. Jung and U. Paik, *ACS Nano.*, 2012, **6**, 8308.
483. S. Brutti, V. Gentili, H. Menard, B. Scrosati and P. G. Bruce, *Adv. Energy Mater.*, 2012, **2**, 322.
484. Y. Wang, S. Q. Liu, K. L. Huang, D. Fang and S. X. Zhuang, *J. Solid State Electrochem.*, 2012, **16**, 723.
485. H. Xiong, M. D. Slater, M. Balasubramanian, C. S. Johnson and T. Rajh, *J. Phys. Chem. Lett.*, 2011, **2**, 2560.
486. G. D. Du, B. Wan, Z. P. Guo, J. N. Shen, Y. Li and H. K. Liu, *Adv. Sci. Lett.*, 2011, **4**, 469.
487. M. G. Choi, Y. G. Lee, S. W. Song and K. M. Kim, *Electrochim. Acta*, 2010, **55**, 5875.
488. Z. Wei, Z. Liu, R. R. Jiang, C. Q. Bian, T. Huang and A. S. Yu, *J. Solid State Electrochem.*, 2010, **14**, 1045.
489. J. Y. Yan, H. H. Song, S. B. Yang and X. H. Chen, *Mater. Chem. Phys.*, 2009, **118**, 367.
490. J. W. Xu, C. H. Ha, B. Cao and W. F. Zhang, *Electrochim. Acta*, 2007, **52**, 8044.
491. G. Armstrong, A. R. Armstrong, J. Canales and P. G. Bruce, *Electrochem. Solid State Lett.*, 2006, **9**, A139.

492. J. Wang, Y. K. Zhou, B. Xiong, Y. Y. Zhao, X. J. Huang and Z. P. Shao, *Electrochim. Acta*, 2013, **88**, 847.
493. Y. Q. Fan, N. Zhang, L. Y. Zhang, H. B. Shao, J. M. Wang, J. Q. Zhang and C. N. Cao, *Electrochim. Acta*, 2013, **94**, 285.
494. X. Q. Meng, J. Y. Yao, F. L. Liu, H. C. He, M. Zhou, P. Xiao and Y. H. Zhang, *J. Alloys Compd.*, 2013, **552**, 392.
495. N. A. Kyeremateng, C. Lebouin, P. Knauth and T. Djenizian, *Electrochim. Acta*, 2103, **88**, 814.
496. L. Yu, Z. Y. Wang, L. Zhang, H. B. Wu and X. W. Lou, *J. Mater. Chem. A*, 2013, **1**, 122.
497. K. Y. Kang, Y. G. Lee, S. Kim, S. R. Seo, J. C. Kim and K. M. Kim, *Mater. Chem. Phys.*, 2012, **137**, 169.
498. J. W. Zhang, X. X. Yan, J. W. Zhang, W. Cai, Z. S. Wu and Z. J. Zhang, *J. Power Sources*, 2012, **198**, 223.
499. S. J. Park, Y. J. Kim and H. Lee, *J. Power Sources.*, 2011, **196**, 5133.
500. J. W. Xu, Y. F. Wang, Z. H. Li and W. Zhang, *J. Power Sources*, 2008, **175**, 903.
501. H. E. Wang, L. J. Xi, R. G. Ma, Z. G. Lu, C. Y. Chung, I. Bello and J. A. Zapien, *J. Solid State Chem.*, 2012, **190**, 104.
502. J. Z. Wang, N. Du, H. Zhang, J. X. Yu and D. R. Yang, *J. Phys. Chem. C*, 2011, **115**, 11302.
503. P. Wu, N. Du, H. Zhang, J. X. Yu, Y. Qi and D. R. Yang, *Nanoscale*, 2011, **2**, 746.
504. L. M. Li, X. M. Yin, S. A. Liu, Y. G. Wang, L. B. Chen and T. H. Wang, *Electrochem. Commun.*, 2010, **12**, 1383.
505. J. F. Ye, H. J. Zhang, R. Yang, X. G. Li and L. M. Qi, *Small*, 2010, **6**, 296.
506. Y. Wang, J. Y. Lee and H. C. Zeng, *Chem. Mater.*, 2005, **17**, 3899.

CHAPTER 6

Measurements of Photovoltaic Cells

HUANG XUEBO AND ZHANG JING*

National Metrology Centre, Agency for Science, Technology and Research
(A*STAR), 1, Science Park Drive, Singapore 118221
*Email: zhang_jing@nmc.a-star.edu.sg

6.1 Basic Concepts and Quantities of Photovoltaic Measurements

The photovoltaic effect was first demonstrated experimentally in 1839 by the French physicist, A. E. Becquerel, while the first solid state photovoltaic (PV) cell was built in 1883. The first solar module was built in 1950s. A PV cell is essentially a solid state electrical device converting light energy directly into electricity, based on the photovoltaic effect.

6.1.1 Solar Cell Efficiency

The overall efficiency of the solar cell is the key evaluating factor of it. Solar cell efficiency is the ratio of the maximum electrical output power of a solar cell to the incident optical power in the form of sunlight under standard testing conditions defined in the IEC 60904-3 (2008) standard,[1] $\eta = \frac{P_{max\,electrical}}{P_{optical}}$. The optical power is quantified as the total irradiance from the Sun incident on the solar cell surface. $P_{optical} = EA_{cell}$, where E is the solar irradiance of sunlight incident on the solar cell surface with an active area of A_{cell}; it is power per unit area on the Earth's surface.

RSC Nanoscience & Nanotechnology No. 32
Nanofabrication and its Application in Renewable Energy
Edited by Gang Zhang and Navin Manjooran

The accurate measurement of the overall efficiency is important for both manufacturers and users. For the PV manufacturers, accurate measurements will help to improve the reliability and quality of the solar cells in applications; it can provide accurate information for the designer in the product design; it will also affect the pricing which may cost huge amount of dollars. For PV users, accurate efficiency is needed for power budgeting.

6.1.2 I-V Characteristics of a PV Cell

A PV cell can be modeled as a diode in parallel with a current source (Figure 6.1). The diode characteristics can be determined by measuring the current versus voltage behavior.

The current in the diode under illumination is $I = I_L - I_0[\exp(\frac{qV}{nkT}) - 1]$. The maximum current is the short-circuit current, $I_{SC} = I_L \propto$ solar irradiance. The maximum voltage happens at the open circuit, which is open circuit voltage, $V_{OC} = \frac{nkT}{q}\ln(\frac{I_L}{I_0} - 1)$. Figure 6.2 shows a typical current *vs.* voltage characteristic of a solar cell as well as its power *vs.* voltage curve measured at the National Metrology Centre, A*STAR, Singapore.

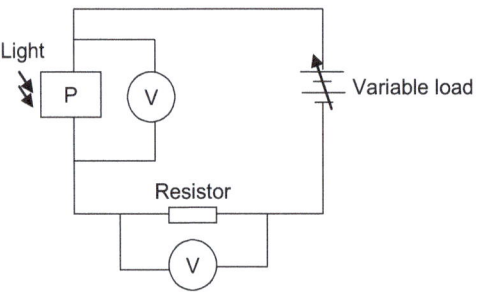

Figure 6.1   The model for PV cell characterization.

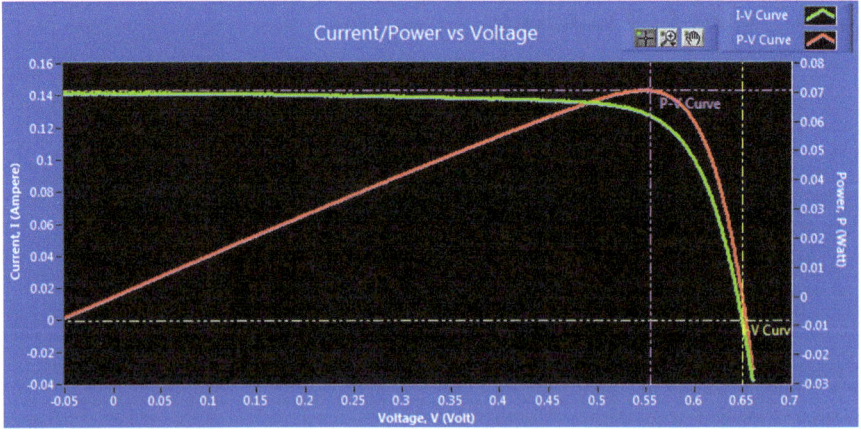

Figure 6.2   A typical I-V characteristic of a PV cell.

6.1.3 Filling Factor

The electrical power generated by the PV cell varies with the load. The electrical power is the product of current and voltage. In the power versus voltage curve, we can find the maximum electrical power, $P_{\text{max electrical}}$. The filling factor is defined as $FF = \frac{P_{\text{max electrical}}}{V_{OC}I_{SC}}$. Hence, $P_{\text{max electrical}} = V_{OC}I_{SC}FF$, and the PV cell efficiency is $\eta = \frac{V_{OC}I_{SC}FF}{P_{\text{optical}}}$.

6.2 Properties of Sunlight and Solar Simulator

6.2.1 Properties of Sunlight

The photovoltaic solar cells convert the energy of sunlight to electrical power. The Sun approximates to a black body with an emission spectrum peaked in the central, yellow-green part of the visible spectrum, with significant power in the ultraviolet and infrared as well. According to Planck's Law, radiation emitted by a black body radiator is a function of wavelength and absolute temperature only. The effective temperature of the Sun is approximately 5800 K. The spectral radiance is $L_{e,\lambda}(\lambda, T) = \frac{c_1}{\pi n^2 \lambda^5} \frac{1}{e^{c_2/n\lambda T}-1}$, where $c_1 = 2\pi hc^2$, $c_2 = hc/k$, h is the Planck constant, c is the speed of light in the vacuum, and k is the Boltzmann constant. Figure 6.3 shows the calculated spectral radiance of the light of the Sun, based on the black body model.

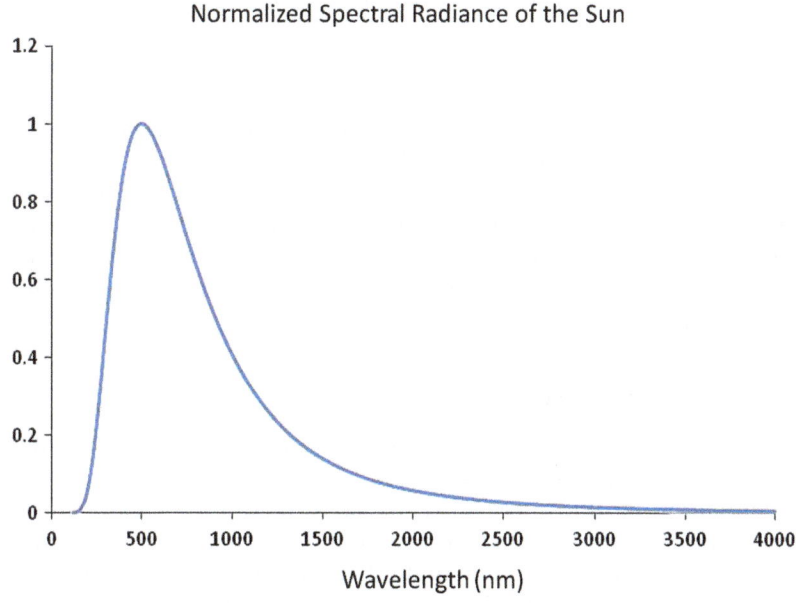

Figure 6.3 The normalized spectral radiance of the light of the Sun.

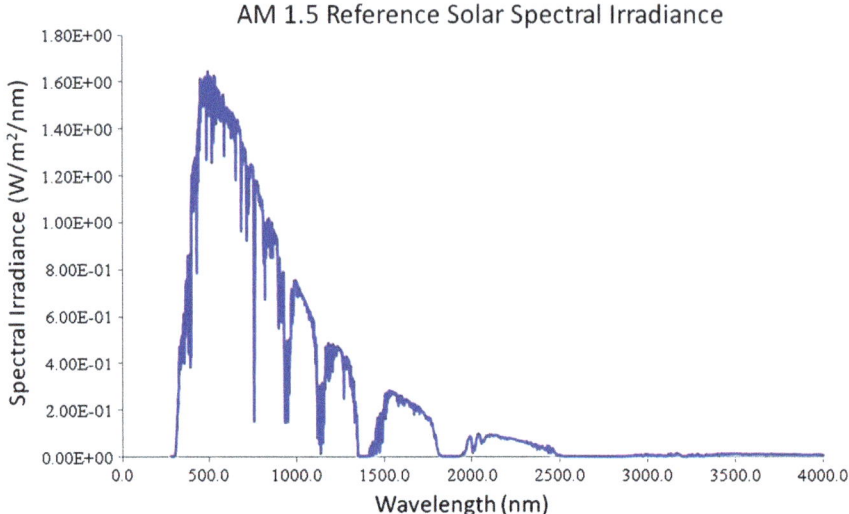

Figure 6.4   AM1.5 Global reference spectrum.

6.2.2   Standard Testing Conditions

Due to the atmosphere absorption and scattering, the sunlight spectral irradiance differs at different locations on the Earth. Latitude and climate will affect the total energy output of the solar cells used in a photovoltaic system. For example, a solar panel can produce more when the Sun is high in the sky and will produce less in cloudy conditions and when the Sun is low in the sky.

To standardize the measurement condition, the solar cell measurement must be under standard test conditions (STC) unless stated otherwise. STC specifies a cell temperature of 25 °C and an irradiance of 1000 W m^{-2} with an air mass 1.5 (AM1.5) spectrum. These conditions correspond to a clear day with sunlight incident upon a 37°-tilted sun-facing surface with the Sun at an angle of 41.81° above the horizon.[2] Under these test conditions a solar cell of 20% efficiency with a 100 cm^2 (0.01 m^2) surface area would produce 2.0 watts of power. Figure 6.4 shows the standard and defined sun radiation spectrum AM1.5 Global.[3]

6.2.3   Solar Simulator

The solar irradiance outdoors is subject to location, weather, and time in a year. In order to measure the solar cell accurately indoors, a controllable solar simulator becomes necessary. A solar simulator is a test facility that provides illumination approximating natural sunlight. It is used for the testing of solar cells, sun screen, and other materials and devices.

The solar simulator is classified according to its spectral content, spatial uniformity and temporal stability. Table 6.1 shows the classification of the

Table 6.1 Solar simulator classification.

Classification	Spectral match to all intervals specified	Irradiance spatial non-uniformity	Temporal instability – short term	Temporal instability – long term
Class A	0.75–1.25	2%	0.5%	2%
Class B	0.6–1.4	5%	2%	5%
Class C	0.4–2.0	10%	10%	10%

Table 6.2 AM 1.5 Global reference solar irradiance distributions.

Wavelength interval (nm)	300–400	400–500	500–600	600–700	700–800	800–900	900–1100	1100–1400
AM1.5G	No spec	18.4%	19.9%	18.4%	14.9%	12.5%	15.9%	No spec

solar simulators.[4] The percentages of integrated irradiance across several wavelength intervals over total irradiance are further specified for the solar simulation spectrum. The percentage is shown below in Table 6.2 for the standard terrestrial spectra of AM1.5G according to IEC standard 60904-3 (2008).

6.3 Characteristics of Solar Cell and Device

The solar cells are similar to photodetetors in terms that they can convert the optical energy or power into electrical current or voltage. The spectral responsivity $S(\lambda)$ of the solar cell is the electric current (or voltage) produced by an optical radiation of unit radiant power at a wavelength falling onto the solar cell surface.

$$S(\lambda) = \frac{I_{SC}}{P_{optical}} = \frac{q \cdot n_e}{\frac{hc}{\lambda} \cdot n_{photon}} = \frac{q \cdot \lambda}{h \cdot c} \frac{n_e}{n_{photon}}$$

$$= 0.80656\lambda \cdot EQE(\lambda)$$

$$= 0.80656\lambda \cdot [1 - R(\lambda)] \cdot IQE(\lambda)$$

$$(6.1)$$

Where $EQE(\lambda)$ is the external quantum efficiency, $IQE(\lambda)$ is the internal quantum efficiency, $R(\lambda)$ is the reflectivity, n_e is the number of electrons produced, and n_{photon} is the number of photons falling onto the solar cell surface. The spectral responsivity will determine the overall efficiency of the solar cell. However, the spectral responsivity is subject to many conditions.

The spectral responsivity is dependent on the incident angle, as the incident angle affects the reflection and hence affects the external quantum efficiency. Therefore, a standard testing condition must be applied to standardize the solar cell measurement. A solar cell will have special variation of its spectral responsivity, which is due to the variation in the fabrication process. The linearity of the spectral responsivity is the deviation

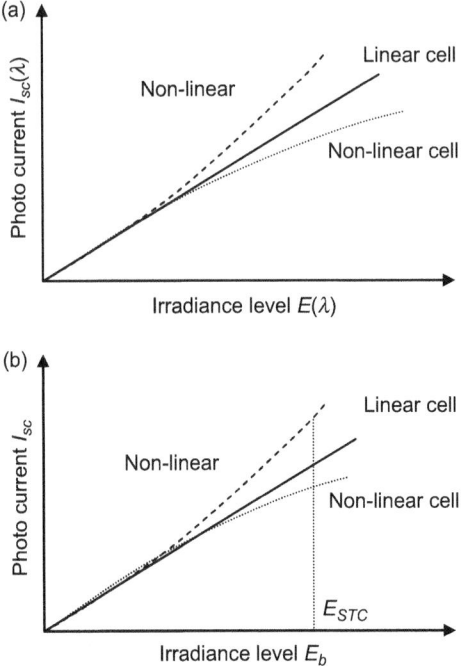

Figure 6.5 (a) Linearity of spectral responsivities to irradiance. (b) Linearity of solar cell response to irradiance.

of the electrical output from being proportional to the input radiant power. The temperature will also affect the spectral responsivity.

6.3.1 Linearity of the Responsivity

The external quantum efficiency of the solar cell is a function of the bias light level due to the nonlinearity of the cell. External quantum efficiency of some types of solar cells may have weak dependence on the bias level (<0.1%). Dye-sensitized solar cells are greatly affected by the bias intensity and chopping frequency. Figures 6.5(a) and 6.5(b) show the spectral responsivities of linear and nonlinear solar cells, and the linear and non-linear responses of the solar cells *vs.* the irradiance level of the sunlight, respectively. Thus, for a nonlinear cell, the spectral responsivity cannot be measured at low irradiance level.

Many silicon based solar cells are found to have a higher responsivity at a higher irradiance level, which is due to the solar cell's spectral responsivity increasing with the increase of irradiance level at certain wavelength ranges.

6.3.2 Temperature Effect

The temperature will affect the external quantum efficiency. Solar panel performance should be evaluated at a temperature of 25 °C according to

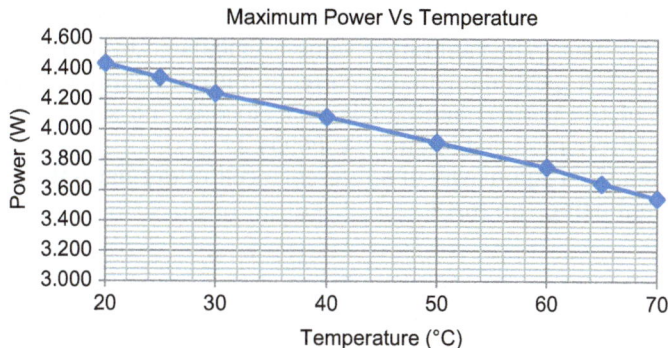

Figure 6.6 Dependency of maximum electrical output power on the temperature of a typical silicon solar cell.

standard testing conditions in the IEC 60904-3 standard. Some solar panels operate at environmental temperatures of up to 70 °C in tropical regions (*e.g.* Singapore), and in the deserts. Based on the diode model of the solar cell, the short-circuit current and open circuit voltage, maximum power and filling factor are linear with temperature.[5]

$$I_2 = I_1 + I_{SC1}\left(\frac{E_2}{E_1} - 1\right) + \alpha(T_2 - T_1) \tag{6.2}$$

$$V_2 = V_1 - R_S(I_2 - I_1) - I_2 K\alpha(T_2 - T_1) + \beta(T_2 - T_1) \tag{6.3}$$

Where α and β are the temperature coefficients, R_S is the series resistance, and K is a curve-shape correlation factor.

Figure 6.6 shows the dependency of maximum electrical output power on the temperature of the solar cell. It is found that output power of a type of silicon polycrystalline solar cell would drop about 17% when it operates at 70 °C comparing to that at 25 °C.

6.4 Calibration Method, Setup and Traceability

6.4.1 Primary Calibration Methods

6.4.1.1 Outdoor Method Under Global Sunlight[6]

The reference solar cell to be calibrated is compared with two reference radiometers under natural sunlight, which are a pyrheliometer measuring direct solar irradiance and a pyranometer with a shade device measuring diffuse solar irradiance under normal incidence conditions. The total solar irradiance is determined by the sum of direct and diffuse irradiance. The pyrheliometer is used in the form of an absolute cavity radiometer traceable to the World Radiometric Reference (WRR).

Under certain conditions the simplified global sunlight method is applicable. The short-circuit current of the reference cell is used to set the solar irradiance at 1000 W m^{-2} and then plotted versus pressure corrected geometric air mass. The calibration value is calculated by a linear square fit at air mass 1.5. A spectral mismatch correction is not required as the spectral irradiance of the natural sunlight is very close to the standard spectrum defined in international standard IEC 60904-3.

The irradiance responsivity of a reference cell under natural sunlight can be calculated using eqn (6.4):

$$s_{STC} = \frac{I_{SC}}{E_{tot}} \frac{\int_{\lambda_1}^{\lambda_2} E_{Ref}(\lambda) S_T(\lambda) d\lambda}{\int_{\lambda_1}^{\lambda_2} E_{Ref}(\lambda) S_R(\lambda) d\lambda} \frac{\int_{\lambda_1}^{\lambda_2} E_S(\lambda) S_R(\lambda) d\lambda}{\int_{\lambda_1}^{\lambda_2} E_S(\lambda) S_T(\lambda) d\lambda} \qquad (6.4)$$

s_{STC}: irradiance responsivity of the device/cell under test (DUT)
I_{SC}: short-circuit current of DUT
$E_S(\lambda)$: solar spectral irradiance (through measurement)
E_{tot}: total reference solar irradiance (integrated from tabulated spectrum in IEC standard)
$S_T(\lambda)$: absolute spectral responsivity of DUT
$E_{Ref}(\lambda)$: reference solar spectral irradiance (tabulated in IEC standard)
$S_R(\lambda)$: spectral responsivity of reference thermal detector

As the spectral responsivity of a reference thermal detector is constant, it can be cancelled out in the equation. The incident irradiance is measured by the thermal detector (*e.g.* absolute cavity radiometer or pyranometer) that is traceable to the WRR scale. The researchers at the National Renewable Energy Laboratory (NREL) of USA use an absolute cavity radiometer for measurements because it is the primary instrument used to calibrate pyranometers.

The pyranometer-based calibration method requires the measurement of spectral irradiance in a spectral range from 300 nm to 2500 nm. Typically irradiance responsivity is an average of many measurements taken over several days.

6.4.1.2   Indoor Method using a Solar Simulator

In the indoor method, the spectral irradiance of a solar simulator is measured by a spectroradiometer, which is calibrated with standard reference lamps. Traceability is based on the absolute spectral irradiance of simulated sunlight and relative spectral responsivity of the reference solar cell to be calibrated. The absolute spectral irradiance are calibrated by standard lamps and the spectral responsivity are calibrated by standard detectors, which are directly traceable to SI units. The calibration value is computed from the measured spectral responsivity of the reference cell, the spectral irradiance of the solar simulator and the reference solar spectral irradiance (IEC 60904-3). The researchers at the National Institute of Advanced Industrial Science and Technology (AIST) of Japan use this method to calibrate reference solar cells.

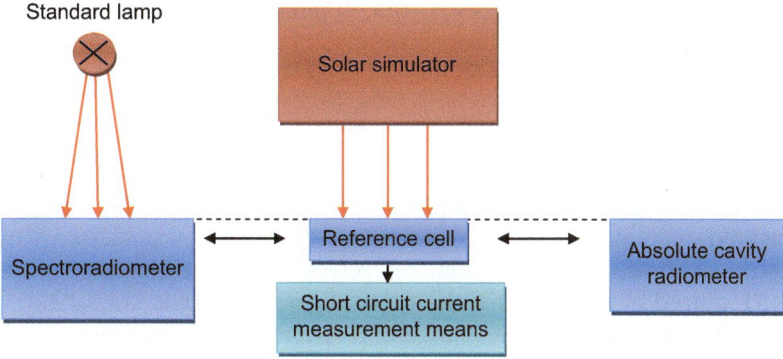

Figure 6.7   Schematic diagram of the solar simulator method.

If the absolute spectral irradiance, $E_S(\lambda)$, of the solar simulator on the test plane is known, irradiance responsivity of a reference cell, s_{STC}, can be calculated by equation (6.5):

$$s_{STC} = I_{SC} \frac{\int_{\lambda_1}^{\lambda_2} E_{Ref}(\lambda)S_T(\lambda)d\lambda}{\int_{\lambda_1}^{\lambda_2} E_{Ref}(\lambda)d\lambda} \frac{1}{\int_{\lambda_1}^{\lambda_2} E_S(\lambda)S_T(\lambda)d\lambda} \qquad (6.5)$$

As the measurement of spectral irradiance of a solar simulator is traceable to the primary spectral irradiance scale – black body spectrum through a group of spectral irradiance lamps – this is a light source-based method for the calibration of the reference solar cell. Figure 6.7[7] shows a schematic diagram of the solar simulator method.

6.4.1.3   *Indoor Method using Differential Spectral Responsivity (DSR) Measurement Facility[8]*

This method is the most straightforward method of determining the short-circuit current with respect to absolute spectral responsivity of a solar cell, $S_T(\lambda)$; the reference solar spectrum, $E_{Ref}(\lambda)$; the reference total irradiance, E_{tot}; and the cell area, A; under standard testing conditions, using the following equation:

$$I_{STC,SC} = \frac{E_{tot}A \int_{\lambda_1}^{\lambda_2} E_{Ref}(\lambda)S_T(\lambda)d\lambda}{\int_{\lambda_1}^{\lambda_2} E_{Ref}(\lambda)d\lambda} \qquad (6.6)$$

For a linear solar cell, its broadband irradiance responsivity, $s = I_{sc}(E)/E$, determined by the ratio of the short-circuit current (SCC), $I_{sc}(E)$, generated by the applied irradiance, E, is a constant, independent of different irradiance levels.

For a nonlinear cell, however, the differential irradiance responsivity (DIR), $\tilde{s}(E_b)$, must be taken into account:

$$\tilde{s}(E_b) = \frac{\partial I_{sc}(E)}{\partial E}\bigg|_{E_b} \tag{6.7}$$

where E_b is a sun-like broadband bias irradiance under which the DIR is measured.

As the calibrated SCC of a solar cell refers to standard testing conditions (STC) required by IEC 60904-3, the spectrally resolved DSR, $\tilde{s}(\lambda, I_{sc}(E_b))$, ratio of the change of SCC to the change of a spectral irradiance generated by a monochromatic probe beam under a bias irradiance, E_b, must be measured and calculated as eqn (6.8).

$$\tilde{s}(\lambda, I_{sc}(E_b)) = \frac{\partial I_{sc}(\lambda, I_{sc}(E_b))}{\partial E(\lambda)}\bigg|_{E_b} \tag{6.8}$$

The broadband irradiance responsivity, $\tilde{s}_{AM1.5}(I_{sc}(E_b))$, as function of SCC of the solar cell under test in accordance with AM1.5 global reference solar spectrum, $E_{AM1.5}(\lambda)$, is calculated from eqn (6.9).

$$\tilde{s}_{AM1.5}(I_{sc}(E_b)) = \frac{\int_0^\infty \tilde{s}(\lambda, I_{sc}(E_b))E_{AM1.5}(\lambda)d\lambda}{\int_0^\infty E_{AM1.5}(\lambda)d\lambda} \tag{6.9}$$

Note that the expressions for both DSR and DIR use the short-circuit current $I_{sc}(E_b)$ instead of the bias light E_b as a parameter as it is easily measurable.

If DIRs at different bias levels are measured, the SCC under STC, I_{STC}, can be decided by finding the upper limit of the following equation (eqn 6.10):

$$E_{STC} = \int_0^{I_{STC}} \frac{dI_{sc}}{\tilde{s}_{AM1.5}(I_{sc})} \tag{6.10}$$

where $E_{STC} = 1000\ W\ m^{-2}$ is the irradiance under STC.

In order to calibrate and characterize solar cells with low uncertainty, NMC designed and constructed a DSR calibration system (Figure 6.8) based on the principle originated from Physikalisch-Technische Bundesanstalt (PTB) of Germany with some special built-in features.[9] A monochromatic probe beam used in the measurement is produced by a light source (1600 W Xe discharge or 1000 W halogen lamp) through a double-grating monochromator and modulated by a mechanical chopper. A specially designed homogeniser was inserted in the collimating optic to ensure the probe beam has spatial uniformity better than 2% over the area of 20 mm × 20 mm.

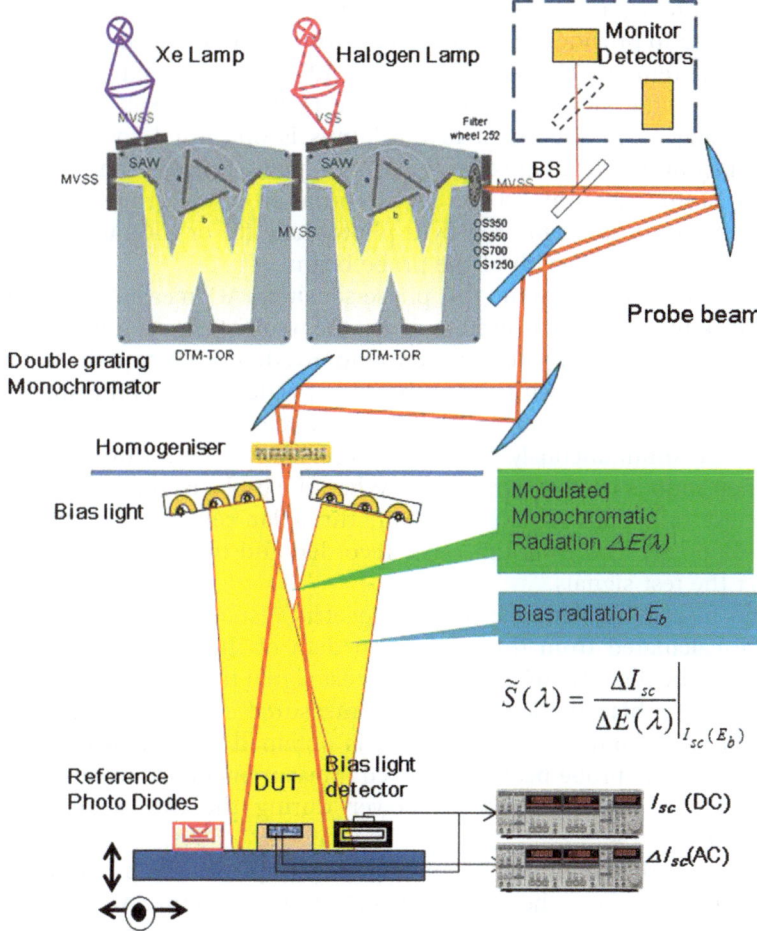

Figure 6.8 Schematic diagram of DSR measurement system at NMC-A*STAR.

A movable stage which can hold a large solar cell (156 mm×156 mm), two WPVS (World Photovoltaic Scale) reference cells (20 mm×20 mm) and two reference photodiodes (Si and InGaAs) with precision apertures of nominal diameter of 4.0 mm is mounted on a motorized X-Y-Z stage, and enables accurate alignment of any of the above components to the probe beam. The temperature of the sample stage is controlled at 25 °C by a closed loop temperature control chiller. The temperature of each solar cell can be individually controlled by additional thermal electrical controllers from 20 °C to 60 °C for the temperature coefficient measurement.

As the probe beam is slightly divergent, by increasing the distance of the sample stage from the probe beam optics the probe beam can be adjusted to overfill solar cells of sizes from 20 mm×20 mm up to 156 mm×156 mm. An

array of 24 halogen lamps (50 W), arranged in a two-layered circular geometry, is used to provide the required bias irradiance on the solar test cell for
DSR measurement. The level of bias is adjustable by changing the com-
binations of the lamps and the distance between the bias light with the
sample stage. The alignment of the lamps is optimized so that the non-
uniformity of bias irradiance on the sample plane is better than 2% over an
area of 20 mm×20 mm. To correct the variation of the probe beam power,
the probe beam is monitored by a photodiode (Si or InGaAs) *via* a beam
splitter in the optical path of the probe beam.

In the detection system, the phase-sensitive AC measuring technique
using lock-in amplifiers is used to detect the weak AC signal under the probe
beam from the strong bias background irradiance. When the reference
photodiode is aligned with the probe beam, which overfills the aperture on
the reference diode, both AC signals from reference and monitor photodiodes are simultaneously recorded at every wavelength during each scan
and their ratio is used as a reference. When the probe beam is aligned to the
solar test cell, both AC signals from the solar cell and the monitoring
photodiode are also simultaneously recorded and their ratios calculated to
correct the test signals.

If no bias is applied, the relative spectral responsivity of the solar cell
can be calculated from the calibration data of the reference photodiode
multiplied by the ratio of the corrected test signal to the corrected reference
signal. With a bias applied, the DSR is measured instead and in the mean-
time the SCC from the solar test cell is obtained directly by a DC source
meter when the probe beam is blocked. Measurement errors caused by the
fluctuation of the probe beam power during the two scans are then
corrected.

The use of the beam homogenizer greatly improves the probe beam uniformity to < 2% over the sample area (20 mm × 20 mm). Any error due to
the area difference between the solar test cell and reference photodiodes is
corrected using data from spatial uniformity measurements of the probe
beam. The two reference photodiodes used in the system are calibrated
against NMC's spectral responsivity scale with an uncertainty typically
$\approx 0.2\%$ in the spectral range from 400 nm to 900 nm and 1–2% below 400 nm
and above 900 nm. The absolute spectral irradiance responsivity (SR) of the
solar cell under test at 650 nm is calibrated by directly comparing the AC
short-circuit currents generated by the DUT and reference Si photodiode using
a calibrated current amplifier.

$$SR(650\text{nm})_{DUT} = \frac{s(650\text{nm})_{ref}}{A} \frac{i_{DUT}}{i_{ref}} \tag{6.11}$$

where A is the aperture area, $s(650 \text{ nm})_{ref}$ is spectral irradiance responsivity
of the reference photodiode at 650 nm, i_{DUT} and i_{ref} are short circuit
currents of DUT and reference Si photodiode respectively. Correction is
made for the slight difference in spectral irradiance received by the DUT and

Figure 6.9   Photo of the motorized sample stage and bias lighting system.

reference photodiode due to their differences in area. Figure 6.9 shows a photo of the motorized sample stage and bias lighting system.

6.4.2   Secondary Calibration Method[10]

If the calibration is carried out by comparing the DUT with a reference solar cell, which is calibrated by primary calibration methods, the method is defined as a secondary calibration method. There are three common methods for secondary calibration of solar cells, which are described here.

6.4.2.1   Measurements in Natural Sunlight

Measurements in natural sunlight are made when the instability of the global solar irradiance is less than $\pm 1\%$ during a measurement. As the measurements refer to the STC, the irradiance on the solar cell to be tested shall be at least 800 W m^{-2}.

6.4.2.2   Measurements in Steady-State Simulated Sunlight

Measurements in steady-state simulated sunlight shall meet the requirements of IEC 60904-9. The spatial uniformity of light distribution in the test plane are measured and periodically checked. The reproducibility of the measurements are verified periodically by successive measurements under the same test conditions.

The solar test cell is connected to a source measurement unit (SMU) through a 4-wire connection. The I–V curve is measured directly by the SMU, which applies a voltage sweep on the test sample and measures the output current from the sample and the voltage across the sample simultaneously under the radiation of the solar simulator. Calibration setup at NMC-A*STAR is shown in Figure 6.10.

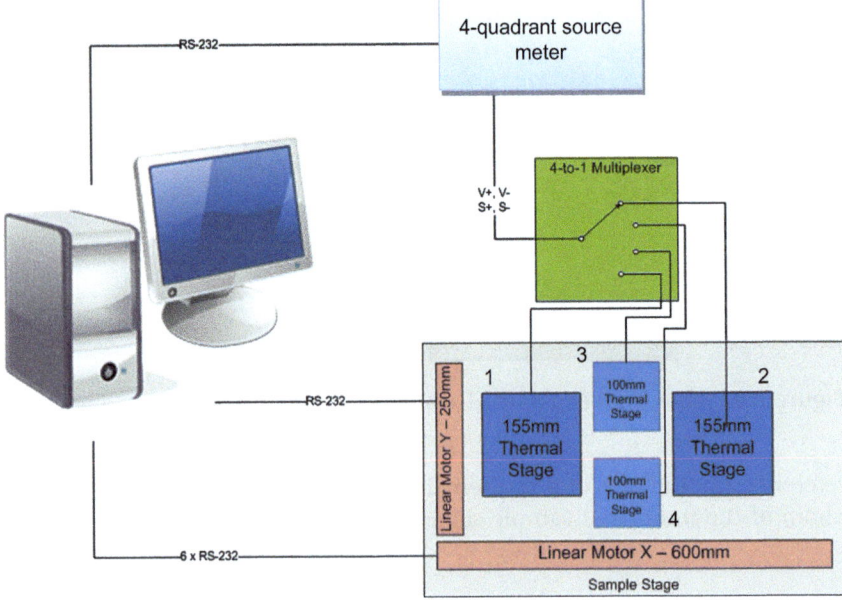

Figure 6.10   Schematic diagram of secondary calibration setup.

6.4.2.3   *Measurement in Pulsed Simulated Sunlight*

Measurement in pulsed simulated sunlight shall meet the requirements of IEC 60904-9. The spatial uniformity of light distribution in the test plane is measured and periodically checked. The reproducibility of the measurements is verified periodically by successive measurements at the same test conditions.

6.4.3   **Measurement Traceability**

6.4.3.1   *The World Radiometric Reference (WRR)*[11,12,13]

The WRR is the reference standard representing the SI unit of irradiance. It was introduced in order to ensure world-wide homogeneity of solar radiation measurements and has been in use since 1980.

The WRR was determined from the weighted mean of the measurements of a group of 15 absolute cavity radiometers, which have been fully characterized. The estimated accuracy of the scale is 0.3%. The World Meteorological Organization (WMO) introduced its mandatory use in 1979.

The 'Physikalisch-Meteorologisches Observatorium Davos'/World Radiation Centre (PMOD/WRC) conducts an International Pyrheliometer Comparison (IPC) in every five years to transfer the WRR to the participating pyrheliometers in order to ensure world-wide homogeneity of solar radiation measurements. It is intended for the calibration of absolute radiometers from the six Regional Radiation Centres.

6.4.3.2 The World Photovoltaic Scale (WPVS)

The WPVS is international reference cell calibration program. The WPVS provides a scale for PV performance measurements that has been established through world-wide round-robin calibration of a group of primary reference cells and is traceable to the Système International (SI) units. Procedures for recalibration of the existing reference cell group have been devised, along with procedures for admittance and calibration of new reference cells. A reference cell package has been designed that meets the unique requirements of the WPVS.

The World Photovoltaic Scale was established after a formal international comparison was carried out during 1993 to 1996. The results of this international comparison were published and the protocol for conducting recalibration was established.[14] Of the four WPVS qualified calibration laboratories, Physikalisch-Technische Bundesanstalt (PTB) in Germany; Japan Quality Assurance Organization (JQA) in Japan, Tianjin Institute of Power Sources (TIPS) in China, and National Renewable Energy Laboratory (NREL) in USA, NREL performed the first recalibration.

The second WPVS comparison was conducted at NREL in October and November of 1998. Twenty WPVS reference cells were calibrated along with six candidate WPVS reference cells. One of NREL's primary Si reference cells with a long calibration history was also calibrated at the same time. The comparison result[15] showed that a small difference when WPVS spectral responsivity and temperature coefficient data are used in the calibrations. The spectral range was extended to 300–4000 nm in the spectral irradiance measurement at NREL. Its effect on the calibration value is also presented.

6.4.3.3 Traceability of Solar Reference Cell Calibration at NMC-A*STAR

The traceability of solar reference cell calibration at NMC-A*STAR is based on a calibration of differential spectral responsivities of the cell comparing with standard detectors directly traceable to SI units. The calibration value is computed from the measured absolute spectral responsivity of the reference cell and the reference solar spectral irradiance distribution. The DSR calibration is conducted by a dual beam method in which the reference cell is illuminated by a modulated monochromatic light and a CW sun-like bias light at different irradiance levels. Electrical measurement of current to voltage characterization of the solar cell is traceable to the SI units through NMC electrical standards.

The cell temperature measurement and the cell area measurement are also traceable to SI units through NMC temperature and length standards respectively as shown in Figure 6.11.

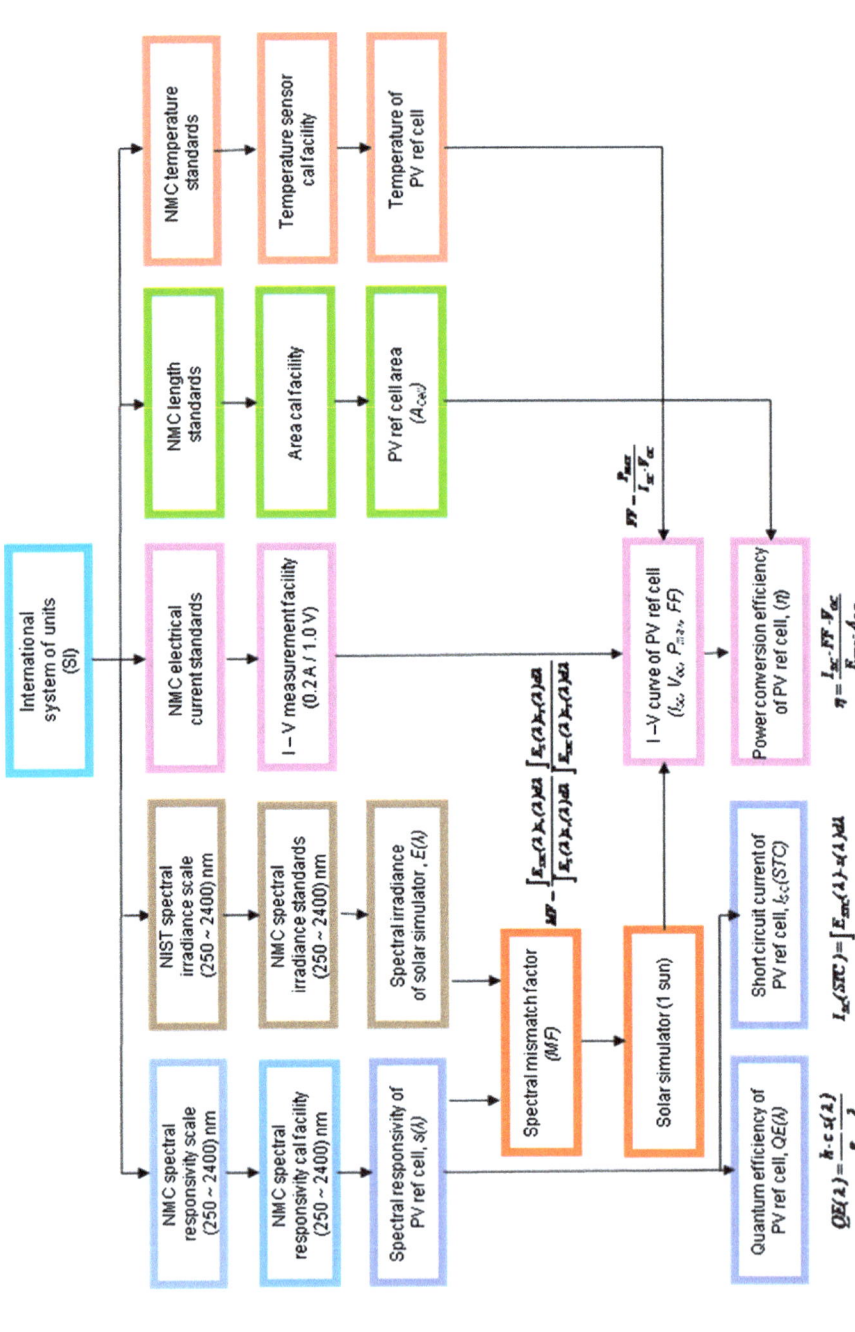

Figure 6.11   Traceability chart of PV reference cell calibration at NMC-A*STAR.

References

1. IEC 60904-3, *Photovoltaic devices – Part 3: measurement principles for terrestrial photovoltaic (PV) solar devices with reference spectral irradiance data*, ed. 2.0, 2008.
2. ASTM G 173-03, *Standard tables for reference solar spectral irradiances: direct normal and hemispherical on 37° tilted surface*, ASTM International, 2003.
3. Solar Spectral Irradiance: Air Mass 1.5. National Renewable Energy Laboratory. http://rredc.nrel.gov/solar/spectra/am1.5/, Retrieved December 2013.
4. IEC 60904-9, *Photovoltaic devices – Part 9: Solar simulator performance requirements*, ed. 2.0, 2007.
5. K. Emery, in *Handbook of photovoltaic science and engineering*, ed. A. Luque and S. Hegedus, John Wiley & Sons Ltd, Chichester, 2003, p. 719.
6. K. Emery, in *Handbook of photovoltaic science and engineering*, ed. A. Luque and S. Hegedus, John Wiley & Sons Ltd, Chichester, 2003, p. 723.
7. IEC 60904-4, *Photovoltaic devices – Part 4: Reference solar devices – Procedure for establishing calibration traceability*, ed.1.0, 2009, p. 21.
8. J. Metzdorf, W. Möller, T. Wittchen and D. Hünerhoff, *Metrologia*, 1991, **28**, 247.
9. G. Xu and X. Huang, *Energy Procedia*, 2012, **25**, 70.
10. IEC 60904-1, *Photovoltaic devices – Part 1: Measurement of photovoltaic current-voltage characteristics*, ed. 2.0, 2006.
11. C. Fröhlich. World Radiometric Reference, *WMO/CIMO Final Report*, WMO No. 490, 97–110. (PMOD/WRC intern: 545a), 1977.
12. C. Fröhlich., *Metrologia*, 1991, **28**, 111.
13. C. Fröhlich, R. Philipona, J. Romero and C. Wehrli, *Optical Engineering*, 1995, **34**, 2757.
14. C. R. Osterwald, S. Anevsky, A. K. Barua, J. Dubard, K. Emery, D. King, J. Metzdorf, F. Nagamine, R. Shimokawa, N. Udayakumar, Y. X. Wang, W. Zaaiman, A. Zastrow and J. Zhang, *NREL Tech. Rep.* NREL/TP-520-23477, 1998.
15. K. Emery. *NREL Tech. Rep.* NREL/TP-520-27942, 2000.

Subject Index

Illustrations and figures are in **bold**.